AF785891

Lecture Notes in Physics

Editorial Board

R. Beig, Wien, Austria
J. Ehlers, Potsdam, Germany
U. Frisch, Nice, France
K. Hepp, Zürich, Switzerland
W. Hillebrandt, Garching, Germany
D. Imboden, Zürich, Switzerland
R. L. Jaffe, Cambridge, MA, USA
R. Kippenhahn, Göttingen, Germany
R. Lipowsky, Golm, Germany
H. v. Löhneysen, Karlsruhe, Germany
I. Ojima, Kyoto, Japan
H. A. Weidenmüller, Heidelberg, Germany
J. Wess, München, Germany
J. Zittartz, Köln, Germany

Springer-Verlag Berlin Heidelberg GmbH

Physics and Astronomy ONLINE LIBRARY

http://www.springer.de/phys/

Editorial Policy

The series *Lecture Notes in Physics* (LNP), founded in 1969, reports new developments in physics research and teaching – quickly, informally but with a high quality. Manuscripts to be considered for publication are topical volumes consisting of a limited number of contributions, carefully edited and closely related to each other. Each contribution should contain at least partly original and previously unpublished material, be written in a clear, pedagogical style and aimed at a broader readership, especially graduate students and nonspecialist researchers wishing to familiarize themselves with the topic concerned. For this reason, traditional proceedings cannot be considered for this series though volumes to appear in this series are often based on material presented at conferences, workshops and schools (in exceptional cases the original papers and/or those not included in the printed book may be added on an accompanying CD ROM, together with the abstracts of posters and other material suitable for publication, e.g. large tables, colour pictures, program codes, etc.).

Acceptance

A project can only be accepted tentatively for publication, by both the editorial board and the publisher, following thorough examination of the material submitted. The book proposal sent to the publisher should consist at least of a preliminary table of contents outlining the structure of the book together with abstracts of all contributions to be included. Final acceptance is issued by the series editor in charge, in consultation with the publisher, only after receiving the complete manuscript. Final acceptance, possibly requiring minor corrections, usually follows the tentative acceptance unless the final manuscript differs significantly from expectations (project outline). In particular, the series editors are entitled to reject individual contributions if they do not meet the high quality standards of this series. The final manuscript must be camera-ready, and should include both an informative introduction and a sufficiently detailed subject index.

Contractual Aspects

Publication in LNP is free of charge. There is no formal contract, no royalties are paid, and no bulk orders are required, although special discounts are offered in this case. The volume editors receive jointly 30 free copies for their personal use and are entitled, as are the contributing authors, to purchase Springer books at a reduced rate. The publisher secures the copyright for each volume. As a rule, no reprints of individual contributions can be supplied.

Manuscript Submission

The manuscript in its final and approved version must be submitted in camera-ready form. The corresponding electronic source files are also required for the production process, in particular the online version. Technical assistance in compiling the final manuscript can be provided by the publisher's production editor(s), especially with regard to the publisher's own Latex macro package which has been specially designed for this series.

Online Version/ LNP Homepage

LNP homepage (list of available titles, aims and scope, editorial contacts etc.):
http://www.springer.de/phys/books/lnpp/

LNP online (abstracts, full-texts, subscriptions etc.):
http://link.springer.de/series/lnpp/

F. C. Lázaro M. J. Arévalo (Eds.)

Binary Stars: Selected Topics on Observations and Physical Processes

Lectures Held at the Astrophysics School XII Organized by the European Astrophysics Doctoral Network (EADN) in La Laguna, Tenerife, Spain, 6-17 September 1999

Springer

Editors

F. Carlos Lázaro
Maria J. Arévalo
Department of Astrophysics
Universidad de la Laguna
Tenerife, Spain

Cover picture: The interbinary x-ray emission in the orbital plane of AR Lac. The observations were made with the Solid-State Imaging Spectrograph (SIS) in the energy range 0.4–1.5 keV. (see Strassmeier et al. in this volume)

Library of Congress Cataloging-in-Publication Data.

Die Deutsche Bibliothek - CIP-Einheitsaufnahme

Binary stars : selected topics on observations and physical processes ; lectures held at the Astrophysics School XII in La Laguna, Tenerife, Spain, 6 - 17 September 1999 / F. C. L=E1zaro ; M. J. Ar=E9valo (ed.). Organized by the European Astrophysics Doctoral Network (EADN).
(Lecture notes in physics ; Vol. 563)
ISBN 978-3-642-07455-4 ISBN 978-3-540-44395-7 (eBook)
DOI 10.1007/978-3-540-44395-7

ISSN 0075-8450
ISBN 978-3-642-07455-4

This work is subject to copyright. All rights are reserved, whether the whole or part of the material is concerned, specifically the rights of translation, reprinting, reuse of illustrations, recitation, broadcasting, reproduction on microfilm or in any other way, and storage in data banks. Duplication of this publication or parts thereof is permitted only under the provisions of the German Copyright Law of September 9, 1965, in its current version, and permission for use must always be obtained from Springer-Verlag Berlin Heidelberg GmbH. Violations are liable for prosecution under the German Copyright Law.

http://www.springer.de
© Springer-Verlag Berlin Heidelberg 2001
Originally published by Springer-Verlag Berlin Heidelberg New York in 2001
Softcover reprint of the hardcover 1st edition 2001
The use of general descriptive names, registered names, trademarks, etc. in this publication does not imply, even in the absence of a specific statement, that such names are exempt from the relevant protective laws and regulations and therefore free for general use.

Typesetting: Camera-ready by the authors/editors
Camera-data conversion by Steingraeber Satztechnik GmbH Heidelberg
Cover design: *design & production*, Heidelberg

Printed on acid-free paper
SPIN: 10787507 55/3141/du - 5 4 3 2 1 0

Preface

The XII Predoctoral School of Astrophysics, organized by the European Astrophysics Doctoral Network (EADN) took place on 6–17 September 1999, and was hosted by the Faculty of Physics of the University of La Laguna (Tenerife, Spain). The topic of the School was chosen because of the important role played by binary stars in modern astrophysics. Today, it is thought that at least half of all known stars in our galaxy form and evolve in binary or multiple systems, and this makes the study of binary stars essential to our understanding of stellar evolution. The physics of binary stars is a very active area, in which recent major advances have been made, both in theoretical modelling and observations. While the study of binary stars is a field with too many branches to be fully covered in a two-week course, for the 1999 EADN School we selected a few topics that gave a rather wide view of the present research.

The school was attended by 41 students from 12 different European countries (Belgium, Croatia, Estonia, France, Germany, Greece, Italy, Ireland, Poland, Portugal, Spain and the United Kingdom). The lectures were delivered in the mornings over two weeks, with two afternoon sessions in which the students had the possibility of presenting their research work in the form of a short talk or poster paper.

The students' accommodation and other organizational costs were partially covered by the *European Union Training and Mobility of Researchers Programme*, Contract No. ERBFMMACT960173, and by a grant from the Gramholm Foundation of Sweden. We acknowledge the support received from the following institutions: the University of La Laguna, the Instituto de Astrofísica de Canarias and the island's local authorities (*Cabildo de Tenerife*) and La Laguna Town Hall). The travel costs of the invited lecturers were partially covered by Iberia airlines, while the Disa Oil Corporation supported the cost of the coffee breaks offered to attendants at the School. We are much indebted to Prof. Tom Ray, from the Dublin Institute for Advanced Studies (Ireland), Coordinator of the EADN Schools, for his advice in the organization of the XII Predoctoral School.

La Laguna
February 2001

F. Carlos Lázaro
Maria J. Arévalo

Contents

List of Contributors

Dr. Antonio Claret
Instituto de Astrofísica de Andalucia, CSIC
C/ Sancho Panza s/n. Apartado 3004
E-18080 Granada, Spain
Claret@iaa.es

Dr. Álvaro Giménez
Laboratorio de Astrofísica Espacial y Física Fundamental, INTA
Apartado 50727
E-28080 Madrid, Spain
ag@laeff.esa.es

Dr. Klaus G. Strassmeier
Institut für Astronomie
Universität Wien
Türkenschanzstr. 17
A-1180 Vienna, Austria
strassmeier@astro.univie.ac.at

Prof. Józef I. Smak
N. Copernicus Astronomical Center
Bartycka 18
00-716 Warszawa, Poland
jis@alfa.camk.edu.pl

Dr. Tom R. Marsh
University of Southampton
Department of Physics and Astronomy
Highfield SO17 1BJ Southampton, United Kingdom
trm@astro.soton.ac.uk

Prof. Bert C. de Loore
Astrophysical Institute.
Vrije Universiteit Brussel
Pleinlaan 2, B-1050 Brussels, Belgium
cdeloore@ruca.ua.ac.be

Dr. Jorge Casares
Instituto de Astrofísica de Canarias
C/ Vía Láctea s/n.
38200 La Laguna, Tenerife, Spain
jcv@ll.iac.es

1 Physical Processes in Close Binary Systems

Antonio Claret[1] and Alvaro Giménez[1,2]

[1] Instituto de Astrofísica de Andalucía, CSIC,
Apartado 3004, E-18080 Granada, Spain
claret@iaa.es

[2] Laboratorio de Astrofísica Espacial y Física Fundamental, INTA,
Apartado 50727, E-28080 Madrid, Spain
ag@laeff.esa.es

Abstract. We review briefly some basic aspects of the structure and evolution of a star distorted by tides and rotation. Proximity effects like, for example gravity-darkening, are discussed using the properties of stellar internal structure. On the other hand, the differential equations of the dynamics of close binary systems are discussed with particular attention to apsidal motion. The tidal evolution equations for late and early-type stars are also included and theoretical predictions are compared, when possible, with observations. Concerning stellar atmospheres, limb-darkening laws are revised using recent models. Furthermore, the influence of an external radiation field on the distribution of the intensities and on the irradiated spectrum is also investigated.

1.1 Introduction

The theory of stellar structure and evolution is certainly one of the most important achievements of modern Astrophysics. It allows us to understand the observational HR diagram, the stellar content of clusters and galaxies, their chemical evolution, the estimation of ages and distances for single stars, or the evolution of close binary systems. A quite good description, both qualitative and quantitative, of stellar structure and evolution is now-a-days available and the comparison of theoretical models with observations has provided important constraints to our knowledge of the Universe. Among the most important ones is the possibility to look to stars into the past and compare stellar structure for different ages. These results have to be carefully evaluated under the light of the adopted input physics for the models.

In order to test the theory, or derive empirical values for adopted parameters, observational data are needed. But the main parameter driving the structure and evolution of a star is its mass, which can only be determined accurately in binary systems. The study of astrophysical processes in detached eclipsing binaries is thus a well-known method to empirically explore stellar structure for conditions different than those in the Sun. The need for double-lined eclipsing binaries comes from the requirement of accurate dimensions for a variety of stars, allowing a reliable comparison of observational data with theoretical predictions. In our aim of deriving new information from the comparison of observations with theoretical models, another condition is to restrict the star sample to members of well-detached binaries. We consider a double-lined eclipsing binary to be "well-detached" when the two components behave like "normal" single stars in all aspects except their dynamical evolution, i.e. they are well inside their Roche critical limits, there is no mass transfer, and previous evolution has not been affected by binarity. The problem from the observational point of view is of course that most well-detached binaries will become interacting when they evolve and their radii become close to their Roche lobe. It is therefore difficult to find evolved stars in binaries still in a well-detached configuration.

In the following pages, we intend to introduce briefly how theoretical models are constructed, which are their main problems and uncertainties, and how they can be better understood under the illuminating information provided by double-lined eclipsing binaries. For this purpose, we have separated this presentation in five sections, including the Introduction. The second section set the scene for theoretical developments indicating the needs for additional information. In the third section, we explore the modifications introduced in stellar structure for binary stars, while their atmospheres are treated in Sect. 4. The dynamical behavior of detached binary systems is discussed in the fifth section. The comparison between theoretical predictions and actual observations are performed in each section, when possible. Finally, on the word *close* in the title of this Chapter. We refer to close binaries as double systems for which the proximity effects caused by deformations are important; not those which interact strongly, exchanging mass, for example.

1.2 Basics of Stellar Structure

1.2.1 Equations of Hydrodynamics

Because of the prevailing physical conditions in their interiors, stars can generally be studied as self-gravitating fluids. We therefore use standard equations of hydrodynamics to specify local physical quantities (like density, ρ, pressure, p, and instantaneous velocity, $\boldsymbol{v}$, or any other thermodynamical quantity that may be needed) as a function of position, $\boldsymbol{x}$, and time, t. For a given point, the position in space corresponds to what is seen by a stationary observer in the Eulerian description while, if it corresponds to that seen by an observer that follows the motion of a reference element, we use the Lagrangian description of the fluid. Both are connected through the Lagrangian derivative,

$$\frac{D}{Dt} \equiv \frac{\partial}{\partial t} + \boldsymbol{v} \cdot \nabla \tag{1.1}$$

which can be applied to any studied quantity of the fluid.

For the study of stellar structure, three well-known equations of hydrodynamics are found to be useful:

1. The equation of continuity,

$$\frac{\partial \rho}{\partial t} + \mathrm{div}(\rho \boldsymbol{v}) = 0 \tag{1.2}$$

which gives the rate of change of mass-density in a given element of flowing gas. If we furthermore define the specific volume as $V = 1/\rho$, measuring the volume taken up by a unit of mass, and use the Lagrangian description of the equation of continuity, we can write,

$$\frac{1}{V}\frac{DV}{Dt} = \mathrm{div}\boldsymbol{v} \tag{1.3}$$

and div$\boldsymbol{v}$ is the rate of expansion of a given volume of gas during its motion.

2. The equation of motion

$$\frac{\partial \boldsymbol{v}}{\partial t} + (\boldsymbol{v}\cdot\nabla)\boldsymbol{v} + 2\boldsymbol{\omega}\times\boldsymbol{v} = -\frac{1}{\rho}\nabla p - \boldsymbol{\omega}\times(\boldsymbol{\omega}\times\boldsymbol{r}) + \boldsymbol{f} \tag{1.4}$$

for a frame of reference rotating at the angular rate $\boldsymbol{\omega}$, where $\boldsymbol{f}$ denotes the external forces per unit mass (which have yet to be specified) other than those corresponding to rotation or pressure (either from the gas or radiation). Let us now consider three possible contributions to $\boldsymbol{f}$:

- The force per unit mass due to gravity, which is nothing else than the gravitational acceleration $\boldsymbol{g}$ and can be expressed through the gradient of a potential Φ_G,

$$\boldsymbol{g} = \nabla\Phi_G \tag{1.5}$$

such that the mass-density obeys Poisson's equation,

$$\nabla^2\Phi_G = -4\pi G\rho \tag{1.6}$$

where G is the gravitational constant and $\Phi_G(\boldsymbol{x}, t)$ is the full gravitational potential, i.e, taking into account self-gravitation terms as well as external contributions. Moreover, we can express the term of the centrifugal force by means of a rotational potential,

$$\Phi_R = -\frac{1}{2}(\boldsymbol{\omega}\times\boldsymbol{r})^2 \tag{1.7}$$

such that $\nabla\Phi_R$ is the centrifugal term and we can combine it with the gravitational term using $\Phi = \Phi_G + \Phi_R$. Of course, Poisson's equation should be re-written as,

$$\nabla^2\Phi = -4\pi G\rho + 2\omega^2 \tag{1.8}$$

In the absence of any further contribution to the force acting on the fluid element, Euler's equation of motion takes the form,

$$\frac{\partial \boldsymbol{v}}{\partial t} + (\boldsymbol{v}\cdot\nabla)\boldsymbol{v} + 2\boldsymbol{\omega}\times\boldsymbol{v} = -\frac{1}{\rho}\nabla p + \nabla\Phi \tag{1.9}$$

- The force due to the presence of a magnetic field with strength $\boldsymbol{B}$,

$$\boldsymbol{f} = -\frac{1}{8\pi c}\nabla B^2 + \frac{1}{4\pi c}(\boldsymbol{B}\cdot\nabla)\boldsymbol{B} \tag{1.10}$$

under the assumption of no displacement currents and setting the magnetic permeability $\mu = 1$.

- The force due to frictions or viscous force can be expressed through div$\hat{\tau}$, where $\hat{\tau}$ denotes the stress tensor of viscous dissipation forces.

3. The equation of energy conservation,

$$\frac{DQ}{Dt} = \frac{DU}{Dt} + \frac{p}{\rho}\mathrm{div}\boldsymbol{v} \tag{1.11}$$

which is derived from the first law of thermodynamics as a Lagrangian description for the rate of change of the heat per unit mass, Q, in terms of that for the total internal energy U.
The heat rate of change will equal that of the internal energy only for incompressible fluids. Moreover, the heating term can be expressed, taking into account all contributions, as

$$\rho\frac{DQ}{Dt} = \Phi_v + \rho\epsilon - \mathrm{div}(\boldsymbol{F} + \boldsymbol{F}_r) \tag{1.12}$$

where $\boldsymbol{F}$ and $\boldsymbol{F}_R$ are, respectively, the heat and radiative flux vectors, while ϵ represents the rate of energy released by thermonuclear reactions and Φ_v denotes the rate at which heat is generated by viscous friction, thus called the dissipation function.
Following Fourier, we can express the heat flux using the thermal conductivity, χ,

$$\boldsymbol{F} = -\chi\nabla T \tag{1.13}$$

and the radiative flux using the opacity of matter,

$$\boldsymbol{F}_R = -\frac{4acT^3}{3\kappa\rho}\nabla T \tag{1.14}$$

where a is the radiation density constant, κ is the opacity, and c is the speed of light.
An additional contribution to (eq. 1.12) should also be considered if transport of energy is possible via turbulent convection of fluid elements. We will see this later. On the other hand, we can express the contribution to the energy of the system due to the displacements of matter shells through the entropy function S.

1.2.2 Equations of Equilibrium

We can obviously study the figures of equilibrium of stellar structure using the equations in the previous section under different external conditions. To start with, let us assume the simplest case, when,

1. Viscosity in the stellar structure is negligible ($\hat{\tau} = 0$).
2. There are no magnetic fields, or they are also negligible ($\boldsymbol{B} = 0$).
3. Stars are in hydrostatic equilibrium and there is no motion of the fluid elements with respect to the rotating reference frame ($\boldsymbol{v} = 0$).

The equation of motion (eq. 1.9) is then reduced to the equation of hydrostatic equilibrium,

$$\nabla p = \rho \nabla \Phi \tag{1.15}$$

where the potential Φ still represents the combination of the gravitational and centrifugal contributions. Moreover, the gravitational term includes self gravitation of the stellar configuration as well as the contribution of external potentials. In other words, this equation is valid for the case of binary stars with rotation and tides. It implies moreover that, even for non-spherical figures of equilibrium, surfaces of equal pressure, density, or mean molecular weight, are defined by equipotential surfaces.

In addition, the equation of continuity takes the form,

$$\frac{\partial \rho}{\partial t} = 0 \tag{1.16}$$

and both, (eq. 1.15) and (eq. 1.16), can be re-written in the more familiar form,

$$\frac{dp}{dm} = -G\frac{m}{4\pi r^4}\rho \tag{1.17}$$

$$\frac{dr}{dm} = \frac{1}{4\pi r^2 \rho} \tag{1.18}$$

if we consider only the contribution of self gravitation for the potential, i.e., non rotating isolated stars, and thus spherical symmetry. In such case, Poisson's equation can be integrated to obtain an explicit form of the gravitational potential.

To determine three quantities (ρ, m, p) as functions of the relative radius r, we need something more than two differential equations. Additional constraints are needed, either in the form of equations relating the three unknown parameters or with new differential equations. In the first case, we should adopt relations $p = p(\rho)$ like that adopted for the set of the polytropic models. Since we know that a realistic equation of state $p = p(X, \rho, T)$ requires an adequate knowledge of the local temperature and chemical composition, additional differential equations are needed to close the problem, taking into account the thermal behavior of the stellar structure.

The simplest expression available for the equation of state of stellar matter considers an ideal gas and radiation pressure in such a way that,

$$p = \frac{k}{\mu m_U}\rho T + aT^4 \tag{1.19}$$

being k the Boltzman's constant, μ the mean molecular weight, and m_U the atomic mass unit.

Using the definition of luminosity as,

$$L(r) = 4\pi r^2 F(r) \tag{1.20}$$

the equation of energy conservation (eq. 1.13) for spherical symmetry can now be written as,

$$\frac{dL}{dm} = \epsilon - T\frac{\partial S}{\partial t} \tag{1.21}$$

which, under equilibrium conditions, neglecting the rate of heat change in comparison with the energy flux, becomes the more familiar,

$$\frac{dL}{dm} = \epsilon \tag{1.22}$$

This later simplification, equivalent to neglecting the time derivative of the entropy function, is permitted as long as we consider variations in time which are slow compared to the Kelvin-Helmhotz time scale.

On the other hand, the equation of energy transport for the radiative flux can be used to determine the temperature gradient, neglecting the contribution of thermal conduction, since for spherical symmetry,

$$\frac{dT}{dm} = -\frac{GmT}{4\pi r^4 p}\nabla \tag{1.23}$$

where

$$\nabla = \nabla_{rad} = -\frac{3\kappa L(r)p}{16\pi acGmT^4} \tag{1.24}$$

The only remaining term is that taking into account the transport of energy through turbulent gas motions, as expected in convective regions where $\nabla = \nabla_{conv}$ which is computed using the theory of convection. This depends on the value of ∇_{cri} that is given by

$$\nabla_{cri} = \nabla_{ad} - \left(\frac{\partial lnT}{\partial ln\mu}\right)_{p,\rho} \frac{\partial ln\mu}{\partial m}\frac{4\pi r^4 p}{Gm} \tag{1.25}$$

The computation of which is the dominant contribution to the energy transport, either radiation or convection, is made through the Schwarzschild criterium, i.e., the lowest temperature gradient.

For the computation of simple stellar structure models we have now 4 differential equations (hydrostatic equilibrium, conservation of mass, conservation of energy, and energy transfer). Establishing physically realistic boundary conditions, we can solve them for four independent quantities provided that any additional parameter involved in the equations is independently known (nuclear reaction rates, opacities, or convective energy transport coefficients).

The above equilibrium equations then take the form given by eqs. 1.17, 1.18, 1.22 and 1.23, where $\nabla = \nabla_{rad}$ if $\nabla_{rad} < \nabla_{cri}$ or $\nabla = \nabla_{conv}$ if $\nabla_{rad} > \nabla_{cri}$ where ∇_{cri} and ∇_{rad} are the critical and radiative gradients, respectively. In the right-hand side of (eq. 1.22) we can include the term $-T\partial S/\partial t$ if we want to take into account the time variation of the total entropy or ϵ_ν for energy loss via neutrinos.

As a consequence, we can compute all basic stellar parameters (radius, density, pressure, temperature, and luminosity) throughout the stellar interior as a function of the relative mass for a given total mass and chemical composition distribution.

The distribution of chemical elements is a function of the initial chemical composition and age. If composition changes are assumed to be only due to nuclear reactions (neglecting e.g., diffusion), the abundance by mass X of hydrogen for example is given through the variation,

$$\frac{dX}{dt} = -r_X \tag{1.26}$$

where r_X is the net destruction rate obtained by summing over the reactions in which hydrogen is involved.

Numerical methods to solve the given set of differential equations with the corresponding internal and external boundary conditions follow two different approaches. In the first method, known as the shooting technique, the four differential equations are integrated from the center to a selected fitting point and from the surface down to the same point. The final solution is obtained by requiring that both integrations agree at the selected matching point. The second method, leading to relaxation techniques, approximate the differential equations by a set of difference equations. In this way, N (the number of shells) equations of the form,

$$\left(\frac{Y_{i+1} - Y_i}{x_{i+1} - x_i}\right) - G_{i+\frac{1}{2}} = 0 \tag{1.27}$$

are used, where $G_{i+\frac{1}{2}}$ are the generic right-hand terms of the differential equations evaluated at the middle of the interval. A method is then applied using differential corrections until some pre-adopted criterium for convergence is achieved. Details about the different numerical methods available can be found in Kippenhahn et al. (1967) and Kippenhahn and Weigert (1990).

1.2.3 Changes in the Adopted Simplifications

Reality is of course much more complicated than the up to now adopted simple assumptions. We will describe in Sect. 3 the modifications introduced by considering the effects of rotation and tides. In this section we will only make some comments about the expected problems for single stars even with negligible distortion, that have not been discussed in the evaluation of stellar structure or were assumed to be well known.

1. Opacity: Due to the complexity involved in the physics, the opacities are computed by specific codes and tabulated as a function of the chemical composition, temperature and density. Given that an evolving model changes its internal distribution of chemical elements due to nuclear burning, several tables are needed to take into account these changes. For example, non-ideal gases have to be considered for partial ionization of heavy elements and absorption by more than just hydrogenic approximations have to be taken into account.

2. Nuclear Reaction rates: Let us consider only the nuclear reactions that are important for energy generation. They are the p-p chain, and the CNO cycle, for hydrogen burning, and the triple-alpha reaction, for helium burning. If nucleosynthesis studies are to be carried out, additional reactions must be included (Fowler et al. 1975, Caughlan et al. 1985, 1988).
   The change with time of a given chemical element X_i is
$$\frac{\partial X_i}{\partial t} = -\sum_k \frac{\epsilon_k}{Q_{ik}} + \sum_j \frac{\epsilon_j}{Q_{ij}} \tag{1.28}$$
   where k represents all reactions that destroy i while j denote the reactions creating it, Q_{ik} and Q_{ij} are the involved energies, while ϵ_{ik} and ϵ_{ij} are the nuclear energy generation rates.
   If we are dealing with a convective region we have
$$\frac{\partial X_i}{\partial t} = \frac{\int_{conv} \left(-\sum_k \frac{\epsilon_k}{Q_{ik}} + \sum_j \frac{\epsilon_j}{Q_{ij}} \right) dm}{\int_{conv} dm} \tag{1.29}$$
3. Equation of state: The total pressure is the sum of the gas pressure and the radiation pressure. The ions and the electrons contribute to the gas pressure. The latter contribution depends on the degree of ionization and degeneracy. For a non ideal gas, as in the case of the opacities, the thermodynamical quantities are given in extense tables computed by specific codes. For example, for the physical conditions prevailing at the stellar cores, relativistic effects should be considered in the evaluation of velocities and pressures due to electrons. A good review on the role of the equation of state in stellar evolution can be found in Christensen-Dalsgaard and Däppen (1992).
4. Chemical composition distribution: Often the chemical profile in the interior of a stellar models depends on the nuclear rates which took place locally and on the time. However, it is possible that the chemical elements may migrate to regions of higher temperature/pressure. Such a diffusion process is slower than the nuclear time but for some cases it may be important. Rotation also may induce changes in the chemical profile (Zahn 1992).
5. The energy transport: The energy in the interior of a star may be transported by radiation, convection and conduction depending on the physical conditions. In the case of convection, there are hotter macroscopic mass elements which move upwards while the cooler ones move in the opposite direction allowing for an exchange of energy. Of course, a region where convection dominates is chemically homogeneous. The switch between radiation/convection follows the Schwarzschild criterium. Such a limit may be exceeded if we consider convective overshooting. This is due in few words to the non-zero velocity of the convective cells when they reach the surface where the Schwarzschild criterium is verified. Penetration in stable radiative regions is unavoidable, but it is only poorly known how relevant the effect is and the way to quantify it. On the other hand, for a gas of electrons which is highly degenerate, another process appear: heat conduction.

6. Mass loss/gain: A star can lose mass during its evolution through stellar winds (slow process) or through catastrophic events. Binary systems also gain/lose mass when their components achieve their Roche lobes. These loss/gain of mass obviously change the structure and future evolution of such stars. Concerning the mass loss/gain in binary systems, see de Loore (1999, this volume) and for rates of mass loss through stellar winds, see Nieuwenhuijzen & de Jager (1990).

1.3 Perturbed Stellar Structure

1.3.1 Stellar Shapes

Stars are, in general, not spherical bodies. Rotation produces a flattening at the poles while tidal effects in close binary systems produce an elongation or oblateness in the direction of the companion star. The real shape of a star which belongs to a binary system will thus be given by the combination of both effects. For those readers interested in more details on the mathematical formulation, see Kopal (1959, 1978).

The equations of hydrodynamics should be able to reproduce the behavior of stellar structure under the effects of rotation and tides. In fact, the equation of hydrostatic equilibrium (eq. 1.15) is valid for stellar configurations with both contributions, and showed that surfaces of equal density, pressure (and chemical composition), correspond to those with equal potential. The problem of obtaining the actual distorted shape of real stars is therefore reduced to the evaluation of the potential due to all forces acting on any point of the configuration.

Let us consider a generic point within the star, $P(r, \theta, \phi)$ (Fig. 1.1) attracted by a stratum comprised between r_o and r_1. The interior and the exterior potential will be given by:

$$V^{int} = G \int_{r_o}^{r_1} \frac{dm'}{\Delta}$$
$$V^{ext} = G \int_{0}^{r_o} \frac{dm'}{\Delta} \qquad (1.30)$$

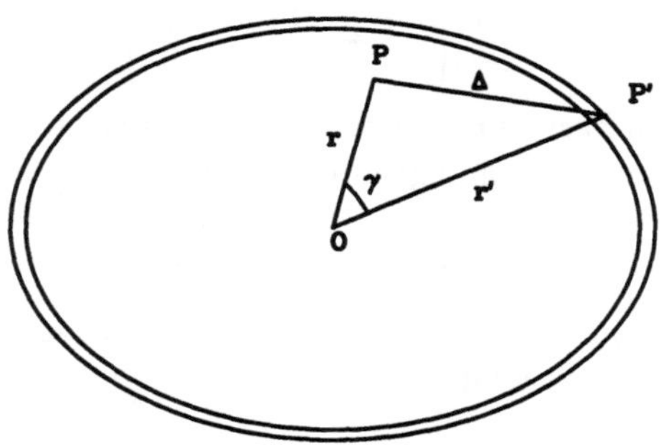

Fig. 1.1. Geometry for the determination of the potential.

where the distance between the mass element dm' and the point P is given by,

$$\Delta = \left(r'^2 + r^2 - 2rr'\cos\gamma\right)^{1/2} \tag{1.31}$$

where the angle γ between the position vectors of the mass element and the point P is,

$$\cos\gamma = \cos\theta\cos\theta' + \sin\theta\sin\theta'\cos(\phi' - \phi) \tag{1.32}$$

Using now Legendre polynomials,

$$\frac{1}{\Delta} = \frac{1}{r}\sum_{n=0}^{\infty}\left(\frac{r'}{r}\right)^n P_n(\cos\gamma) \quad \text{for } r' < r \tag{1.33}$$

$$\frac{1}{\Delta} = \frac{1}{r'}\sum_{n=0}^{\infty}\left(\frac{r}{r'}\right)^n P_n(\cos\gamma) \quad \text{for } r' > r \tag{1.34}$$

respectively, for internal and external points, the potential becomes,

$$V = \sum_{n=0}^{\infty}(r')^n V_n^{int} + \sum_{n=0}^{\infty}(r')^{-n-1}V_n^{ext} \tag{1.35}$$

where,

$$V_n^{int} = G\int_r^R\int_0^{\pi}\int_0^{2\pi}\rho r^{1-n}P_n(\cos\gamma)dr\sin\theta d\theta d\phi \qquad r' < r \tag{1.36}$$

$$V_n^{ext} = G\int_0^r\int_0^{\pi}\int_0^{2\pi}\rho r^{2+n}P_n(\cos\gamma)dr\sin\theta d\theta d\phi \qquad r' > r \tag{1.37}$$

for the potential on P of mass elements internal (external) to it in the stellar configuration.

The radius r will be a function $r(a, \theta, \phi)$ representing the equation of an equipotential surface, where a is the average radius of the equipotential. We can then write,

$$r = a\left[1 + \sum_{j=0}^{\infty}Y_j(a, \theta, \phi)\right] \tag{1.38}$$

where Y_j are spherical harmonic functions with respect to the center of mass. If $Y_j(a, \theta, \phi)$ can be factored using Legendre polynomials and an amplitude function as

$$Y_j(a, \theta, \phi) = f_j(a)P_j(\theta, \phi) \tag{1.39}$$

and we can neglect terms including Y_j^2, it is possible to use a well-known condition of orthogonality for Legendre polynomials,

$$\int_0^{\pi}\int_0^{2\pi}P_n(\cos\gamma)Y_j(a, \theta, \phi)\sin\theta d\theta d\phi = \frac{4\pi}{2j+1}Y_j(a, \theta', \phi') \tag{1.40}$$

for $j = n$, but it is null if $j \neq n$.

This allows us to change variable from r to the more convenient a, as well as the simplification of the involved integrals, to finally get,

$$V = 4\pi G \int_a^R \rho a da + \sum_{j=1}^{\infty} a^j \frac{4\pi G}{2j+1} \int_a^R \rho \frac{\partial(a^{2-j}Y_j)}{\partial a} da + \frac{Gm(a)}{r} +$$
$$+ \sum_{j=2}^{\infty} \frac{4\pi G}{(2j+1)a^{j+1}} \int_0^a \rho \frac{\partial(a^{j+3}Y_j)}{\partial a} da \qquad (1.41)$$

where we have used the mass internal to the distance a as given by,

$$m(a) = 4\pi \int_0^a \rho a^2 da \qquad (1.42)$$

and the surface value R corresponds to the first root of $\rho(a) = 0$.

The terms of the potential corresponding to $n = 0$ were extracted from the series and the term for $n = 1$ of the external potential vanishes due to symmetry (the center of mass coincides with the center of coordinates).

Let us assume now that the external disturbing potential can also be expressed in terms of Legendre polynomials as,

$$V' = \sum_{j=0}^{\infty} c_j a^j P_j(\theta, \phi) \qquad (1.43)$$

where c_j depend on the nature of the distortion. This is just a general solution of Laplace equation $\nabla^2 V' = constant$.

Since the total potential, $V + V'$, should be constant for surfaces of equal pressure or density, the coefficients corresponding to individual values of j must all cancel, i.e. from (eq. 1.41) using (eq. 1.38) and (eq. 1.42),

$$Y_j^i \int_0^a \rho a^2 da - \frac{1}{(2j+1)a^j} \int_0^a \rho \frac{\partial(a^{j+3}Y_j)}{\partial a} da - \frac{a^{j+1}}{2j+1} \int_a^R \rho \frac{\partial(a^{2-j}Y_j)}{\partial a} da =$$
$$\frac{c_j}{4\pi G} a^{j+1} P_j(\theta, \phi) \qquad (1.44)$$

which is known as Clairaut's equation. If we are dealing with equipotential surfaces, then there will be n equations, as the one above, describing the distorted stellar surfaces (due to V' through c_j).

Further derivation with respect to a of Clairaut's equation gives,

$$\left(\frac{jY_j}{a^{j+1}} + \frac{1}{a^j} \frac{\partial Y_j}{\partial a} \right) \int_0^a \rho a^2 da - \int_a^R \rho \frac{\partial(a^{2-j}Y_j)}{\partial a} da = \frac{c_j}{4\pi G}(2j+1) P_j(\theta, \phi) \qquad (1.45)$$

where, for $a = R$ the second integral becomes zero while the first one is $M/(4\pi)$. Using logarithmic derivatives we can now define,

$$\eta_j(a) = \frac{a}{Y_j} \frac{\partial Y_j}{\partial a} \qquad (1.46)$$

and the surface distortion parameter is defined as

$$\Delta_j = \frac{2j+1}{j+\eta_j(R)} \tag{1.47}$$

all of which allows us to finally write, for the spherical harmonics at the stellar surface,

$$Y_j(R,\theta,\phi) = c_j \Delta_j \frac{R^{j+1}}{GM} P_j(\theta,\phi) \tag{1.48}$$

describing the shape of the stars, although not explicitly because the harmonics are also found in the distortion parameter and, of course, c_j is still to be specified.

1.3.2 The Internal Structure

Let us now evaluate how the internal constitution of the star affects its external shape in a close binary system. A second derivation of Clairaut's equation with respect to a gives,

$$\left(a^2 \frac{\partial^2 Y_j}{\partial a^2} - j(j+1)Y_j\right) \frac{3}{a^2} \int_0^a \rho a^2 da + 6\rho \left(a \frac{\partial Y_j}{\partial a} + Y_j\right) = 0 \tag{1.49}$$

and, using the logarithmic derivatives (eq. 1.46), we can transform the second order differential equation into a first order one, which is the Radau's equation (or Clairaut-Radau),

$$a \frac{d\eta_j}{da} + 6 \frac{\rho(a)}{\overline{\rho}(a)} (\eta_j + 1) + \eta_j(\eta_j - 1) = j(j+1), \qquad (j = 2, 3, 4, \ldots) \tag{1.50}$$

with $\eta_j(0) = j - 2$ as a boundary condition, being $\overline{\rho}(a)$ the average density internal to the distance a as given by $m(a)$.

The solution of Radau's equation, for a particular density distribution, gives values of $\eta_j(R)$, and allows us to know how the internal structure $(\rho/\overline{\rho})$ affects the distorted configuration through (eq. 1.47) and (eq. 1.48). The solution is generally obtained for a given density distribution through a fourth-order Runge-Kutta numerical integration of (eq. 1.50). An example based on more realistic models is shown in Fig. 1.2.

For practical reasons, since the dependence of the dynamical behavior of the binary system is a function of $\Delta_j - 1$, the apsidal motion constants are also introduced by means of

$$k_j = \frac{1}{2}(\Delta_j - 1) = \frac{j + 1 - \eta_j(R)}{2\,(j + \eta_j(R))} \tag{1.51}$$

where, again, η_j describe the derivatives of the deviations of the equipotential surfaces from spherical symmetry and R is the radius of the configuration. The apsidal motion constants can be directly compared with observational values as we will show in Section 5. A value of $\eta_j(R) = j+1$ is obtained for configurations

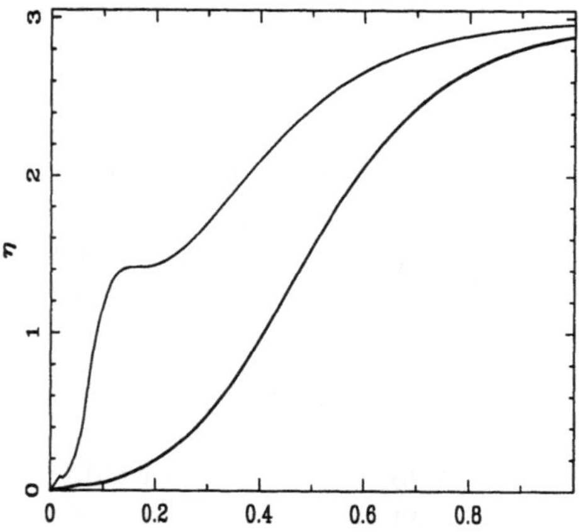

Fig. 1.2. The integrated values de $\eta(r)$ for a 10 $M_\odot$ model. The thick line denotes a ZAMS model while the thin one represents a model at the TAMS.

with infinite concentration of mass towards the center while for a homogeneous mass distribution is $\eta_j(R) = j - 2$. Thus the apsidal motion parameters, k_2, for these two extreme examples, following (eq. 1.51), will be between 0 and 0.75 (observational data show that real stars are in fact highly concentrated in mass with values of k_2 around 0.01).

1.3.3 The Perturbing Potential

We can now evaluate the significance of the distortion through the coefficients c_j in (eq. 1.43). Using the obtained derivatives of Clairaut's equation, it is possible to show that the perturbing potential, at a distance $r > R$ can be written as,

$$V'(r) = \frac{GM}{r} + \sum_{j=2}^{\infty} 2k_j \frac{R^{2j+1}}{r^{j+1}} c_j P_j(\theta, \phi) \tag{1.52}$$

Let us assume that now that, at large distances, the perturbing potential is well reproduced by that of a mass point, i.e., we can neglect the second term in the right-hand side of (eq. 1.52) and, using again Legendre polynomials,

$$V'(r) = \frac{GM}{r} = \frac{GM}{d} \sum_{j=2}^{\infty} \left(\frac{R}{d}\right)^j P_j(\lambda) \tag{1.53}$$

where d is the distance between the components, $\lambda = \cos\phi\cos\theta$ is measured over the surface of the perturbed configuration with radius R, and the total mass M corresponds to the perturbing star, since

$$r^2 = d^2 + R^2 - 2dR\lambda \tag{1.54}$$

A simple comparison of (eq. 1.53) with (eq. 1.43) shows that, c_j will be given by

$$c_j = \frac{GM}{d^{j+1}} \tag{1.55}$$

It is then easy to see that neglecting square values of the spherical harmonics implies that only terms up to $j = 4$ should be considered. On the other hand, the assumption about perturbation due to mass points is consistent with the same approximation. We thus have for the spherical harmonics representing tidal distortion,

$$Y_j^{tid} = \Delta_j \frac{M_2}{M_1} \left(\frac{R}{d}\right)^{j+1} P_j(\lambda) \tag{1.56}$$

On the other hand, if the configuration rotates rigidly with angular velocity ω around the z axis the corresponding potential (eq. 1.7) will be:

$$V' = -\frac{1}{3}\omega^2 r^2 P_2(\cos\theta) \tag{1.57}$$

as can be easily checked through the solution of Poisson's equation (eq. 1.6). Comparison of (eq. 1.57) with (eq. 1.43) immediately leads to,

$$c_2 = -\frac{1}{3}\omega^2 \tag{1.58}$$

and zero for $j \neq 2$. The corresponding rotational spherical harmonic takes then the form,

$$Y_2^{rot} = -\frac{1}{3}\Delta_2 \frac{R^3\omega^2}{GM} P_2(\cos\theta) = -\frac{1}{3}\Delta_2 (1+q) \left(\frac{R}{d}\right)^3 P_2(\cos\theta) \tag{1.59}$$

where we have used for the second part Kepler's third law and rotational synchronization with the orbital motion.

1.3.4 Application to the Stellar Surfaces

As a direct application of the mentioned equations, let us compute the projected stellar surface over the plane of the sky. Defining 3 points in the distorted stellar surface: x, in the direction of the perturbing star, z, towards the poles, and y, in the perpendicular direction to the previous points, we have a three-axis ellipsoid given by,

$$x = a\left(1 + \frac{1}{6}(1+7q)\Delta_2 \left(\frac{R}{d}\right)^3\right) \tag{1.60}$$

$$z = a\left(1 - \frac{1}{6}(2+5q)\Delta_2 \left(\frac{R}{d}\right)^3\right) \tag{1.61}$$

$$y = a\left(1 + \frac{1}{6}(1-2q)\Delta_2 \left(\frac{R}{d}\right)^3\right) \tag{1.62}$$

which correspond to the surface values of,

$$r = a\left(1 + Y_2^{rot} + Y_2^{tid}\right) \tag{1.63}$$

if only terms $j = 2$ are considered and rotational synchronization is adopted. The projection of the defined three-axis ellipsoid over the plane of the sky is obviously given, in terms of the orbital inclination i and the orbital phase β, by

$$S = \pi(x^2 z^2 \cos^2 i + z^2 y^2 \cos^2 \beta \sin^2 i + x^2 y^2 \sin^2 \beta \sin^2 i)^{1/2} \tag{1.64}$$

This would certainly allow us the computation of theoretical light curves for eclipsing binaries if the emerging flux is constant over the projected surfaces and we can treat eclipses by means of a suitable function giving the portion of star eclipsed in terms of orbital phase and inclination. We will come back to this point in next Section.

Another interesting effect of the distorted stellar shapes is gravity-darkening. In 1924, von Zeipel showed that in a distorted star the emerging flux is proportional to the local gravity, that is,

$$T_{eff}^4 \propto g^{\beta_1} \tag{1.65}$$

with $\beta_1 = 1$ for envelopes in radiative equilibrium. This can be immediately seen as a result of (eq. 1.14). Later, Lucy (1967) demonstrated that, for stars with convective envelopes, the same effect may be present but with an average value of $\beta_1 = 0.32$.

The local gravity can be approximated by,

$$g = \frac{d\Phi}{dr} \tag{1.66}$$

while the average value is given by,

$$g_o = G\frac{M}{r^2} \tag{1.67}$$

at the stellar surface. Then, for $j = 2$, it is easily found that,

$$\frac{g - g_o}{g_o} = -\sum_j \left(1 + \eta_2(R)\right) Y_2(\theta, \phi) = \left(1 - \frac{5}{\Delta_2}\right)\left(\frac{r}{a} - 1\right) \tag{1.68}$$

and a rotationally distorted star, for example, will show larger temperatures in the poles. We will return to this subject in Section 4.

1.3.5 Application to the Structure of Rotating Stars

We have seen the effects of perturbations in the shape of the stars, and how the stellar interior drives them. An extensive source on the subject can be found in Tassoul (1978). Now we would like to show some ways to obtain stellar structure models taking into account rotational perturbations in the basic set of differential equations discussed in Sect. 2.

Using only a first order approach (Kippenhahn and Thomas 1970) , we can use the total potential for rotation at any distance r within the interior of the star for $j = 2$, i.e.,

$$\Phi = \frac{Gm_\phi}{r} + \frac{1}{2}\omega^2 r^2 \sin^2\theta - \frac{4\pi G}{5r^3} P_2(\cos\theta) \int_0^a \rho \frac{d}{da}\left(a^5 f_2\right) da \tag{1.69}$$

where, m_ϕ stands for the enclosed within the equipotential surface, and in agreement with (eq. 1.38), (eq. 1.39), and (eq. 1.57),

$$r = a\,(1 - f_2 P_2(\cos\theta)) \tag{1.70}$$

$$f_2 = \frac{5\omega^2 a^3}{3Gm_\phi(2+\eta_2)} \tag{1.71}$$

where ω is the angular velocity, $\mathrm{P}_2(\cos\theta)$ is the second Legendre polynomial, a the radius of the level surface, η_2 is the logarithmic derivative of the spherical harmonic defined through Radau's equation, and the remaining symbols have their usual meaning.

The differential equations of stellar structure are then changed as follows:

$$\frac{\partial r_\phi}{\partial m_\phi} = \frac{1}{4\pi\rho r_\phi^2} \tag{1.72}$$

$$\frac{\partial p_\phi}{\partial m_\phi} = -\frac{Gm_\phi}{4\pi r_\phi^4} f_P \tag{1.73}$$

$$\frac{\partial L_\phi}{\partial m_\phi} = \epsilon - T\frac{\partial S}{\partial t} \tag{1.74}$$

$$\frac{\partial \ln T_\phi}{\partial \ln P_\phi} = \frac{3\kappa L_\phi p_\phi f_T}{16\pi ac Gm_\phi T_\phi^4 f_P} \tag{1.75}$$

where we have considered the entropy S of the stellar structure and the Schwarzschild criterion is given by:

$$\frac{\partial \ln T_\phi}{\partial \ln P_\phi} = min[\nabla_{ad}, \nabla_{rad}\frac{f_T}{f_P}] \tag{1.76}$$

where ∇_{ad} and ∇_{rad} are the spherical adiabatic and radiative gradients. Some modes of convective motions may be changed by rotation but we do not take into account here these effects and the conventional criterion of stability is used. The quantities f_P and f_T are given by

$$f_P = 4\pi r_\phi^4 \frac{1}{Gm_\phi S_\phi \overline{\overline{g^{-1}}}} \tag{1.77}$$

and,

$$f_T = \left(\frac{4\pi r_\phi^2}{S_\phi} \right)^2 \frac{1}{\overline{g}\overline{g^{-1}}} \tag{1.78}$$

where the mean values of g and g^{-1} are taken over equipotentials and S_ϕ is the integrated value of the distorted surface.

To compute f_P and f_T one has to know the relationship between a and r_ϕ. The radius of a sphere with equivalent volume is related to the radius of the level surface by

$$r_\phi^3 = a^3 \left(1 + \frac{3}{5} f_2^2 - \frac{2}{35} f_2^3 \right) \tag{1.79}$$

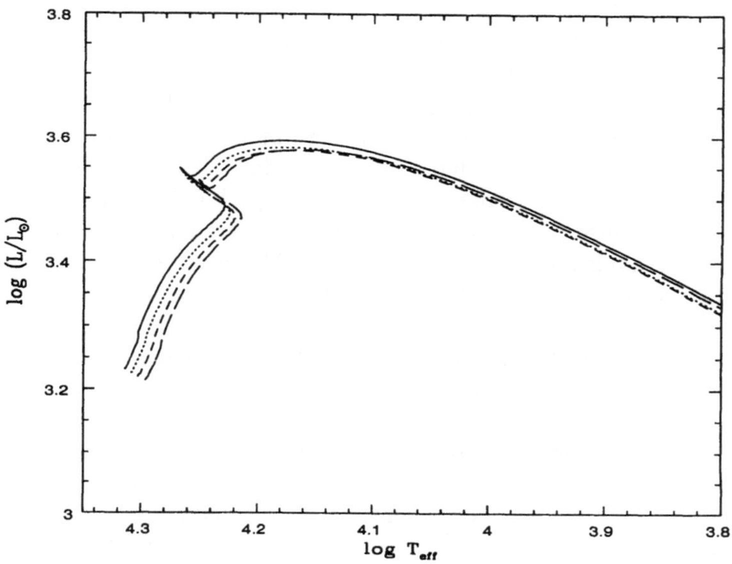

Fig. 1.3. Standard and rotating models (7 M$_\odot$). Continuous line indicates a model without rotation, long dashed rotating model with Ω_i=8.5×10$^{-5}s^{-1}$, small dashed line denotes model with Ω_i=7×10$^{-5}s^{-1}$ and dotted line represents a model with Ω_i=5×10$^{-5}s^{-1}$.

The value of a is then obtained through iteration of (eq. 1.79). Figure 1.3 shows an example of a model with and without rotation for the case of local conservation of the angular momentum (Claret 1999). The effect of luminosity and effective temperature lowering can be clearly noticed. However, the distribution of angular momentum should be determined through integration of a fourth order differential equation and the assumption of local conservation is only an approximation (Zahn 1992). Moreover, the above equations are valid only for the case of conservative potential. In the case of local conservation, as the models evolve the potential is no longer conservative and the density and the temperature are not constant on the equipotential (Meynet and Maeder 1997). For not excessively distorted configurations we do not expect large differences if a more elaborated treatment is used. Another limitation is concerned with the transport of angular momentum due to the rotationally-induced instabilities. Another way to treat the stellar configuration under the effect of rotation and tides, as well as the potential in the interstellar region between the component stars is derived from the approximation to the solution of the restricted three-body problem named after Roche (Kopal 1959, Tassoul 1978).

1.4 Stellar Atmospheres in Close Binary Systems

1.4.1 Introduction

Eclipsing binaries provide the royal road to the study of surface inhomogeneities in stellar surfaces. Eclipses are nothing else than a scan of the surface of a star by the shadow of it mate that occur once per orbit and star. We have seen that the study of the light curves of detached eclipsing binaries would have been almost straightforward if variations of light were only due to surface distortions and mutual eclipses. In fact it is not so because of the highly non linear problem involved which requires the solution through the frequency domain for a very simplified case or an inverse problem in general. This is done by the optimization of a χ^2 function derived from the comparison of actual observations with a synthetic light curve. The search for the optimum combination of parameters is done, either through a direct exploration of the phase space within physically realistic boundaries, or differential corrections.

We are not going to provide details about the problem here but it is certainly possible to check the accuracy of the distortion of the component stars plus the flux distribution over the surface of the two components to some level and favorable cases. It is obviously always important to avoid systematic errors in the synthetic light curve models in order to obtain accurate and realistic physical parameters from the light curves, i.e. relative radii, luminosity ratio, orbital inclination, etc.

The flux distribution over the surface of a star is known to be non-uniform even for single non-rotating stars. This is due to the effect known as limb darkening. But in rotating configurations, we have seen that an additional contribution is due to the effect of gravity-darkening which is also affected by mutual tides in close binaries. Finally, we have to consider in binary stars the contribution of mutual irradiation of the stellar atmospheres adding to their computation an external source of energy. Let us have a look to all of them.

1.4.2 Limb-Darkening

Limb-darkening coefficients are an important tool to study the light curves of eclipsing binaries, but also to measure stellar angular diameters, to investigate line profiles perturbed by rotation or, more recently, to study some effects of gravitational micro-lensing (Alcock et al. 1997). Of course, limb-darkening is a well known effect of the semi-transparency of the upper atmospheric layers of the stars. It is clearly visible in the Sun and expected to be present in all stars irrespective of temperature, binarity, or rotational distortion. We are mainly interested in the modifications introduced by the effects of a close companion.

Many efforts have been dedicated to compute and analyze limb-darkening coefficients for stellar atmospheres and their use in synthetic light curves in the past (Grygar 1965, Shul'berg 1973, Manduca et al. 1977) However, the basic tool used in these investigations was plane parallel model atmospheres developed for stars generally hotter than the Sun. The more simple fitting to the intensity at any point of the stellar surface is given by the well-known linear expression:

$$I(\mu) = I(1)(1 - u(1 - \mu)) \tag{1.80}$$

while the quadratic approximation can be written as,

$$I(\mu) = I(1)(1 - a(1 - \mu) - b(1 - \mu)^2) \tag{1.81}$$

and, more recently, a square root approximation was proposed (Díaz-Cordovés & Giménez 1992) as given by,

$$I(\mu) = I(1)(1 - c(1 - \mu) - d(1 - \sqrt{\mu})) \tag{1.82}$$

where u is the linear limb-darkening coefficient, a and b the quadratic coefficients, c and d the square root ones, $\mu = \cos\gamma$ (γ being the angle between the line of sight and the emergent flux), and $I(1)$ is the monochromatic specific intensity at the center of the disk.

The mathematical method used to derive limb-darkening coefficients from the model atmospheres is still a matter of discussion. Some authors prefer the least-squares fit while others adopt some kind of weighted average approach based on the conservation of the total flux (i.e, the flux computed using the integration of the intensity values should be equal to the flux using the adopted limb-darkening law). For at least the case of eclipsing binaries, we certainly prefer the least-squares method as much better representative of the actual flux distribution over the stellar surface, and this is what we basically need for the study of scans of the stellar disk. Flux conservation then should be a consequence of the good quality of the fitting and adopted law (Claret and Giménez 1990, Wade and Ruciński 1985, Van Hamme 1993, Díaz-Cordovés et al. 1995, Claret et al. 1995b)

For close binaries, an additional complication is added to the theoretical values of limb-darkening coefficients. This is the mutual irradiation of the atmospheres of the component stars which was certainly not taken into account in the models for isolated stars. Irradiation may change drastically, under certain conditions, the distribution of intensities along the stellar disk. Qualitatively, an irradiated stellar atmosphere will present a more uniform distribution of brightness. We will come back to this point later.

The least-squares method was used to fit the computed specific distribution to the three approximations given in (eq. 1.80), (eq. 1.81), and (eq. 1.82). We adopted model atmospheres by Kurucz (ATLAS 1993) with solar abundances and a mixing-length parameter $\alpha = 1.25$. These models were computed for 1221 wavelengths at 17 values of μ (1.0-0.01). The range of effective temperatures considered goes from 3500 up to 50000 K while the logarithm of the surface gravity varies between 0.0 and 5.0 (Fig. 1.4). From the observational point of view, it is interesting to perform the calculations of limb-darkening coefficients for the most frequently used photometric systems (i.e., uvby, UBV, and its extension RIJHK).

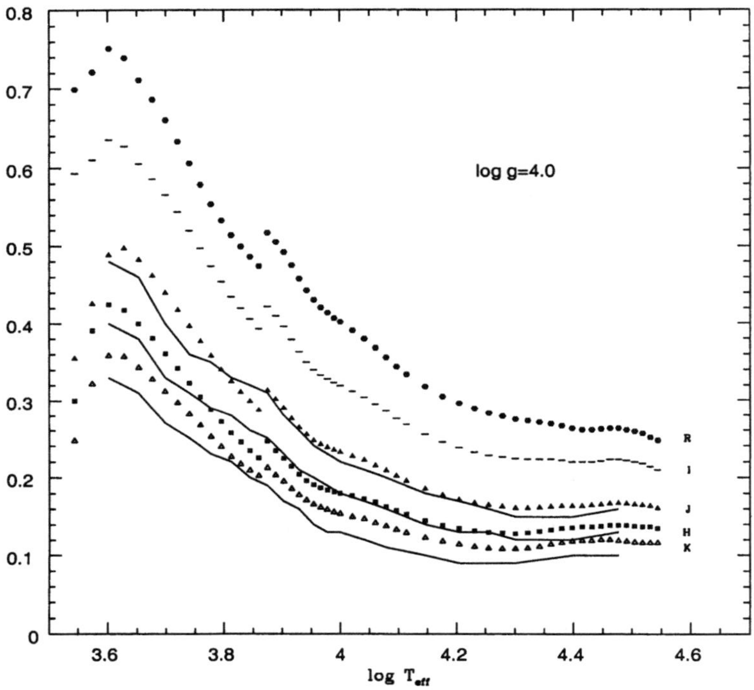

Fig. 1.4. The linear limb-darkening coefficients as a function of log T_{eff} for the filters R,I,J,H,K with Al-Naimiy's results included (solid lines). Case log g=4.0.

A discontinuity in the values of u around $\log T_{eff} = 3.87$ and a change in the general trend, for the cooler stars, around $\log T_{eff} = 3.6$, are evident (Díaz-Cordovés et al. 1995, Claret et al. 1995b). The discontinuity is certainly an effect of the theoretical model atmospheres most probably related to the treatment of convection or, at least the change from models dominated by radiative atmospheres to those with an important contribution of convection. The change of trend around $T_{eff} = 4000$ K is linked to deep convective envelopes. But to explore this behavior in more detail further models are needed in the cooler end of the main sequence, i.e., models for M-type dwarfs. New models, down to $T_{eff} = 2000$ K have been therefore used to cover such a gap (Claret 1998a). These data are important in order to study stars in the lower part of Main-Sequence as well as brown dwarfs. The results show that the change of trend is confirmed but followed soon by a reversal back to the normal behavior, i.e., higher values of limb-darkening coefficient for cooler atmospheres, around $\log T_{eff} = 3.52$ (Fig. 1.5).

Even though linear fittings are known to be unrealistic for a detailed description of the stellar flux distribution, the analysis of eclipsing binaries does not require for the time being the use of non-linear expressions. Future accurate light curve may show a very different situation. In any case, we can compare the results derived for cool stars using the model atmosphere code PHOENIX (Allard and Hauschildt 1995, Allard et al. 1997, Hauschildt et al. 1997a, 1997b) with those calculated using the intensities generated by ATLAS. For effective temperatures - $\log T_{eff} \geq 3.8$ - the PHOENIX limb-darkening coefficients compare well with the ATLAS results but they are systematically larger for all photometric bands. Such a behavior may be caused by abundances and mixing-length differences used in both codes.

For coolest models this tendency is maintained, with the exception of Strömgren u and v, and Johnson U. For these filters the tendency is the opposite and the linear coefficients computed following the ATLAS code is larger achieving a difference of about 25%. The worst situation is for the U band. In this range of effective temperatures the results using ATLAS are less accurate due to the missing of molecules.

1.4.3 Irradiated Atmospheres

It is well known that when the two components of an eclipsing binary system are sufficient close to each other, their mutual irradiation is very important in the analysis of the light curves. This is one of the proximity effects observed in the light curves and treated through the reflection coefficient or, its equivalent, the albedo. For the phase dependence of the effect, generally a Lambertian reflection is considered as the simplest approach to the real problem but over a distorted stellar surface. Most authors have adopted an albedo of 0.5 for stars with convective envelopes and 1.0 for those with radiative atmospheres. We summarize the main theoretical and observational aspects of the irradiation as follows:

- Dugan (1908) and Stebbins (1911) ⇒ first observations
- Eddington (1926) and Milne (1927) ⇒ first theoretical interpretation
- Chandrasekhar (1945, 1947, 1950) ⇒ exact solution for the grey case
- Hosokawa (1959) ⇒ albedo of 0.50 for secondaries of Algol-type
- Rucinscki (1969) ⇒ albedo is 0.50 for star with deep convective envelopes
- Miner (1966), Mc Namara (1967), Hall (1967) ⇒ general conclusion that secondaries of classical Algols present strong ultraviolet excess and lower metallicity than their mates

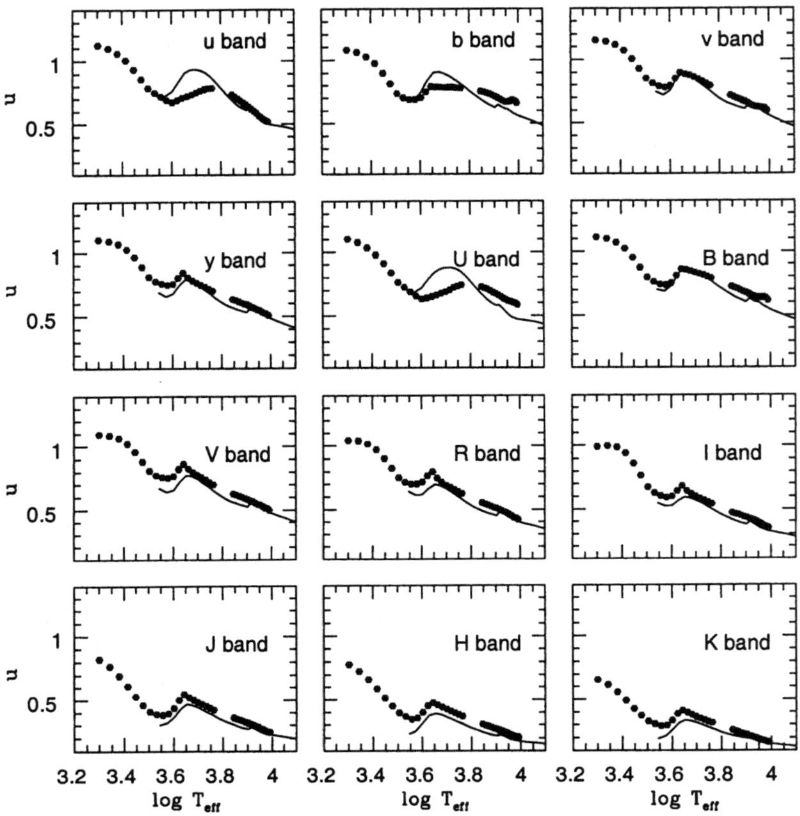

Fig. 1.5. The linear limb-darkening u computed following PHOENIX intensities (full hexagons) and those computed following ATLAS results (full lines) versus effective temperatures for log g = 5.0.

- Under the spectroscopic point of view Naftilan (1975a, 1975b, 1976), Bond (1972), Kondo and Okazaki (1980) ⇒ metal deficiency in these secondaries.

On the other hand, it has also been observationally found that, stars with irradiated atmospheres generally present ultraviolet excesses and low metal contents when compared to their companions. This of course has been discussed in the light of late-type stars with enhanced chromospheric activity due to tidally induced fast rotation. In this case, it is well known that active chromospheres produce both an ultraviolet excess and an apparent low metallicity because of the filling of absorption lines with chromospheric emissions (Hall 1989, Giménez et al. 1991). But, in close binaries, irradiation of the atmospheres may be an important if not the dominant contribution to the observed effects (Claret and Giménez 1992; Smak 1999, this volume).

Chandrasekhar (1945, 1947, 1950) was the first to to treat the external irradiation and presented an exact solution for the grey atmospheres. Later and using an ATLAS version Buerger (1969, 1972) investigated the case where the two stars, source and irradiated, are hot. Kirbiyik and Smith (1976) studied the circulation currents due to irradiation. More recently Claret and Giménez (1992) investigated the specific case of classical Algols (see this paper for more extensive references on the numerical methods used in the irradiation investigation).

The relation between the effective temperature of the models and their albedoes can be written in the form:

$$W_{bol}F_r^* = \sigma(T_H^4 - T_{eff}^4) \tag{1.83}$$

where $F_r^* = \sigma T^{*4} r^2 \nu_j$ is a radial external flux, T^* is the effective temperature of the irradiating star, r is the apparent radius of the irradiating star, ν_j is the cosine of the incidence angle, W_{bol} is the bolometric albedo, T_{eff} is the effective temperature of the non-irradiated model and T_H is the effective temperature of the irradiated model. The theoretical albedoes computed with this approach have been found to be in good agreement with the observational values. Let us have a look to the actual theoretical predictions and their possible observational effects.

Overluminosity of the Secondaries in Classical Algols. As already seen, the albedo is smaller than 1 for convective atmospheres which are expected to be valid for the secondary mass losing components of classical Algol-type binaries. These atmospheres reflect W_{bol} of the incident flux giving rise to the mentioned reflection effect. The amount $(1 - W_{bol})F^*$ is absorbed and the atmosphere should re-emit this excess of energy sooner or later. The net effect is that the stars increase their temperature and luminosity. In this way, it will appear to be overluminous and its position in the HR diagram will be displaced toward the left when compared to their normal expected positions for subgiants. The amount of the mentioned shifts depends on the conditions of the irradiation. Figure 1.6 shows the behavior of the change in the effective temperature caused by irradiation (δT) as a function of the relative incident flux. This should explain, at least partially, the observed overluminosity of the secondaries of classical Algols (Claret and Giménez 1992).

Metal deficiency in the secondaries of classical Algols. When a model atmosphere is compared to an observed spectrum the effects of irradiation are not normally taken into account. By inspecting Fig. 1.7 it can be seen that, for some regions of the spectrum, standard spectra with less metals reproduce the irradiated models better than those with the original chemical composition. This means that systematic errors may be introduced in the interpretation of

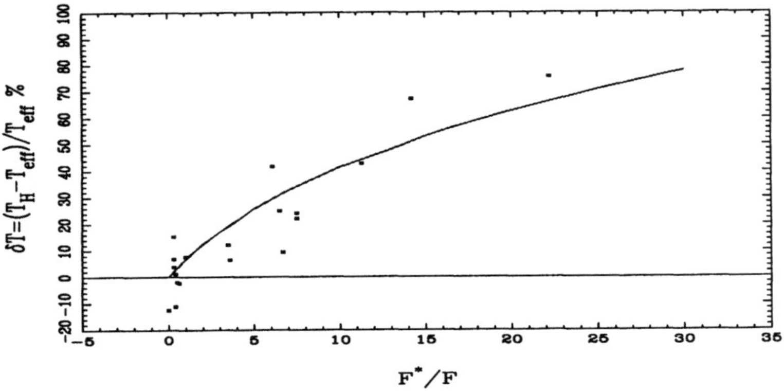

Fig. 1.6. Theoretical shift in the effective temperature due to irradiation as a function of the relative flux (continuous line). The observational data are from García (1990).

atmospheres of close binary systems. Observational data seems to actually indicate the apparent metal deficiency of the cooler components of Algol-type stars (Claret and Giménez 1992).

Limb-darkening of irradiated atmospheres. Irradiated models show limb-darkening coefficients very different from those for non-irradiated models with the same effective temperature and $\log g$. This means that the use of coefficients computed from non-irradiated model atmospheres is not the correct procedure in the case of some types of binaries like classical Algols (Claret and Giménez 1990; see Fig. 1.8). It is certainly more complicated to introduce the effect of irradiation in the prediction of theoretical coefficients for limb-darkening since we have to take into account a number of additional parameters such as the relative distance between the two stars and realistic values for the effective temperatures, consistent with the parameters of the system under study.

1.4.4 Gravity-Darkening

An approximated expression of the gravity-darkening law was derived before using the parameters which characterize the distortion of a star in a binary system. In fact, this indicates that the gravity-darkening is a structural problem rather than an atmospheric one. In spite of this fact, in such a derivation we only treated the geometric parameters rather than the physical ones. It showed though that gravity darkening is a typical consequence of distorted stars, either rotationally, tidally, or both.

Gravity darkening is certainly an important, but not always sufficiently studied, parameter in the analysis of the light curves of eclipsing binaries and in the investigation of the effects of rotation in clusters (Pérez Hernandez et al. 1999). The exponent β_1 should therefore be well known under different physical circumstances for such studies. In a stellar envelope the energy is often transported by

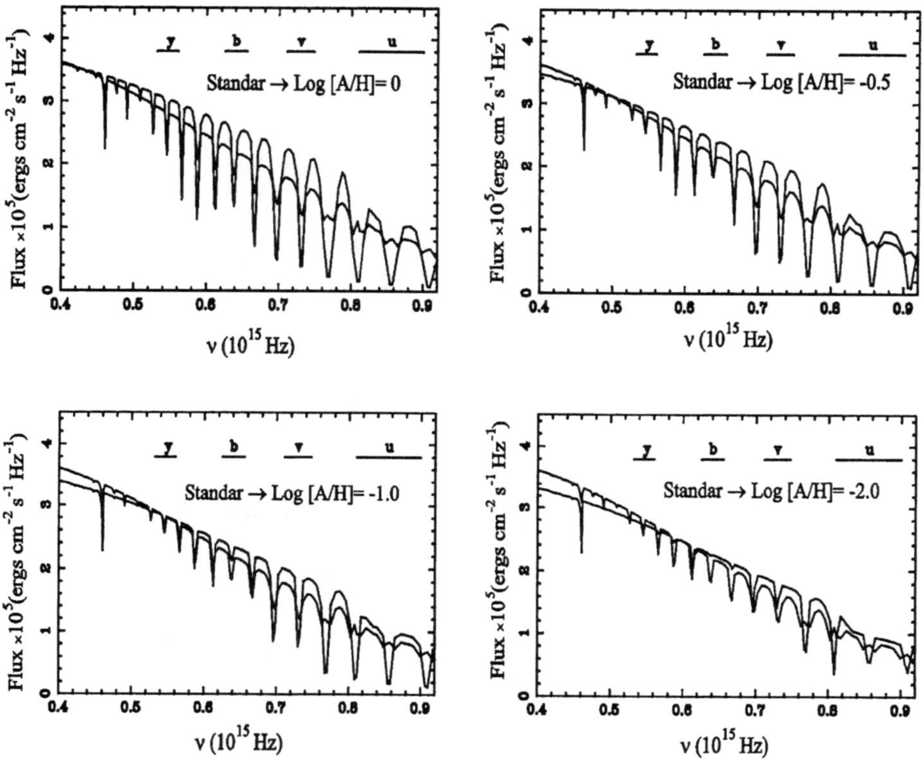

Fig. 1.7. Comparison between irradiated and standard spectra. The thick line represents the irradiated spectra computed using typical values of classical Algols: T^*=12000 K, T_{eff}=5300 K, T_H=5890 K, r=0.22, log g=3.44, ν=0.67 and log [A/H]=0.0. The thin line corresponds to the standard one with the same effective temperature and log g but adopting different values of log [A/H]. The Strömgren filters are indicated for reference.

convection and/or radiation. One does not expect that the values of β_1 would be the same for the two cases. In fact, for stars in strict radiative equilibrium (pseudo-barotrope), von Zeipel demonstrated that the variation of brightness over the surface is proportional to the effective gravity. In mathematical form,

$$F = -\frac{4acT^3}{3\kappa\rho}\frac{dT}{d\Phi}g^{\beta_1} \tag{1.84}$$

where $\beta_1 = 1.0$, g is the local gravity, Φ is the total potential and the remaining variables have their usual meaning.

From the point of view of convective envelopes, Lucy could find that the same relation (eq. 1.84) was valid but using $\beta_1 = 0.32$ (average value). This jump of β_1 was, and still is, very uncomfortable from the observational and theoretical points of view. The two processes of energy transport can even exist simultaneously in a given stellar envelope. Therefore, one would expect a less

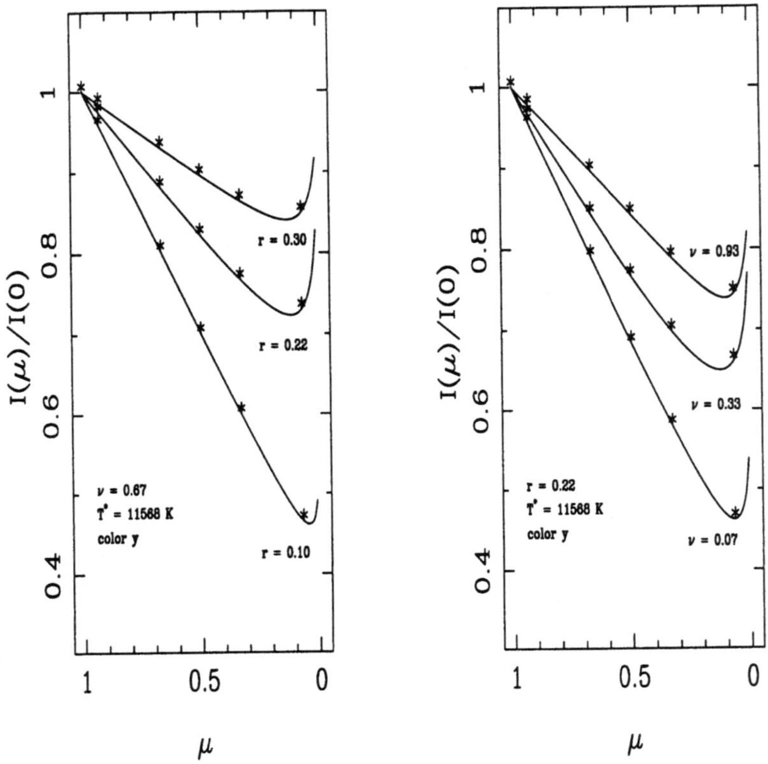

Fig. 1.8. Effect of the distance of the irradiating source and of the angle of incidence of the irradiating source on the emergent intensity distribution. Asterisks indicate the six μ points used previously to this work (no limb-brightening is detected). Note that only with the introduction of 12 additional μ points (continuous line) it is possible to discriminate if there is limb-darkening or limb-brightening (Claret 2000b).

abrupt change. In order to advance a little further, a more sophisticated method, based on the properties of the stellar interior models, was developed (Claret 1995, 1998b, 2000a). In this way the gravity-darkening exponent is not only a function of the effective temperature as until now accepted but also of the mass, radius and evolutionary status.

An interesting behavior of β_1 is expected for moderately massive models whose tracks cross over the boundary between radiative and convective equilibrium. A 2 $M_\odot$ model was selected to represent this situation. In Fig. 1.9 one can see how β_1 depends on the effective temperature while on the upper right corner of the same figure we plot the usual HR Diagram. As expected, the gravity-darkening exponent for Main-Sequence models is 1.0 confirming the prediction by von Zeipel and it begins to decrease until the point indicated by letter B which is the boundary between both competing regimes of transport of energy. During contraction (B-C) the effective temperature increases, which is reflected by an increase of β_1. After C the effective temperature decreases as well as the

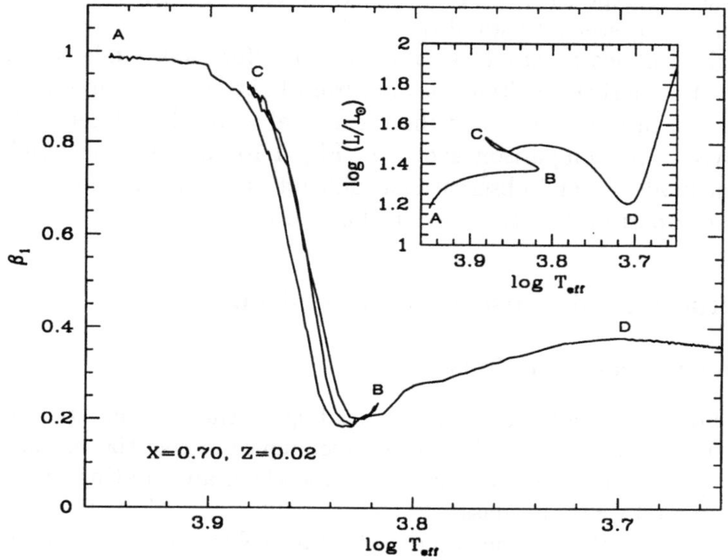

Fig. 1.9. The gravity-darkening exponent β_1 for a 2 $M_{\odot}$ model. In the upper right corner the corresponding HR Diagram is shown.

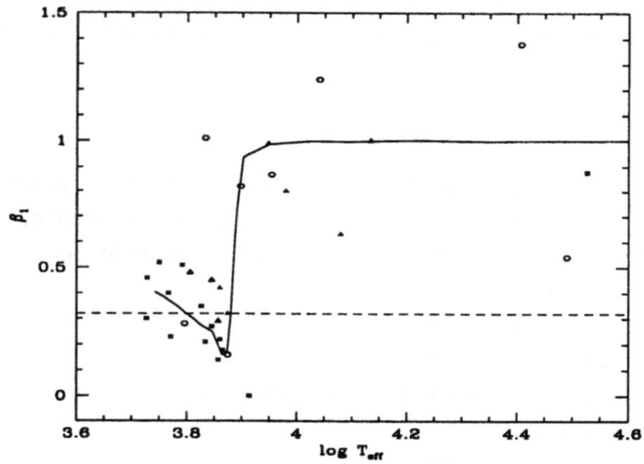

Fig. 1.10. Observational values of the gravity-darkening exponent (Rafert & Twigg 1980) as a function of the effective temperatures. Open circles represent detached systems, full triangles semi-detached systems, full squares denotes contact systems while open triangles denotes other systems. The full line represent the theoretical β_1 for homogeneous models.

gravity-darkening exponent. For very deep convective envelopes, for stars at the red giant phase, β_1 seems to stabilize at 0.3.

Although it is not a easy task, a significant effort was made by Rafert & Twigg (1980) to derive gravity-darkening from light curves analysis. Their results are shown in Fig. 1.10 where the theoretical values by Claret (1998b) are also displayed. The comparison shown in Fig. 1.10 can be considered within the inherent modeling and observational difficulties. Evolutive effects may also change the comparison (see the previous figure).

1.5 Dynamics of Close Binary Systems

1.5.1 The Secular Equations

We have obtained already the equation leading to the internal structure constants for theoretical models. Now, using techniques of celestial mechanics, we shall derived them for two real distorted stars which are rotating and orbiting around a common center of mass.

It is well known that, if the component stars behave like mass points, there is no mass exchange or loss, and the system is isolated, their dynamical behavior is given by the solution of the equations:

$$\frac{d^2\boldsymbol{r}}{dt^2} + \frac{G(m_1+m_2)}{r^3}\boldsymbol{r} = 0 \tag{1.85}$$

where r is the radius vector. The solution for the angular variation of the radial component is,

$$\frac{d^2u}{d\phi^2} + u = \frac{GM}{j^2} \tag{1.86}$$

where $M = m_1 + m_2$ and j stands for the specific angular momentum $r^2 d\phi/dt$ and can be easily expressed by means of Kepler's laws. Moreover, three second-order differential equations define six constants of integration which are known as the orbital parameters:

Ω - the longitude of the ascending node
i - the inclination of the orbit
a - the semi-major axis of the relative orbit
e - the orbital eccentricity
ω - the longitude of the periastron
T_0 - the time of passage through periastron.

Note that the orbital period is not one of them because of its implicit definition through the other parameters. It is also well known that, working with the above equations, the relative motion of one star around its mate is given by the ellipse equation:

$$r = \frac{a(1-e^2)}{1+e\cos v} \tag{1.87}$$

where v is the true anomaly.

Let us now consider distorted stars, which can not be reduced to mass points, but keeping the assumptions about no mass exchange/loss and system isolation from external gravitational potentials. Then, if the perturbation function can be written in such a form that the total potential,

$$V = G\frac{m_1 m_2}{r} + \frac{m_1 m_2}{m_1 + m_2} R_{12} \tag{1.88}$$

the center of mass will move with constant velocity and the dynamical equations of the system will be:

$$\frac{d^2x}{dt^2} + \frac{G(m_1+m_2)}{r^3}x = \frac{\partial R_{12}}{\partial x} \tag{1.89}$$

$$\frac{d^2y}{dt^2} + \frac{G(m_1+m_2)}{r^3}y = \frac{\partial R_{12}}{\partial y} \tag{1.90}$$

$$\frac{d^2z}{dt^2} + \frac{G(m_1+m_2)}{r^3}z = \frac{\partial R_{12}}{\partial z} \tag{1.91}$$

which can not be solved in a closed form.

For a non negligible, but small, disturbing function R_{12}, we can use the well-known perturbation method leading to Lagrange equations which allow us to write the same solution of the unperturbed problem but showing some variation with time of the six orbital parameters.

Let us now re-define some of the orbital "constants" in such a way that,

$$\varpi \equiv \omega + \Omega \tag{1.92}$$

$$u \equiv \omega + v \tag{1.93}$$

$$\epsilon \equiv \varpi - nT_o \tag{1.94}$$

We have then:

$$\frac{1}{An}\frac{d\Omega}{dt} = \frac{1}{\sqrt{1-e^2}\sin i}\frac{\partial R_{12}}{\partial i} \tag{1.95}$$

$$\frac{1}{An}\frac{di}{dt} = \frac{1}{\sqrt{1-e^2}\sin i}\frac{\partial R_{12}}{\partial \Omega} - \frac{\tan\frac{i}{2}}{\sqrt{1-e^2}}\left(\frac{\partial R_{12}}{\partial \omega} + \frac{\partial R_{12}}{\partial \epsilon}\right) \tag{1.96}$$

$$\frac{1}{An}\frac{dA}{dt} = 2A\frac{\partial R_{12}}{\partial \epsilon} \tag{1.97}$$

$$\frac{1}{An}\frac{de}{dt} = -\sqrt{1-e^2}\left(\frac{1-\sqrt{1-e^2}}{e}\frac{\partial R_{12}}{\partial \epsilon} + \frac{1}{e}\frac{\partial R_{12}}{\partial \omega}\right) \tag{1.98}$$

$$\frac{1}{An}\frac{d\varpi}{dt} = \frac{\tan\frac{i}{2}}{\sqrt{1-e^2}}\frac{\partial R_{12}}{\partial i} + \frac{\sqrt{1-e^2}}{e}\frac{\partial R_{12}}{\partial e} \tag{1.99}$$

$$\frac{1}{An}\frac{d\epsilon}{dt} = \frac{\tan\frac{i}{2}}{\sqrt{1-e^2}}\frac{\partial R_{12}}{\partial i} - \sqrt{1-e^2}\frac{1-\sqrt{1-e^2}}{e}\frac{\partial R_{12}}{\partial e} - 2A\frac{\partial R_{12}}{\partial A} \tag{1.100}$$

If we now consider R to represent the component acting in the direction of the radius vector of R_{12}, S the perpendicular to the radius vector in the orbital plane and W that perpendicular to the orbital plane, the equations above can be re-written as:

$$\frac{d\Omega}{dt} = \frac{W \sin u}{nA^2\sqrt{1-e^2}\sin i} \tag{1.101}$$

$$\frac{di}{dt} = \frac{W \cos u}{nA^2\sqrt{1-e^2}} \tag{1.102}$$

$$\frac{dA}{dt} = \frac{2}{n\sqrt{1-e^2}}\left(\frac{R\sin v + SA(1-e^2)}{r}\right) \tag{1.103}$$

$$\frac{de}{dt} = \frac{\sqrt{1-e^2}}{nA}(R\sin v + (\cos v + \cos E)S) \tag{1.104}$$

$$\frac{d\varpi}{dt} = \frac{\sqrt{1-e^2}}{nAe}\left(-R\cos v + \left[1 + \frac{r}{A(1-e^2)}\right] S\sin v\right) + \frac{W\sin u}{n\sqrt{1-e^2}A^2}\tan\frac{i}{2} \tag{1.105}$$

$$\frac{d\epsilon}{dt} = -\frac{1}{nA}\left(\frac{2r}{A} - \frac{1-e^2}{e}\cos v\right)R - \frac{1-e^2}{nAe}\left[1 + \frac{r}{(1-e^2)}\right] S\sin v \tag{1.106}$$

Now the definition of the orbital period will be different depending on the orbital parameter adopted to measure it. If we adopt the simplifications for perturbed configurations of Sect. 2, the perturbing function R_{12} can be expressed as

$$R_{12} = R_1 + R_2 \tag{1.107}$$

and, $R_i = R_{i,rot} + R_{i,tid}$ being the individual terms for rotation and tides,

$$R_i^{rot} = -\frac{2\mu\omega_i^2 R_i^5 k_{2i}}{3Gm_i r^3} P_2(\nu_i^{'}) \tag{1.108}$$

and

$$R_i^{tid} = \frac{\mu m_{3-i}}{m_i}\sum_{j=2}^{4}\frac{k_{ji}R_i^{2j+1}}{r^{j+1}r_\epsilon^{j+1}} P_j(\lambda_i^{''}) \tag{1.109}$$

where $\mu = \mathrm{G}(\mathrm{m}_1 + \mathrm{m}_2)$, ω_i is the angular velocity of the component i, R_i is the mean radius and k_{ji} are the internal structure constants while $\nu_i^{'}$ is the cosine of the angle between the radius-vector and the rotation axis and $\lambda_i^{''}$ is the cosine of the angle between the direction of the tidal distortion and the radius-vector.

In the case of viscous stars, producing tidal lag, the radius-vector in the direction of the lagging tide is,

$$r_\epsilon = \frac{A(1-e^2)}{1 + e\cos(v-\epsilon)} \tag{1.110}$$

otherwise, $r = r_\epsilon$.

Using the contributions for rotation and tides, we can now obtain the perturbation function in the direction of the above defined directions. After the corresponding differentiations, the components of the rotational contribution are expressed as,

$$R = \frac{2\mu}{G}\sum_{i=1}^{2}\frac{k_{2i}\omega_i^2 R_i^5}{m_i r^4}P_2(\nu_1^{''}) \tag{1.111}$$

$$S = -\frac{2\mu}{G}\sum_{i=1}^{2}\frac{k_{2i}\omega_i^2 R_i^5}{m_i r^4}\nu_1^{''}\nu_2^{''} \tag{1.112}$$

$$W = \frac{2\mu}{G}\sum_{i=1}^{2}\frac{k_{2i}\omega_i^2 R_i^5}{m_i r^4}\nu_1^{''}\nu_3^{''} \tag{1.113}$$

where the $\nu^{''}$ are the partial derivatives of ν' with respect to the three angles described before.

For tidal distortion, if there is no lag, the components S and W become zero due to symmetry and the component R is

$$[R\cos\nu] = -2\mu\sum_{i=1}^{2}\frac{m_{3-i}}{m_i}\sum_{j=2}^{4}(j+1)k_{ji}X_o^{-(2j+3),1}\frac{R_i^{2j+1}}{A^{2j+3}} \tag{1.114}$$

where we have averaged the R function over the orbit using the well known expressions of Celestial Mechanics, defining the zero-order Hansen coefficients,

$$\frac{1}{P}\int_0^P\left(\frac{r}{A}\right)^n\cos(mvt)dt = \binom{-n-2}{m}\left(\frac{e}{2}\right)^m(1-e^2)^{\frac{n+3}{2}}$$
$$F\left(\frac{m+n+2}{2},\frac{m+n+3}{2},m+1,e^2\right) \tag{1.115}$$

where the hypergeometric functions have positive values for $m+n+2 \leq 0$ or zero otherwise. Similarly,

$$\frac{1}{P}\int_0^P\left(\frac{r}{A}\right)^n\sin(mvt)dt = 0 \tag{1.116}$$

for any integral value of m and n.

We can now see that the orbital parameter variations are restricted to the orbital plane if rotation is coplanar. No lagging tides have only contributions to R and the rotational contributions to W vanish for rotational axes perpendicular to the orbital plane so that (1.101) and (1.102) become,

$$\frac{d\Omega}{dt} = 0 \tag{1.117}$$

$$\frac{di}{dt} = 0 \tag{1.118}$$

Furthermore, for the parameters in the orbital plane, we reduce the equations of Lagrange to,

$$\frac{dA}{dt} = \frac{2}{n\sqrt{1-e^2}} \left(\frac{Rersinv + SA(1-e^2)}{r} \right) \tag{1.119}$$

$$\frac{de}{dt} = \frac{\sqrt{1-e^2}}{nA} (R \sin v) \tag{1.120}$$

$$\frac{d\varpi}{dt} = \frac{\sqrt{1-e^2}}{nAe} (-R \cos v) \tag{1.121}$$

$$\frac{d\epsilon}{dt} = -\frac{1}{nA} \left(\frac{2r}{A} - \frac{1-e^2}{e} \cos v \right) R \tag{1.122}$$

It is therefore evident that, in this case, also the orbital eccentricity and separation will remain constant because of the $\sin v$ dependence.

Taking into account the properties of the coefficients, with no dissipative forces, even for no coplanar rotation,

$$\frac{dA}{dt} = \frac{de}{dt} = 0 \tag{1.123}$$

Finally, the rate of variation with time of the longitude of periastron is

$$\frac{d\varpi}{dt} = \sum_{i=1}^{2} \frac{\omega_i^2 k_{2i}}{2\pi \overline{\rho}_i} a_i^2 \left(1 - \frac{3}{2} \sin^2(\theta_i + i) - \frac{1}{2} \sin 2(\theta_i + i) \tan \frac{i}{2} \right)$$
$$\frac{3n}{2(1-e^2)} \tag{1.124}$$

where $a_i = R_i/A$. Note the dependence on the angular arguments θ_i y i. If the equatorial plane does coincides with the orbital plane they are null. However, if this is not the case, there may be a recession of the line of apses.

Concerning the tidal contribution, the Lagrange equations give the following expression:

$$\frac{d\varpi}{dt} = 2\mu\sqrt{1-e^2} \frac{1}{neA^3} \sum_{i=1}^{2} \frac{m_{3-i}}{m_i} \sum_{j=2}^{4} (j+1) k_{ji} X_o^{-(2j+3),1} a_i^{2j+1} \tag{1.125}$$

The total rate of the advance of the periastron is the sum of both contributions. An useful parameter, the rate per cycle or orbital period, will be given by,

$$\frac{\Delta\varpi}{2\pi} = \frac{P}{U} = \frac{1}{n} \left((\frac{d\varpi}{dt})_{rot} + (\frac{d\varpi}{dt})_{tidal} \right) \tag{1.126}$$

where U is the period of revolution of the line of apses.

With the exception of (k_{ji}), all parameters in the above equation can be independently measured and thus the weighted average of the internal structure constants can be empirically derived for individual binary systems.

1.5.2 Relativistic Effects

There are two assumptions in what we have discussed in the previous Sections: that the binary system is isolated from external potentials and that Newton's law of Gravitation is valid. The first one has to be modified for binaries with a third body or moving through a resisting medium, while the second has to be revised under the light of the more realistic theory of General Relativity (Giménez 1985). In this latter case, the displacement does not depend on the rotational and tidal distortions and should be added to the classical Newtonian term. In fact it is found that the change in position of the periastron per orbit is given by (Levi-Civita 1937),

$$\delta\omega = \frac{6\pi GM}{ac^2(1-e^2)} \tag{1.127}$$

where M denotes the total mass of the system and if U' denotes the period of revolution of the line of the apses, we have,

$$\frac{P}{U'} = 6.35 \times 10^{-6} \frac{m_1 + m_2}{A(1-e^2)} \tag{1.128}$$

provided that the semi-major axis and masses are given in solar units.

It is interesting to note that the expression for the relativistic apsidal motion can also be derived using the Lagrange perturbation equations if the appropriate correction to the Newtonian potential, due to relativistic effects, is introduced as R_{12}.

1.5.3 Effects of a Third Body and Interstellar Medium

The other possible corrections come from the fact that the binary system may not be completely isolated. The presence of a third component, with period P' perturbs the orbit of a close binary system, with period P and one of the sensitive elements is the longitude of the periastron. The induced apsidal motion, U', for coplanar orbits and small eccentricities can be approximated by (Martynov 1948),

$$\frac{P}{U'} = \frac{3}{8}\frac{m_3}{m_1+m_2+m_3}\left(\frac{P}{P'}\right)^2 + \frac{225}{32}\frac{m_3^2}{(m_1+m_2+m_3)^2}\left(\frac{P}{P'}\right)^3 \tag{1.129}$$

In general, if we consider that the two orbits are eccentric e, e' and showing an inclination angle I we have,

$$\frac{P}{U'} = 2a\left(1 - \frac{e^2}{2} + \frac{3}{2}e'^2 - 2\tan^2 I\right) + 50a^2 \tag{1.130}$$

where,

$$a = \frac{3}{8}\frac{m_3}{m_1+m_2+m_3}\left(\frac{P}{P'}\right)^2 (1-e^2)^{-3/2} \tag{1.131}$$

On the other hand, the line of the nodes also precesses with a period U''

$$\frac{P}{U''} = 2a\left(1 + 2e^2 + \frac{3}{2}e'^2 - \frac{1}{2}\tan^2 I\right) - 2a^2 \quad (1.132)$$

Another effect that may change the rate of advance of periastron is that of a viscous medium. The resistance itself has no influence on that rate but the mutual attraction is changed. Indeed, there appears a recession of the apsidal motion given by (Hadjidemetriou, 1967)

$$\frac{P}{U'''} = -\frac{GP^2\sigma}{2\pi} \quad (1.133)$$

where U''' is the period of revolution of the line of the apside due to this effect and σ stands for the density of the medium. Average interstellar densities though imply that this effect should be in general a negligible contribution.

1.5.4 Apsidal Motion: Comparison with Observations

Classical systems. During several decades a discrepancy between theory and observation was reported concerning apsidal motion: the theoretical models seems to be less mass concentrated than real stars (Schwarszchild 1958, Kopal 1965, Giménez 1981). As the apsidal motion period depends on the relative radii as r^{-5} only systems with accurate absolute dimensions can be used to compare with the observations. Furthermore, the stellar models available some time ago were not accurate enough to carried out a significant comparison. Many attempts were then made in order to decrease the above commented discrepancy (see Claret and Giménez 1993 for an extense list of contributions of many authors to the field).

Stellar masses and radii are now-a-days available in an amount statistically significant (Andersen 1991) and some of them present apsidal motion. On the other hand, stellar models also improved due mainly to new opacities. The two conditions simultaneously mean that a realistic comparison between observation and theory is possible.

In order to carried out such a task, let us first introduce the appropriate equations for the apsidal motion test.

$$\frac{P}{U} = c_{21}k_{21} + c_{22}k_{22} \quad (1.134)$$

where k_{2i} is the apsidal motion constant for the component i.

$$c_{2i} = \left[\left(\frac{\Omega_i}{\omega_K}\right)^2 \left(1 + \frac{m_{3-i}}{m_i}\right) f(e) + \frac{15 m_{3-i}}{m_i} g(e)\right] \left(\frac{R_i}{A}\right)^5 \quad (1.135)$$

where the auxiliary functions f(e) and g(e) are given by

$$f(e) = (1 - e^2)^{-2} \quad (1.136)$$

$$g(e) = \frac{(8 + 12e^2 + e^4) f(e)^{2.5}}{8} \tag{1.137}$$

In the above expressions m is the mass of the component, R_i the stellar radius, A the semi-major axis, Ω_i is the angular velocity of the component i, e is the eccentricity and ω_K is the keplerian angular velocity.

The first term in (eq. 1.135) is the rotational contribution while the second one concerns to tidal distortions. The mean value for the internal constant structure that will be compared with those derived from theoretical models is given by:

$$\overline{k}_{2obs} = \frac{1}{360\,(c_{21} + c_{22})} \dot{\omega}_{obs} \tag{1.138}$$

if $\dot{\omega}_{obs}$, the rate of variation of the periastron position, is given in degrees/cycle. This last value should be corrected by the relativistic contribution (Section 5.2). The theoretical mean value is

$$\overline{k}_{2theo} = \frac{c_{21} k_{21theo} + c_{22} k_{22theo}}{c_{21} + c_{22}} \tag{1.139}$$

where k_{21theo} and k_{22theo} are the theoretical apsidal motion constants for the primary and secondary, respectively. Such values are inferred from a previous comparison with the absolute dimensions.

Before to perform the apsidal motion test, as a first step we have to check if the models are able to reproduce some basic stellar parameters. The constraints are

- a good agreement between the observed effective temperatures of each of the component stars and the theoretical values given by the stellar models
- a common age for the two components

The first of these requirements is presented in Fig. 1.11 where we plot the observed and theoretical effective temperature for a sample of stars showing apsidal motion. Figure 1.12 shows the evolutionary ages for both components of each system. Of course, slight variations in the chemical composition may improve even more the aspect of these pictures. Finally, Fig. 1.13 displays the comparison between the observed and predicted apsidal motion rates. The old systematic trend, models less mass concentrated than real stars, is no longer detected.

Relativistic systems. The eclipsing binaries which present apsidal motion are also useful to test the predictions of the General Relativity - GR - for the periastron advance (Giménez 1985, Giménez & Scaltriti 1982). It was shown in the previous section that using new opacity calculations, core overshooting, rotation, improved orbital elements and recent apsidal motion rates the theoretical predictions are in good agreement with observations. The old problem, that real stars seemed to be more mass concentrated than predicted by theory, was solved or at least minored. As we did not know a priori which was the cause for these

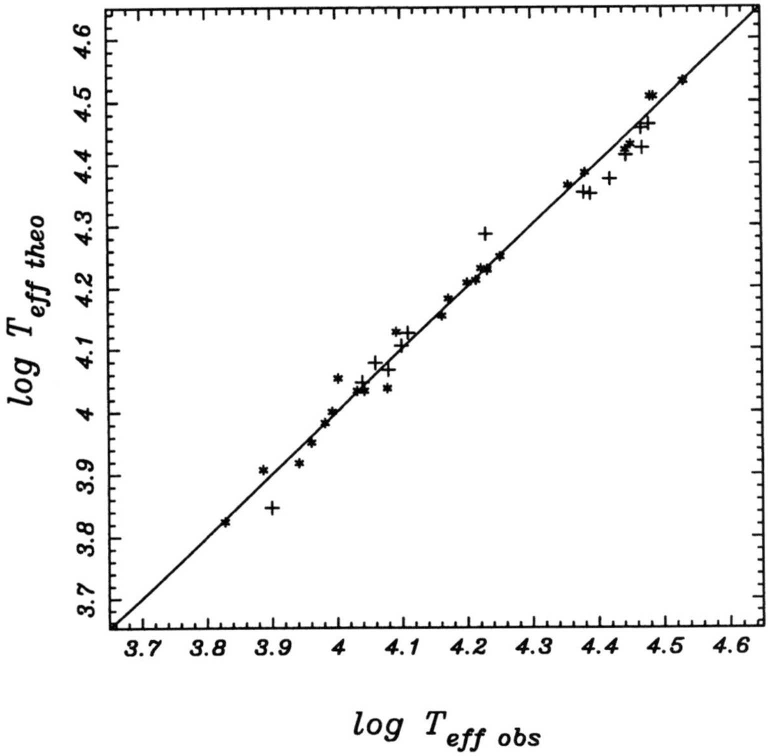

Fig. 1.11. Comparison between observed and theoretical effective temperatures. Models with core overshooting and mass loss

discrepancies we have separated the systems presenting high relativistic contributions in order to avoid these disagreements with the theory to be attributed to relativistic effects. As the name indicates, the relativistic systems are those for which the advance of the periastron predicted by the GR is comparable with the classical contribution. Moreover, the separation in two classes is justified given that the observations for some relativistic systems, DI Her and AS Cam, do not seem to be in agreement with GR predictions. An alternative theory for the periastron shift was presented by Moffat (1984, 1989) and, in principle, it was able to explain such systems. However, such theory presents some problems which will be discussed.

About 10 systems with significant relativistic contribution were selected to test the validity of the current stellar models and the predictions of GR (Claret 1998c). These systems were submitted to the same tests as described in the previous Section. The results can be seen in Fig. 1.14. An analysis of this figure indicates that the predictions by the General Relativity and the new stellar models are able to explain the shift in the periastron position. On the other

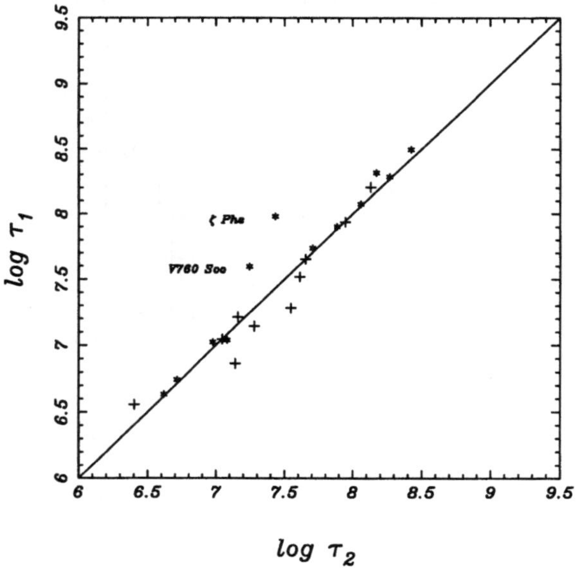

Fig. 1.12. Comparison between evolutionary ages derived for each of the component stars. Models with core overshooting and mass loss.

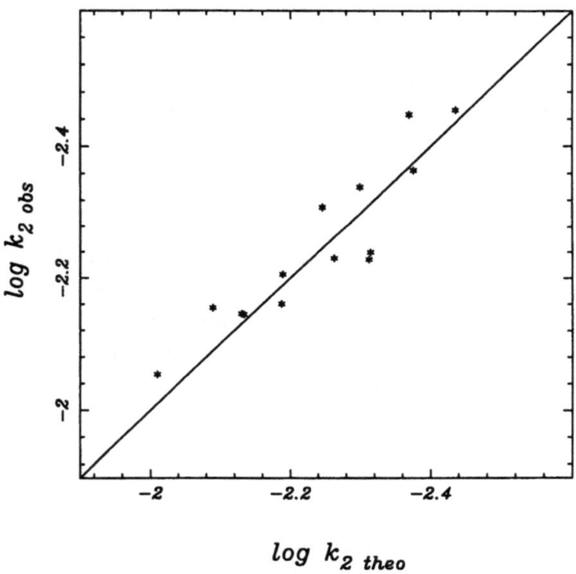

Fig. 1.13. Observed versus predicted average values of log k_2 using models with core overshooting, mass loss corrected for the effect of rotation.

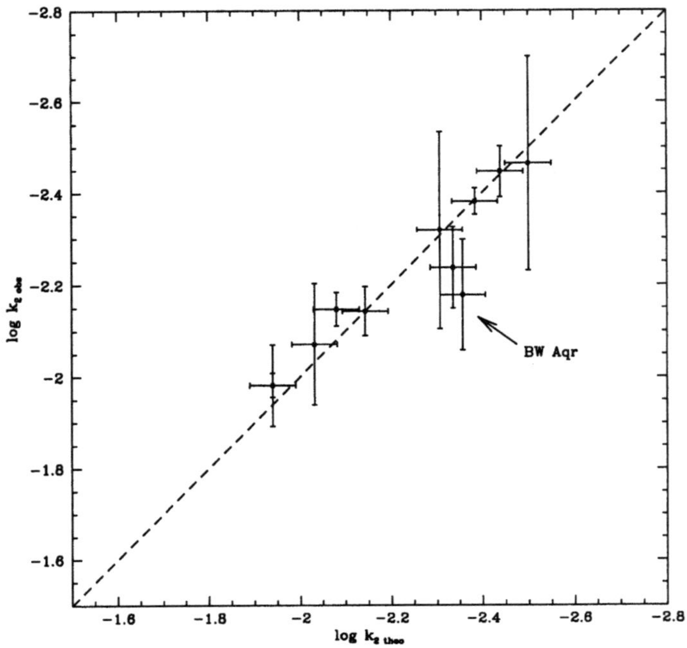

Fig. 1.14. Theoretical and observed apsidal motion constants for relativistic systems. The shift in the periastron was computed correcting $\dot{\omega}_{obs}$ by the prediction of the General Relativity

hand, the two versions of the non-symmetrical theory of gravitation by Moffat are not supported by the observations.

1.5.5 Tidal Evolution

As previously seen, we assume that the stars are in hydrostatic equilibrium. In binaries, however, the angular velocities Ω_1 and Ω_2 may be different from the orbital velocity ω and the bulges due to tides are not in phase with the line joining both stars. It can be shown that for low viscosity the angle α (Fig. 1.15) is related with the tidal frequency σ_{lm} through

$$\alpha = \frac{R^3}{t_f G m} \sigma_{lm} \tag{1.140}$$

where t_f is the friction time.

Introducing (eq. 1.140) in the secular equations we shall have the temporal derivatives of the eccentricity, angular velocity and of the semi-major axis. These equations are coupled but if we assume that the system is near to synchronism and low eccentricity, the equations are decoupled. We can define the time scales for synchronization and circularization,

$$\frac{1}{\tau_{sync}} = \frac{6k_2}{t_f} q^2 \frac{mR^2}{I} \left(\frac{R}{A}\right)^6 \tag{1.141}$$

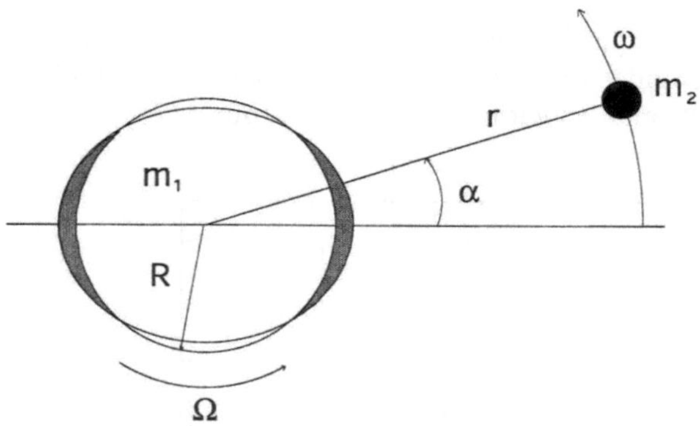

Fig. 1.15. Tides induced in the primary component.

$$\frac{1}{\tau_{cir}} = \frac{21k_2}{t_f} q(q+1)\left(\frac{R}{A}\right)^8 \tag{1.142}$$

where k_2 is the apsidal motion constant, I is the moment of inertia, q the mass ratio, A the semi major axis and R is the stellar radius. These equations hold for stars with convective envelopes where turbulent dissipation is efficient (Zahn 1966). The convective friction time t_f is given by

$$t_f = \left[\frac{mR^2}{L}\right]^{1/3} \tag{1.143}$$

where L is the star luminosity.

For more massive stars, with convective cores and envelopes in radiative equilibrium, the dissipation mechanism is due to radiative damping. The g-modes can resonate with the periodic tidal potential. If there is a departure from synchronism the gravity waves produced by the convective core will be damped in the surface zone since the dissipation time is shorter than the tidal period. The times scales are given by (Zahn 1975)

$$\frac{1}{\tau_{sync}} = 5\left(\frac{Gm}{R^3}\right)^{1/2} q^2(1+q)^{5/6} E_2 \frac{mR^2}{I}\left(\frac{R}{A}\right)^{17/2} \tag{1.144}$$

$$\frac{1}{\tau_{circ}} = \frac{21}{2}\left(\frac{Gm}{R^3}\right)^{1/2} q(1+q)^{11/6} E_2 \left(\frac{R}{A}\right)^{21/2} \tag{1.145}$$

where E_2 is related with the dynamic tidal contribution to the total perturbed potential. E_2 is similar to k_2 but it is much more dependent on the stellar structure than the apsidal motion constant. This parameter, E_2, is computed numerically given the complexity of the differential equations which must be solved (Claret and Cunha 1997).

All astrophysical parameters required by the set of differential equations, including evolutionary age, were obtained by comparing the theoretical models mentioned directly for each star of the sample. After the integration of the above differential equations we are able to compare the theory with the levels of synchronism and circularization. It is shown in Fig. 1.16 the observed and theoretical rotational velocities at the periastron.

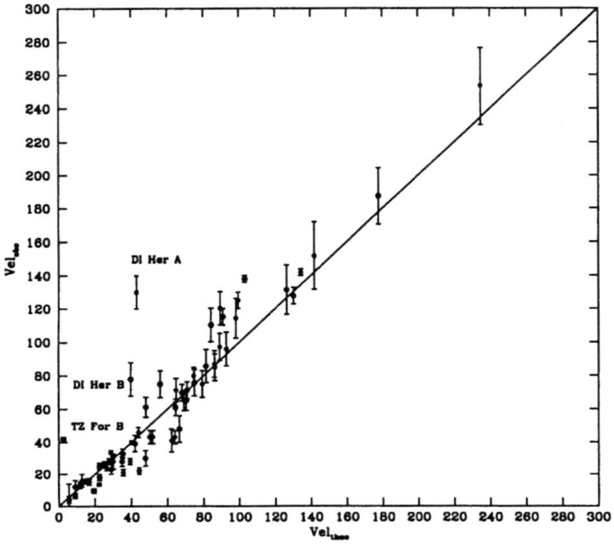

Fig. 1.16. Observed and theoretical rotational velocities assuming synchronization at periastron. Full hexagons denote primaries and open ones represent the secondaries. Note the position of the components of DI Her and TZ For B. (Claret & Cunha 1997).

Analyzing that figure we can consider that the stars compiled by Andersen (1991) are, on average, pseudo-synchronized as indicated by Giménez & Andersen (1983) and more recently by Claret & Cunha (1997). Figures 1.17 and 1.18 show the critical times t_{cri} compatible with Fig. 1.16, even for DI Her given its age this system has not time to synchronize/circularize.

Synchronization levels do not give as conclusive results as in the case of the circularization of orbits: while for the latter case we can say if a system is eccentric or not, a similar reasoning can not be applied directly to synchronization, since the interior can rotate at a different angular velocity than the envelope does. The results of the integration of eqs. 1.142 and 1.145 are shown in Fig. 1.19; the observed eccentricities versus log (t/t_{cri}) where t is the age of the system. The results are satisfactory, especially for the more eccentric orbits, since all these systems are on the left of the zero point, i.e., their ages are smaller, and in some systems equal, to their predicted circularization times.

An interesting case to be analyzed is that of TZ For. TZ Fornacis is an eclipsing binary system whose components have masses of 2.05 and 1.95 $M_\odot$ and a period of 75.7 days. The radius of the more massive component is 8.32 $R_\odot$ while the radius of the secondary is 3.96 $R_\odot$. The mean errors are smaller than 3 %. The primary seems to be synchronized but the secondary does not. In fact, it rotates about 10 times faster than the primary and about 16 times faster than the theoretical value. By comparing the stars with theoretical models, we infer that the primary is burning helium in the core while the secondary is near

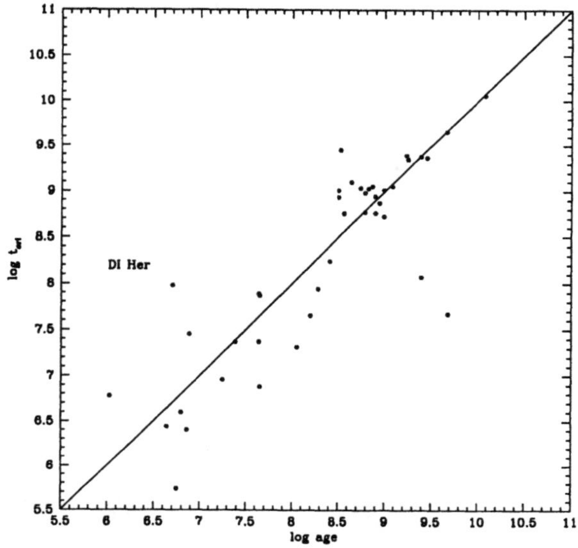

Fig. 1.17. Critical times for synchronization using the turbulent dissipation and radiative damping mechanisms versus the age. Case of primaries. The position of DI Her is indicated.

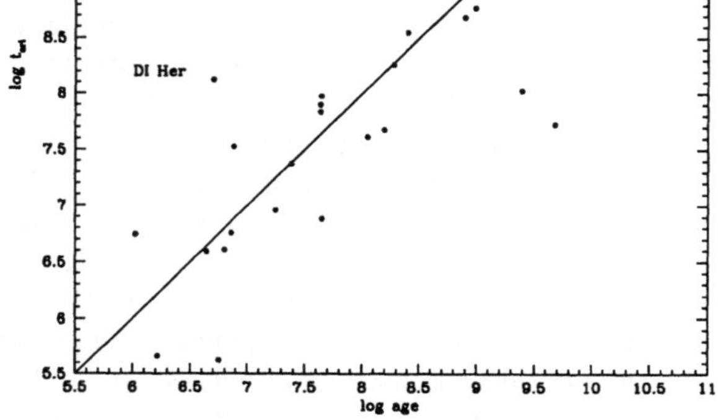

Fig. 1.18. Same remarks as in the previous figure. Case of the secondaries.

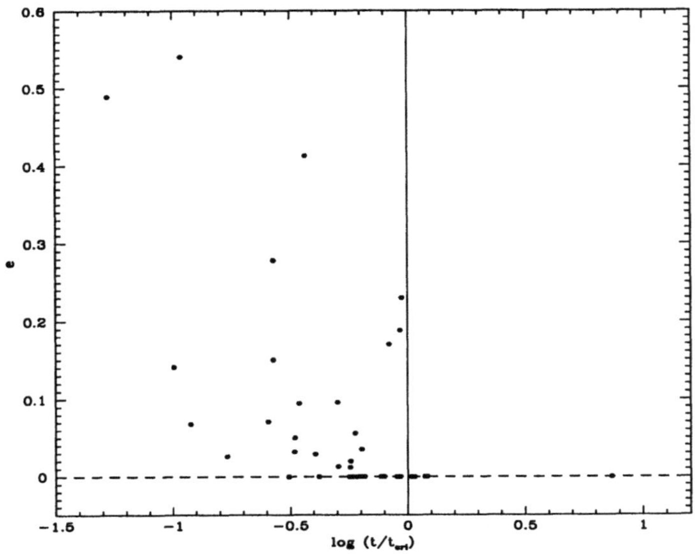

Fig. 1.19. Comparison between the evolutionary ages and critical times for circularization using turbulent dissipation and radiative damping mechanisms.

the "red hook"(see Fig. 1.1 by Claret & Giménez 1995). The diagnosis diagram shown in Fig. 1.20 is very useful useful since it distinguishes the synchronous component from the asynchronous one. In this way, it was have shown how both components of TZ Fornacis probably evolved, the circularization as well as the synchronization for the primary were achieved, while the secondary is still presenting high degree of asynchronism (Claret and Cunha 1997).

Another braking mechanism is based on the work of Tassoul (1987, 1988). Following this description, the stars in a close binary system tend to synchronize and circularize the orbit due to the distortions which cause large scale hydrodynamical currents. Regardless of the dissipation mechanism used, we would like to stress the fact that the differential equations which govern the orbital parameters of a binary system must be integrated instead of using the corresponding time scales. Moreover, they were assumed several approximations: constant periods during the integration of the differential equations, little departure from synchronism, small eccentricities, etc. The integration of the differential equations which control the orbital parameters depends strongly upon the initial conditions. Besides the limitations and approximations quoted above, the most crucial limitation was probably not to consider the pre main-sequence evolution.

Some especial systems. Some binary systems are of especial interest concerning stellar evolution and dynamical behavior. Among them, we select the case of EK Cep and HV 2274. The first one is interesting because it provides more

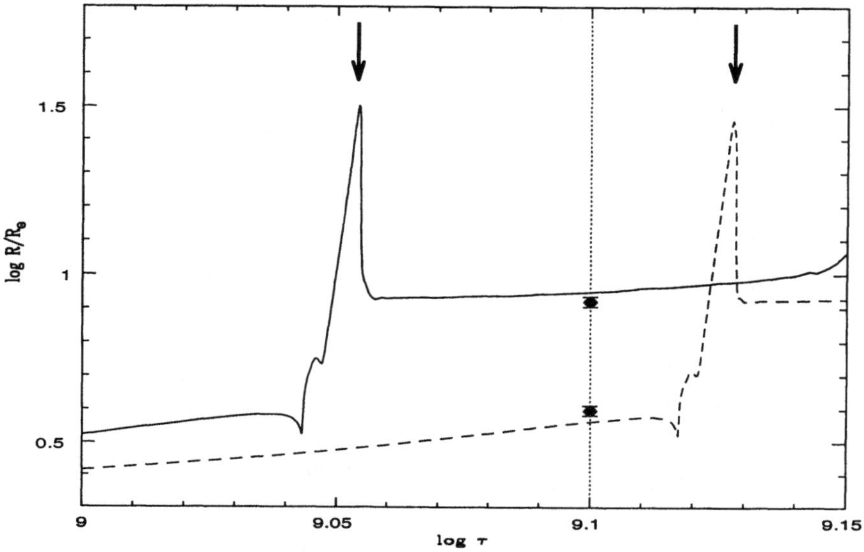

Fig. 1.20. The evolution of the radii of TZ For. The continuous line represents the primary while the dashed one denotes the secondary. The left side arrow indicates the point where circularization and synchronization of the primary are achieved. The other arrow indicates the time where the secondary would synchronize.

constraints than usually double-lined eclipsing binaries do. On the contrary, the second system do not provide very accurate dimensions but, its attraction is that it is an extragalactic eclipsing binary situated in the Large Magellanic Cloud.

Let us analyze first the case of EK Cep (Claret et al. 1995a). Really, it is a very good candidate to test evolutionary models since we have observational data concerning

- accurate absolute dimensions (Popper 1987)
- good determination of the apsidal motion rate (Giménez and Margrave 1985)
- determination of surface lithium abundance (Martín and Rebolo 1993)
- the secondary component is suspected to be in a pre main-sequence phase

Figure 1.21 shows the behavior of the radii of the components of EK Cep.

The models fits very well the absolute dimensions and confirms the secondary as a pre main-sequence star. With respect to the apsidal motion rates predicted values of k_2 are in excellent agreement with the observational data. Concerning lithium burning, the models indicate no depletion for the primary while they predict a depletion of 0.1 dex, which is consistent with the abundance derived by Martín and Rebolo (1993).

On the other hand, the system HV 2274, located in the LMC, observed by Watson et al. (1992) only provides the light curve - no information on the masses

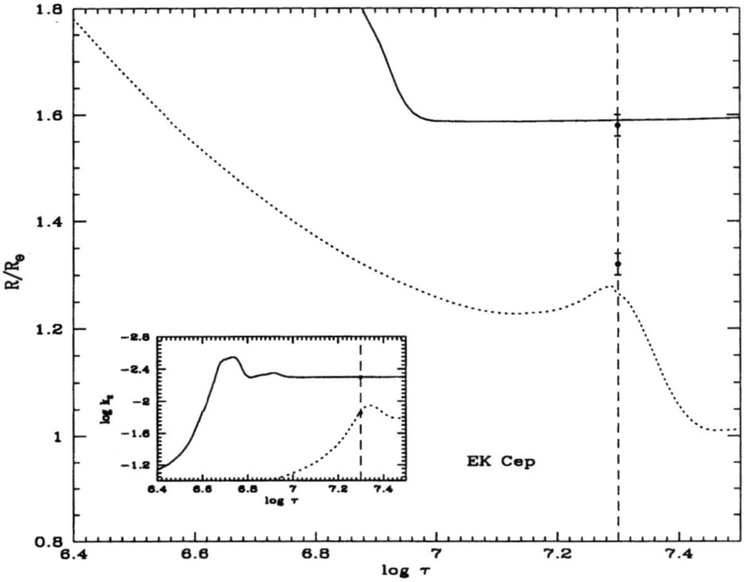

Fig. 1.21. The evolution of the radii and apsidal motion constant for EK Cep.

- and an estimation of the apsidal motion period around 123 years. On the base of theoretical evolutionary models computed for the LMC and the observed apsidal motion rate, it was predicted the masses and radii for both components of this system (Fig. 1.22). The masses were constrained to be between 10 and 12 $M_{\odot}$.

Recently, using ground-based photometry, spectrophotometry on board of Hubble Space telescope, Guinan et al. (1998) were able to determine the absolute dimensions for HV 2274 which are in good agreement with the theoretical predictions. Stellar models computed for the precise observed masses give even a better agreement for the masses and radii as well as for the apsidal motion rate. Of course, the importance of the case of HV 2274 is in the fact that this system is the first extragalactic binary where it has been possible to investigate the surface astrophysical parameters as well as tidal interactions using the techniques of apsidal motion.

Acknowledgements

We wish to thank C. Lázaro and M. J. Arévalo from Universidad de La Laguna for the warm hospitality.

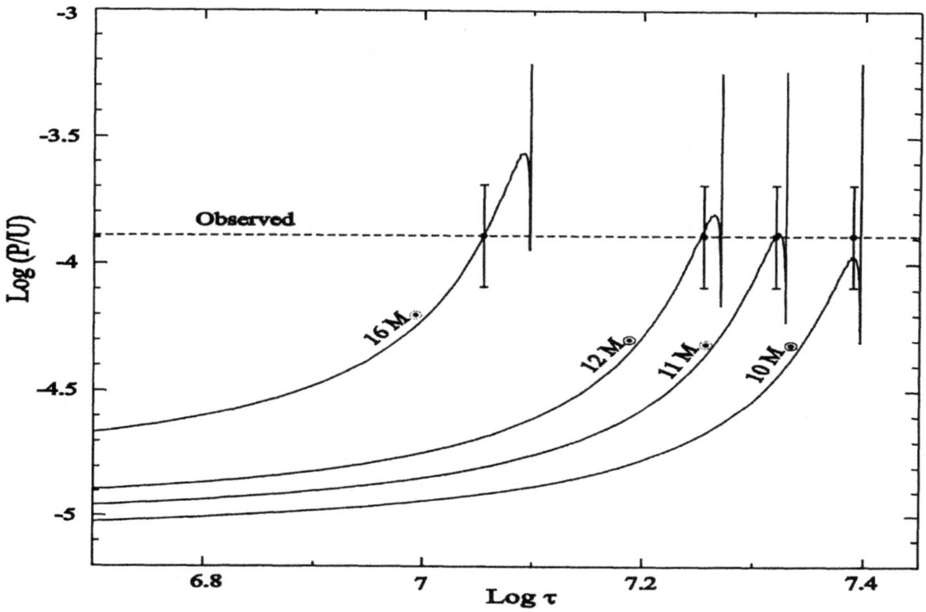

Fig. 1.22. The theoretical P/U for HV 2274 as a function of the time (Claret 1996).

References

1. Alcock, C. et al. 1997, ApJ, 491, 436
2. Allard, F., Hauschildt, P. H. 1995, ApJ, 445, 433
3. Allard, F., Hauschildt, P. H., Alexander, D. R., Starrfield, S. 1997, ARA&A, 35, 137
4. Al-Naimy, H.M. 1978, Astrophys. Space Sci., 53, 181
5. Andersen, J. 1991, A&AR, 3, 91
6. Bond, H.E. 1972, PASP, 84, 839
7. Bueger, P. 1969, ApJ, 158, 1151
8. Bueger, P. 1972, ApJ, 177, 657
9. Caughlan, G. R., Fowler, W. A., Harris, M. J., Zimmermann, B. A. 1985, Atomic Data Nucl. Data Tables, 32, 197
10. Caughlan, G. R., Fowler, W. A. 1988, Atomic Data Nucl. Data Tables, 40, 283
11. Chandrasekhar, S. 1945, ApJ, 101, 348
12. Chandrasekhar, S. 1947, ApJ, 106, 143
13. Chandrasekhar, S. 1950, *Radiative Transfer*, Clarendon Press, Oxford
14. Christensen-Dalsgaard, J., Däppen, W. 1992, A&AR, 4, 267
15. Claret, A. 1995, A&AS, 109, 441
16. Claret, A. 1996, A&A, 315, 415
17. Claret, A. 1998a, A&A, 335, 647
18. Claret, A. 1998b, A&AS, 131, 395
19. Claret, A. 1998c, A&A, 327, 11
20. Claret, A. 1999, A&A, 350, 56
21. Claret, A. 2000a, A&A, 359, 289

22. Claret, A. 2000b, A&A, submitted
23. Claret, A., Cunha, N. C. S. 1997, A&A, 318, 1997
24. Claret, A., Giménez, A. 1990, A&A, 230, 412
25. Claret, A., Giménez, A. 1992, A&A, 256, 572
26. Claret, A., Giménez, A. 1993, A&A, 277, 487
27. Claret, A., Giménez, A. 1995, A&A, 296, 180
28. Claret, A., Giménez, A., Martín, E. L. 1995a, A&A, 302, 741
29. Claret, A., Díaz-Cordovés, J., Giménez, A. 1995b, A&AS, 114, 247
30. de Loore, C. W. H. 1999, this volume
31. Díaz-Cordovés, J., Giménez, A. 1992, A&A, 259, 227
32. Díaz-Cordovés, J., Claret, A., Giménez, A. 1995, A&AS, 110, 329
33. Dugan, R.S. 1908, *P.A.A.S.A.*, 1, 311
34. Eddington, A.S. 1926, MNRAS, 86, 320
35. Fowler, W. A., Caughlan, G. R., Zimmerman, B. A. 1975, *Ann. Rev. Astron. Astrophys.*, 13, 69
36. García, J. M. 1990, PhD Thesis, Universidad Complutense de Madrid
37. Giménez, A. 1981, PhD Thesis, Universidad de Granada
38. Giménez, A. 1985, ApJ, 297, 405
39. Giménez, A., Scaltritri, F. 1982, A&A, 115, 321
40. Giménez, A., Andersen, A. 1983, Comptes Rendus de la Veme Jouneé de l'Observatoire de Strasbourg, p. 155
41. Giménez, A., Margrave, T. E. 1985, AJ, 90, 358
42. Giménez, A. et al. 1991, A&A, 248, 563
43. Grygar, J. 1965, *Bull. Astron. Inst. Czech.*, **16**, 195
44. Guinan, E. F. et al. 1998, ApJ, 509, L21
45. Hadjidemetriou, W. M. 1967, Adv. Astron. Astrophys., **5**, 131
46. Hall, D.S. 1967, AJ, 72, 301
47. Hall, D.S. 1989, Space Science Review, 50, 219
48. Hauschildt, P. H., Baron, E., Allard, F. 1997a, ApJ, 483, 390
49. Hauschildt, P. H., Allard, F., Alexander, D. R., Baron, E. 1997b, ApJ, 488, 428
50. Hosokawa, Y. 1959, *Sendai Astronomiaj Raportj*, 70
51. Kippenhahn, R., Weigert, A., Hofmeister, E. 1967, *Methods for Calculating Stellar Evolution in Computational Physics*, Academic Press Inc. New York, Vol. 7
52. Kippenhahn, R., Weigert, A. 1990, *Stellar Structure and Evolution*, Springer-Verlag, Berlin
53. Kippenhahn, R., Thomas, R. C. 1970, in A. Slettebak (ed.), *Stellar Rotation*, D. Reidel, Publ. Co., Dordrecht, Holland, p. 20
54. Kirbiyik, H., Smith, R.C. 1976, MNRAS., 70, 103
55. Kondo, M., Okasaki, A. 1980, *Close Binary Systems: Observations and Interpretation*, (M.J. Plavec, D.M. Poper y R.K. Ulrich, eds.), p. 221
56. Kopal, Z. 1959, *Close Binary Systems*, Chapman and Hall, London
57. Kopal, Z. 1965, Advances in Astronomy and Astrophysics, 3, 89
58. Kopal, Z. 1978, *Dynamics of Close Binary Systems*, Reidel, Dordrecht, Holland
59. Kurucz, R.L. 1993 in *Light Curve Modeling of Eclipsing Binary Stars*, (E.F. Milone, ed.), Springer-Verlag, New York, p. 93
60. Levi-Civita, T. 1937, *Amer. J. Math.*, 59, 225
61. Lucy, L. B. 1967, Zeitschritft für Astrophysik, 65, 89
62. Mc Namara, D.H. 1967, ApJ, 149, 723
63. Manduca, A.; Bell, R.A.; Gustafsson, B. 1977, A&A, 61, 809
64. Martín, E. L., Rebolo, R. 1993 A&A, 274, 274

65. Martynov, D. Ya. 1948, *Izv. Engelhardt Obs. Kazar*, No. 25
66. Meynet, G, Maeder, A. 1997, A&A, 321, 465
67. Milne, E.A. 1927, MNRAS, 87, 43
68. Miner, E.D. 1966, ApJ, 144, 1101
69. Moffat, J. W. 1984, ApJ, 287, L77
70. Moffat, J. W. 1989, Phys. Rev. D, 39, 474
71. Naftilan, S.A. 1975a, PASP, 87, 321
72. Naftilan, S.A. 1975b, Bull. Amer. Astron. Soc., 7, 476
73. Naftilan, S.A. 1976, ApJ, 206, 785
74. Nieuwenhuijzen, H., de Jager, C. 1990 A&A, 231, 134
75. Pérez Hernández, F, Claret, A., Hernández, M. M., Michel, E. 1999, A&A, 346, 586
76. Popper, D. M. 1987, ApJ, 313, L81
77. Rafert, J.B., Twigg, L.W. 1980, MNRAS, 193, 79
78. Ruciński,S.M. 1969, *Acta Astron.*, 19, 125
79. Schwarzschild, M. 1958, *Structure and Evolution of the Stars*, Dover Publ., New York
80. Shul'berg, A. M. 1973, *Eclipsing Variable Stars*, (V. P. Tsesevich, ed.), Israel Trans. Publ., p. 72
81. Smak, J. 1999, this volume
82. Stebbins, J. 1911, ApJ, 33, 395
83. Tassoul, J. L. 1978, *Theory of Rotating Stars*, Princeton Series in Astrophysics, Princeton University Press, Princeton
84. Tassoul, J. L. 1987, ApJ, 322, 856
85. Tassoul, J. L. 1988, ApJ, 324, L71
86. Van Hamme, W. 1993, AJ, 106, 2096
87. von Zeipel, H. 1924, MNRAS, 84, 665
88. Wade, R.A., Ruciński, S.M. 1985, A&AS, 60, 471
89. Watson, R. D., West, S. R. D., Tobin, W., Gilmore, A. C. 1992, MNRAS, 258, 527
90. Zahn, J.-P. 1966, Ann. Astrophys., 29, 489
91. Zahn, J.-P. 1975, A&A, 41, 329
92. Zahn, J.-P. 1992, A&A, 265, 115

2 Magnetic Activity in Binary Stars

Klaus G. Strassmeier

Institute for Astronomy, University of Vienna, A-1180 Wien, Austria
strassmeier@astro.univie.ac.at

Abstract. "Activity" is a term that came originally from solar astrophysics, where astronomers refer to the quiet Sun and the active Sun depending on whether our star is in a state of sunspot minimum or maximum, respectively. More physically, we mean with activity all phenomena inside and outside of a star that are related to its magnetic field, e.g. the rise of flux tubes in a stellar convection zone and their appearance as spots and plages on the stellar surface or, magnetically induced particle acceleration and its braking effect on the stellar rotation, or simply the interplay between a magnetic field and its surrounding plasma. Bear three things in mind, a magnetic field plays a key role in many astrophysical processes, its mathematical treatment can be somewhat complicated and, therefore, is always the first process that is neglected ... The stellar magnetic field is intimately related to stellar rotation and the quest to understand stellar magnetic activity is thus always a quest to understand stellar rotation. In this lecture, we will focus on the many aspects of stellar rotation and compare the situation of binary stars with their single star counterparts whenever appropriate. I will concentrate on the available observations and their interpretations even though that fundamental progress will also come from the detailed inclusion of magnetic fields in, e.g., radiation hydrodynamics. We emphasize how important considerations of magnetic activity will become once the new generation of large telescopes is turned to late-type binary stars. Parts of the topics in this lecture are adapted and updated from my reference [111] which appeared in german language. I have tried, however, to avoid topics that were already covered in the excellent review by Guinan & Giménez [36]

2.1 Plasma + Magnetic Field + Velocity Field = Magnetic Activity

2.1.1 Ingredient 1: An Astrophysical Plasma

An astrophysical plasma must be electrically neutral to a very high degree because otherwise it would quickly recombine given the good mixing due to the high particle velocities. A plasma's properties are always the sum of the properties of its particles, i.e., its electrons, ions, and neutrons. If the energy per particle is the same, then the plasma is in equilibrium and the velocity distribution of its particles follows a Maxwellian distribution (describes the number of particles per volume with velocities $v + dv$). Less massive particles must have higher average velocities to carry the same energy than a heavy particle and thus dominate the kinetic energy of the plasma while the more massive particles have lower velocities and mainly contribute to the density of the plasma. The kinetic energy of a plasma is then

$$\frac{3}{2}kT_e = \frac{1}{2}m_e\bar{v_e}^2 \tag{2.1}$$

and the plasma density

$$\rho = \rho_{\mathrm{ion}} + \rho_e + \rho_n \quad \text{with} \quad \rho_x = n_x m_x \ , \tag{2.2}$$

where n is the number of particles of species x, and m their mass. The plasma pressure is given by the average momentum from all particles. If sufficient collisions between particles take place, we can assume that on average all particles contribute the same amount of momentum in all three spatial directions. The resulting plasma pressure will be

$$p_i = \frac{1}{3}nmv^2 \quad \text{for} \quad i = 1, 2, 3 \ . \tag{2.3}$$

The electrical conductivity of the plasma is independent of the energy that a particle can carry and depends solely on the collision rate. The more collisions between charged particles, the better will be the charge transport. A simple formula relates conductivity, σ, in electro motoric units (e.m.u.) to the plasma temperature, T, in K

$$\sigma = 1.53 \times 10^{-13} \, \frac{T^{\frac{3}{2}}}{Z} \ , \tag{2.4}$$

where Z is the atomic number. E.g., the conductivity in the solar convection zone varies between $10^6 - 10^3$ A/Vm (SI units).

2.1.2 Ingredient 2: A Magnetic Field

The Maxwell equations decribe the connection between magnetic and electric field, the charge density, and the current while the hydrodynamic equations relate pressure, density, and temperature to the plasma velocity. Combining these two sets of equations, we include the induction equation

$$\frac{\partial \mathbf{B}}{\partial t} = \underbrace{\nabla \times (\mathbf{v} \times \mathbf{B})}_{\text{induction}} + \underbrace{\frac{1}{4\pi\sigma}\nabla^2\mathbf{B}}_{\text{diffusion}}, \tag{2.5}$$

which describes the evolution of the magnetic field with time. The two terms marked can sometimes be approximated by their integral values

$$\nabla \times (\mathbf{v} \times \mathbf{B}) \approx \frac{vB}{\ell} \tag{2.6}$$

$$\frac{1}{4\pi\sigma}\nabla^2\mathbf{B} \approx \frac{B}{4\pi\sigma\ell^2} \ . \tag{2.7}$$

The quantity $4\pi\sigma\ell^2$ is the diffusion time scale, i.e. the time for a magnetic field of scale ℓ to change by its own magnitude through Ohmic decay, and ℓ/v is the advective time scale (e.g. the rise time of a flux tube through the convection zone). The ratio of these two timescales, and thus the ratio of the two terms in (2.5), defines the magnetic Reynoldsnumber ($R_{\mathrm{m}} = \tau_{\mathrm{diff}}/\tau_{\mathrm{adv}}$).

A most important ingredient for our understanding of the solar-stellar connection is the interaction of the magnetic field with the astrophysical plasma. The magnetic field produces the pressure $B^2/8\pi$ (in cgs units) perpendicular to the field lines and a tension $B^2/4\pi$ along the field lines but since charged particles move freely parallel to the field, the plasma is not affected by the magnetic tension. For example, in a young sunspot horizontal equilibrium is reached when the difference of the gas pressure outside and inside the spot equals its magnetic pressure

$$p_{\mathrm{outside}} = p_{\mathrm{inside}} + \frac{B^2}{8\pi} \tag{2.8}$$

while the vertical equilibrium is basically given by the hydrostatic equilibrium

$$\frac{dp}{dr} = -\rho g \ . \tag{2.9}$$

The plasma in a sunspot reaches a stable equilibrium by lowering its vertical position (the so-called Wilson depression) because the pressure in its surrounding increases with depth. Of course, spots can sink because they are cooler than their surrounding. If the magnetic pressure exceeds the gas pressure a current must flow along the field lines and the last of the four Maxwell equations becomes

$$\nabla \times \mathbf{B} = \mathbf{j} = \alpha \mathbf{B} \tag{2.10}$$

where $\mathbf{B}$ is the average magnetic field, $\mathbf{j}$ is the current and α a scalar quantity ($\alpha = 0$ is the special case of a potential field). The magnetic energy is proportional to α^2. The appearance of an additional current in the direction of the average field is called the α effect (see also next section). Because the current itself does not produce a force, the field is a force-free field. Of course there is an upper limit for $\mathbf{B}$ in places where the magnetic forces are so large that the plasma motion is effectively brought to a halt. The simplest form of such a dynamical limit is equipartition, i.e. when the kinetic-energy density equals the magnetic-energy density. The corresponding field strength is

$$B_{\mathrm{eq}} = \sqrt{\rho\mu}\ v \ , \tag{2.11}$$

where ρ is the density of the plasma, μ the magnetic permeability, and v the plasma velocity. A solar granule exhibits a typical velocity of 2 km s^{-1} at the solar surface ($\rho \approx 3 \times 10^{-4}$ kg m^{-3}) and thus allows an equipartition field strength of 0.04 Tesla.

It is assumed that the magnetic field structures on the stellar surface are the product of magnetic flux that was released somewhere in the stellar interior. The origin of this assumption is an effect called magnetic buoyancy. Magnetic structures that have the same temperature than their surrounding and are in pressure equilibrium still experience a net upward lift (Parker [80]). The characteristic lift velocity is the Alfvén velocity from (2.11). Inserting the density from a solar convection-zone model and an expected field strength of $\approx$10 T, the Alfvén velocity at the bottom of the convection zone is of the order of 60 $\mathrm{m\,s^{-1}}$. It would take a magnetic structure about one month to cross the solar convection zone. Therefore, magnetic flux can not be stored in the convection zone for much longer than this and certainly not over the entire 11-year cycle. It needs a sufficiently stable region where it can be stored and thereby strengthened by a dynamo process. The current thinking is that the solar magnetic field is generated in the overshoot region below the convection zone from where it rises to the surface (Fig. 2.1).

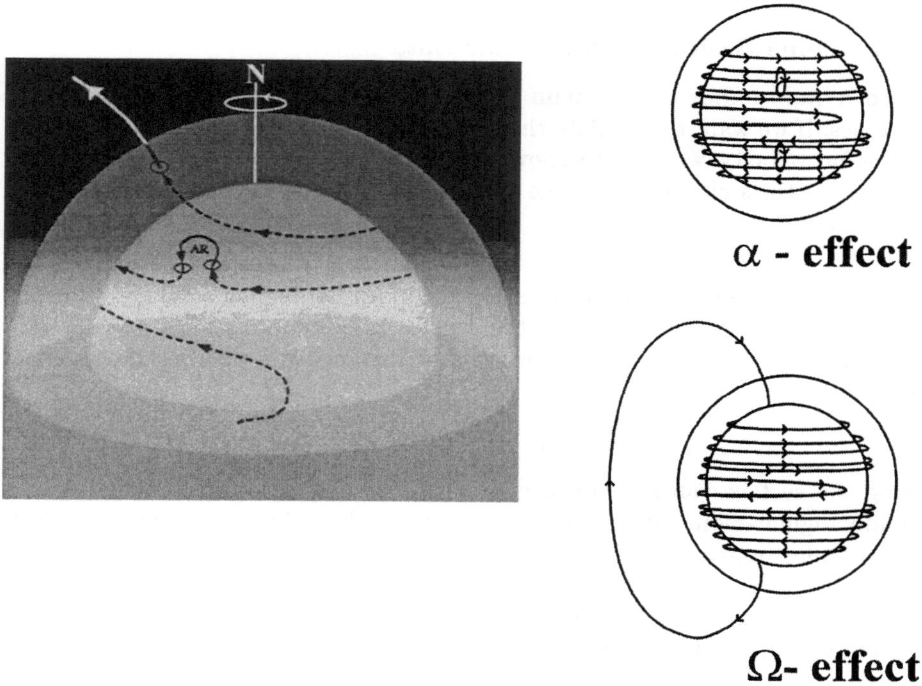

Fig. 2.1. Schematic plots of the α effect (upper right) and the Ω effect (lower right). The left graph shows a pseudo three-dimensional look inside the Sun and highlights a single magnetic field line. The two shaded spheres indicate the surface and the bottom of the convection zone, respectively (so do the thick circle and the thin circle in the small panels). AR indicates a bipolar active region. The axis of rotation is also indicated. (After David H. Hathaway, Solar Physics Branch, Marshall Space Flight Center, NASA).

2.1.3 Ingredient 3: A Velocity Field

The stellar plasma of a rotating star has a preferred sense of motion due to the direction of the stellar rotation. The radial gradient of the angular velocity is called differential rotation. The combination of these two velocity fields convert an embedded poloidal magnetic field into a toroidal (azimuthal) field and back. The toroidal component is generated from the shear of the poloidal field (Ω effect) and the poloidal field itself is generated by the deformation of the toroidal field due to plasma motions in the convective layer (α effect). Figure 2.1 is a schematic plot to illustrate the magnetic geometry. Both effects may be spatially connected (classical $\alpha\Omega$ dynamo) or may be separated (e.g. overshoot-layer $\alpha\Omega$ dynamo).

2.2 Stellar Rotation in the Hertzsprung-Russell Diagram

2.2.1 Some Basics and New Solar Data

When a stars moves off the main sequence, its radius and moment of inertia increases. If we want to correlate these parameters with the (observable) surface rotational velocity, we need to know how the angular momentum is distributed throughout the stellar interior. The angular momentum, L, in units of [$\mathrm{kg\,m^2s^{-1}}$] is generally written as

$$\boldsymbol{L} = I\boldsymbol{\omega} \tag{2.12}$$

where I is the moment of inertia and $\omega = v/R$ the angular velocity (v is the velocity usually in $\mathrm{km\,s^{-1}}$ and R the surface radius in km). The angular momentum of a spherical star of density ρ at distance r from the core is given by

$$L = \frac{8\pi}{3} \int \rho(r)\omega(r)r^4 \, dr \, . \tag{2.13}$$

Using the surface values, we can express the formal equation above also as a pseudo angular-momentum equation when we replace the integral with a single, dimensionless number k (k is the gyration radius, it parameterizes the deviations of density from a homogeneous sphere and is tabulated in some basic physics books; some authors prefer to write $K \equiv k^2$. For most stars, $k^2 \approx 1/13$ is a good approximation). Equation (6.1) then reads

$$L = k^2MRv = k^2MR^2\omega. \tag{2.14}$$

The Sun's radius of gyration is 0.25, and its total angular momentum is approximately 1.7×10^{41} $\mathrm{kg\,m^2s^{-1}}$. When we compare this to the angular momentum of the nine planets in the solar system, around 3×10^{43} $\mathrm{kg\,m^2s^{-1}}$, i.e. approximately a factor 100 higher, we see immediately that rotation in binary stars will be dominated by the angular momentum of the orbital motion rather than by the rotation of the individual components. If we put another Sun into our solar system instead of the nine planets, the orbital angular momentum raises to something in the range of 10^{45} to 10^{48} $\mathrm{kg\,m^2s^{-1}}$, i.e., 10^4 to 10^7 times higher than the solar rotational momentum. The energy that is contained in the (normal) stellar rotation is also relatively small, about 10^8 times smaller compared to the nuclear energy reservoir of the Sun. These numbers significantly change when a star evolves and becomes a giant.

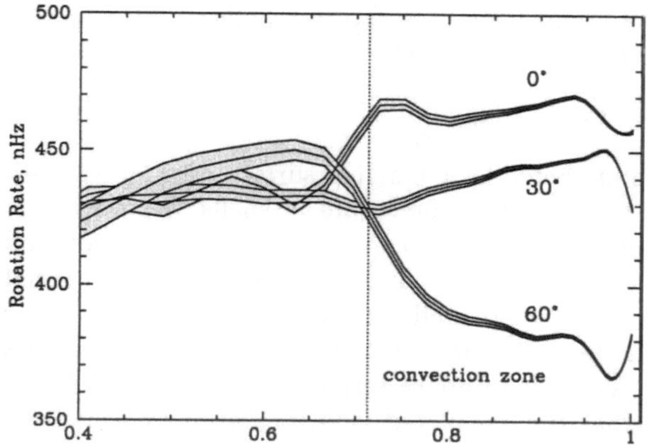

Fig. 2.2. The internal rotation profile of the Sun for three latitudes, 0°, 30°, and 60° between 0.4 and 1 r/R. Data from SOHO MDI. The vertical dotted line indicates the bottom of the convection zone. (After Kosovichev et al. [57]).

Only for the Sun, we have the capability to "look inside a star" by applying seismological inversion techniques to observed radial velocity and brightness variations (there are various proceedings on this topic, see e.g. Weiss & Baglin [137]). Figure 2.2 shows a radial cut through the upper 60% of the Sun. This data were obtained with SOHO's Michelson Doppler Imager (MDI) and reveal rigid rotation down to the radiative zone and, thus, in Eq. (6.1) $\omega(r)$ ≈const. Figure 2.2 also clearly demonstrates that the latitude-dependent surface differential rotation is maintained until close to the bottom of the convection zone. The discontinuity at this depth, e.g. for a latitude of 0° in Fig. 2.2, is now used as good evidence that it is the true location of the solar dynamo. However, this might not be the case for giant stars, or for contracting pre-main-sequence stars, where the evolution of the stellar interior redistributes its mass and thus possibly also redistributes the interior angular momentum.

Historically, Kraft [58], [59] separated between two cases (see also Gray [33]): First, angular momentum is not exchanged radially but is conserved in "shells" and its total amount is conserved with time as the star evolves:

$$\begin{aligned}(I\omega)_{\text{surface today}} &= (I\omega)_{\text{surface ZAMS}}\\ k^2 M R v_{\text{shell}} &= k_0^2 M R_0 v_0\\ v_{\text{shell}} &= v_0 \frac{R_0}{R}.\end{aligned} \tag{2.15}$$

Here, k and M are constant because we assume that angular momentum is conserved within each shell. R_0 denotes the stellar radius on the zero-age-main-sequence (ZAMS) and R is the current radius ($R > R_0$). The second case is pretty much the opposite, that is, angular momentum is completely redistributed throughout the stellar interior and, as a consequence, the star rotates rigidly. Then, the radius of gyration is not a constant anymore and we get

$$v_{\text{rigid}} = v_0 \left(\frac{k_0}{k}\right)^2 \frac{R_0}{R} \tag{2.16}$$

and, because evolved stars have denser cores and larger envelopes as compared to dwarf stars, i.e. $k < k_0$, we should see the relation

$$v_{\mathrm{rigid}} > v_{\mathrm{shell}}. \tag{2.17}$$

On average, a rigidly rotating star should have a higher surface rotation velocity than a star that has conserved its angular momentum in shells. Stars in open clusters of known age are the ideal tracers for a comparison of above prediction with the actual case (cluster stars are chemically homogeneous and their original angular momentum came from the same pre-stellar cloud and likely was of similar amount). If one adopts a simple theoretical model for the change of R with mass and time, i.e. $R(M,t)$, and considers it in above series of equations, Kraft [59] came to the surprising result that the calculated velocities even exceeded the upper limits of the observed velocities. He correctly concluded that the cluster stars observed must have undergone a significant loss of angular momentum.

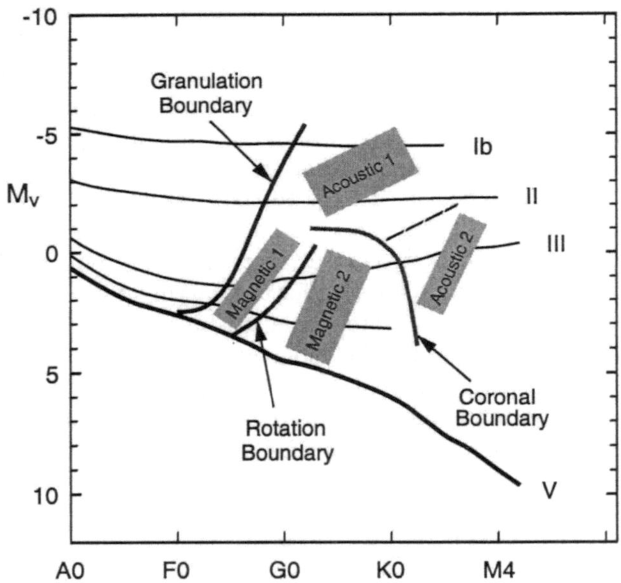

Fig. 2.3. Boundary lines in the H-R diagram. The magnetic and acoustic regions are indicated as described in the text. (After Gray [34]).

2.2.2 Rotation Regimes in the H-R Diagram

Which regions in the Hertzsprung-Russell diagram are influenced by rotation? Figure 2.3 shows four different sections. When more massive stars evolve across the granulation boundary, i.e. develop a convective envelope, their rotation will have already slowed down below what is needed to drive a dynamo. These stars will fall into the domain labeled "Acoustic 1". The less massive stars, say up to 2 $M_\odot$, that thus have shallow convection zones, are not yet in the evolutionary status where their moments of inertia significantly increase nor experience significant magnetic braking, and thus still maintain most of their rotation at this time. These stars already drive a weak dynamo and the corresponding region is labeled "Magnetic 1". Once these stars evolve past the rotation boundary, i.e. where strong magnetic braking takes away much of their angular momentum, their surface rotation slows down and so should their magnetic activity. However, in this region we find the most active stars of all. It is the regime of the tidally locked components of the RS Canum Venaticorum binaries ("Magnetic 2") and of some abnormally fast-rotating, apparently single giants (see the contribution by F. C. Fekel in Strassmeier et al. [119]). The region labeled "Acoustic 2", that appears separated from the others by the coronal-boundary line, is currently being rediscussed. This is because many new ROSAT all-sky survey detections of giants above the coronal line were reported (Hünsch et al. [48]). Nevertheless, beyond the coronal-dividing line stars should have slowed down even more due to their further increase of the moment of inertia during regular evolution and should exhibit no (coronal) X-ray emission.

2.3 Angular Momentum Loss by Magnetic Braking

2.3.1 Principle

The principles of magnetic braking were already layed out by Schatzman [98] back in the early fifties. But once again, let's take our Sun as an example. Consider a particle of the solar wind, an electron or a proton, that leaves the solar surface radially and travels into space. Right at the solar surface its tangential velocity component equals the rotational surface velocity of the Sun, i.e. $v_{\mathbf{surface}} \approx 2$ km s^{-1}. When it reaches the Earth's orbit, it should have slowed down to $R_\odot / a v_{\mathbf{surface}} \approx 10$ m s^{-1} ($a = 1$ AU) if it conserved its angular momentum. However, the HELIOS spacecrafts measured velocities of the order of 1–10 km s^{-1} (Pizzo et al. [87]), that is a factor 100–1000 faster than anticipated. The cause is that the (charged) particles are travelling along open field lines and not just radially outwards. To do so, the energy that is contained in the magnetic field per unit volume must be significantly larger than the particle's kinetic energy and thus its trajectory is dominated by the direction of the magnetic field lines rather than by the gravitational field. If some of the stellar magnetic-field lines are open and reconnect, e.g. with the interstellar medium, a stellar wind exerts a torque on the star and effectively removes angular momentum from it and thus brakes its surface velocity. The magnetic field lines will be distorted opposite to the direction of the stellar rotation (see Fig. 2.4) and do not leave the star radially but will be bent. Their curvature causes a (tangential) counteracting force on the surrounding stellar plasma (a bented field line always wants to straighten because $\nabla \cdot \boldsymbol{B} = 0$ must be fulfilled). If the field is anchored strictly at the poles – we assume that the magnetic poles coincide with the rotation poles (which is, e.g., not true for the Ap stars) – then the amount of dissipated angular momentum is very small and braking almost negligible. However, if the field is anchored near or at the equator (e.g. in active regions) then the braking is strongest.

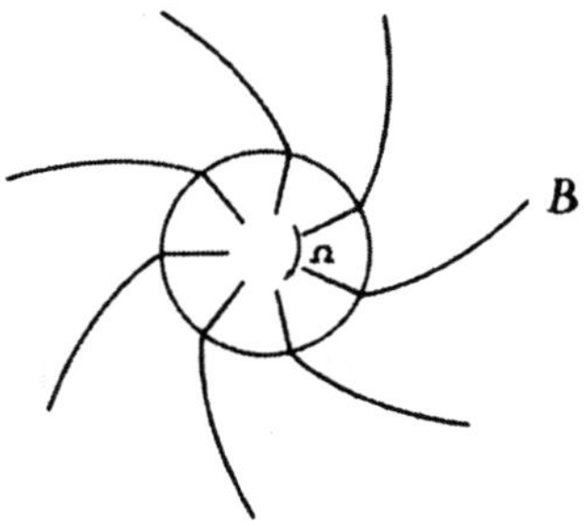

Fig. 2.4. Magnetic braking due to mass loss along open field lines. Pole-on view, the direction of rotation is indicated. (After Kippenhahn & Möllenhoff [55], Bibliographisches Institut Mannheim/Wien/Zürich.)

Naturally, this raises the question of where are the field lines anchored in the star and what part of the star will be braked? A hint comes from observations: we know that the (surface rotational) velocity dispersion on the ZAMS is very large, even for stars in the 1–2 solar-mass range, but that it is very small at the age of the Sun (4.5 Gyr) and already at the age of the Hyades (600 Myr). But even the very young (70 Myr) α-Persei-G stars are rotating on average faster than the – barely older – Pleiades G stars (100 Myr). Compare the individual diagrams in Fig. 2.5. Therefore, magnetic braking must be very effective, at least for the observable surface layer. Significant angular-momentum transport in the convective envelope of these stars must have taken place and theorists believe that the transport mechanism is the turbulent convection induced by the torque of the magnetic field (e.g. Gough [30]). A purely hydrodynamical mechanism, namely the evolutionary change of viscosity of the stellar plasma, may be sufficient to explain the solar spin down and the observed rigid rotation profile of today's Sun (Pinsonneault et al. [82]), but not the rapid-braking phase of young cluster stars. It seems that the entire convection zone participates in the braking.

2.3.2 Two Types of Magnetic Field

The likely mechanism to generate the magnetic field[1] in a solar-like star is Parker's $\alpha\Omega$ dynamo (Parker [80]). It is the main ingredient in Babcock's picture of the solar magnetic cycle. There is good evidence that it operates in a shear layer between the radiative core and the convection zone (see also Sect. 2.7 and Fig. 2.21). The differential rotation between the radiative core and the convective envelope winds up the field and causes a deformation (a "shear") of the

[1] For some basic equations see, e.g., the review on stellar coronae by Schmitt [99] during the Brussels EADN lectures in 1996.

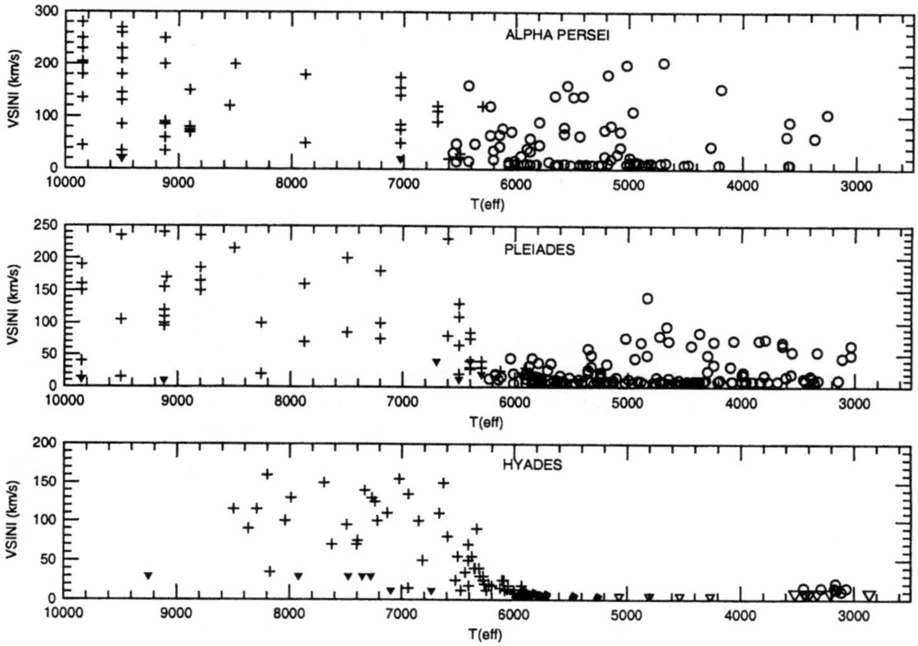

Fig. 2.5. A comparison of observed rotational velocities ($v \sin i$) as a function of effective temperature (or equivalently mass since these are main-sequence stars) for stars in three open clusters. Top panel: α Persei stars (age $\approx$70 Myr), middle panel: Pleiades (100 Myr) and, bottom panel: Hyades (600 Myr). Notice that the Hyades cluster has no rapidly-rotating stars cooler than $\approx$6000 K ($\approx$G0) anymore. (After John Stauffer, Smithsonian Astrophysical Observatory).

poloidal field which, in turn, generates an additional toroidal field component and thus creates a Lorentz force which counteracts the shear due to the poloidal field.

Mestel & Weiss [74] suggested that there is a magnetic field even in those parts of the stellar interior where the energy transport is dominated by radiation and that it may interact with the dynamo-generated field. For the late-type stars of question, i.e. the 0.6–1.6 solar-mass range on the main sequence and up to $\approx$3 solar masses for evolved stars, the radiative zone covers the inner 50–80 % of the stellar radius below a convective envelope. If such an inner field exists, its natural explanation is that it is left over from the contraction of the pre-stellar cloud, i.e. it is the frozen-in interstellar field (sometimes called the primordial field).

Numerical simulations show a strong dependence of the surface rotation upon the radial extent of the inner primordial field (e.g. Charbonneau & MacGregor [14]). Only if the two fields are separated braking can be achieved in roughly 10^7 years, i.e. the correct order that is observed in open clusters. The simulations also show that the rotational dependence is strongest in the time interval between arrival on the ZAMS and an age of 2×10^8 years and converges for all configurations only above an age of approximately 1 Gyr (the age of the ZAMS arrival of the Sun was computed to be 4×10^7 years). In order to agree with observations, the coupling between a primordial inner field and a dynamo-generated outer field must be absent or at least very weak.

2.3.3 Magnetic Activity Depends on the Rotational History

It seems that today's magnetic activity of a main-sequence star is already predetermined in its youth, i.e., at times when the pre-stellar-cloud material is still there and the central object still gains its luminosity mainly from contraction. The link of the central object with its circumstellar environment sets the star's original surface rotational velocity. Therefore, magnetic activity of a star on the ZAMS will be determined by the rotational history when it still was a pre-main-sequence (pms) star. Did the star have a disk? And if so, how long did it live and how efficient did it remove angular momentum from the star? During the very early stages in stellar life (after $\approx10^4$ years) the cloud of which the star has formed is still distributed around it like a cocon (these stars are called "class 0" protostars). During later stages, at an age of $\approx10^5$ years, the left-over cloud material will be concentrated in an accretion disk (these stars are called "class I" protostars or classical T Tauri stars). After all of the disk material has been accreted and/or blown away by a stellar wind, the star is free to rotate. Stars at this stage, roughly at 10^6 years, are called the weak-lined T Tauri stars or "class II" protostars.

In a review by Bouvier [10], it is suggested that pms stars with long-lived disks (classical T Tauri stars) have slowed down their surface due to disk locking but not their interior. From a rotational point of view, their core has decoupled from the envelope and possibly wound up the frozen-in field of the pre-stellar cloud (see also Bouvier et al. [11]). If so, then this is the time when the shear between the core and the envelope causes the poloidal field to change into some toroidal components which effectively brakes the core and possibly accelerates the envelope. We do not know enough yet about the process of core-envelope decoupling in order to fully understand rotational evolution but it seems that differential rotation (surface and internal) is the key to magnetic activity.

2.4 Stellar Rotation in Binary Stars

2.4.1 Magnetic Braking in Close Binaries?

The effect of magnetic braking can also be applied to binaries with magnetically active components like the RS CVn- or BY Dra stars (for a summary of such stars see the catalog of chromospherically active binary stars (CABS catalog; Strassmeier et al. [121]). However, tidal coupling in late-type close binaries is such a powerful mechanism that it alone will be sufficient to force the stellar rotation to synchronize to the orbital revolution. Also, there may not be many open field lines in an active binary that could account for the braking effect. At the moment there is no consistent numerical treatment of magnetic braking in binaries because it is strongly dependent upon the field geometry and we do not know enough about the actual field distributions in close binaries. Also, it is quite common that both stellar components are active once the orbital period is short enough to initiate dynamo action (several days for dwarf components and up to several weeks for giant components). Then, even stars with very shallow convection zones can become magnetically active. An imaginative – but purely hypothetical – picture was put forward by the japanese astronomers Uchida & Sakurai [131] in the early eighties. It nicely demonstrates the complexity of the (joint) stellar magnetic field in a close RS CVn-binary system (Fig. 2.6).

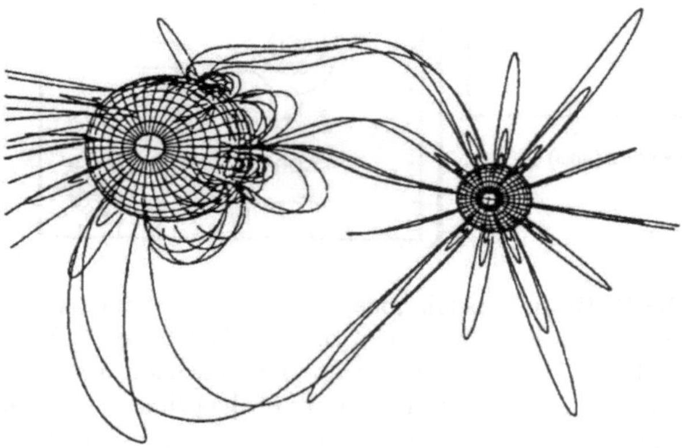

Fig. 2.6. Simulation of a possible field geometry of a RS CVn-type binary. Viewing angle is approximately "pole on". (After Uchida & Sakurai [131]).

2.4.2 Evidence for Inter-binary Material

There is some observational evidence for Uchida & Sakurai's picture from recent VLBI and X-ray observations. Lestrade [67] presented evidence of intra-binary radio emission in the two active systems UX Ari (G2V+K0IV; see Fig. 2.7) and σ^2 CrB (F6V+G0V). It is known that radio emission from these stars stems from the gyrosynchrotron process associated with large-scale magnetic fields and that the interaction between magnetic loops attached to the surfaces of the two stellar components may reconnect and thereby cause the electron acceleration that is needed for radio emission.

Further evidence of intra-binary activity comes from coronal X-ray maps of the eclipsing binary AR Lacertae (G2IV+K0IV). After initial semi-successful attempts in the eighties, Siarkowski et al. [105] presented even 3-D maps of the X-ray emission in and around AR Lac. Their map from June 1993 (see Fig. 2.8) showed that the bulk of the emission comes from in between the binary. However, only about half of the total emission is either rotationally modulated or eclipsed and thus subjected to the mapping. The other half must come either from a halo

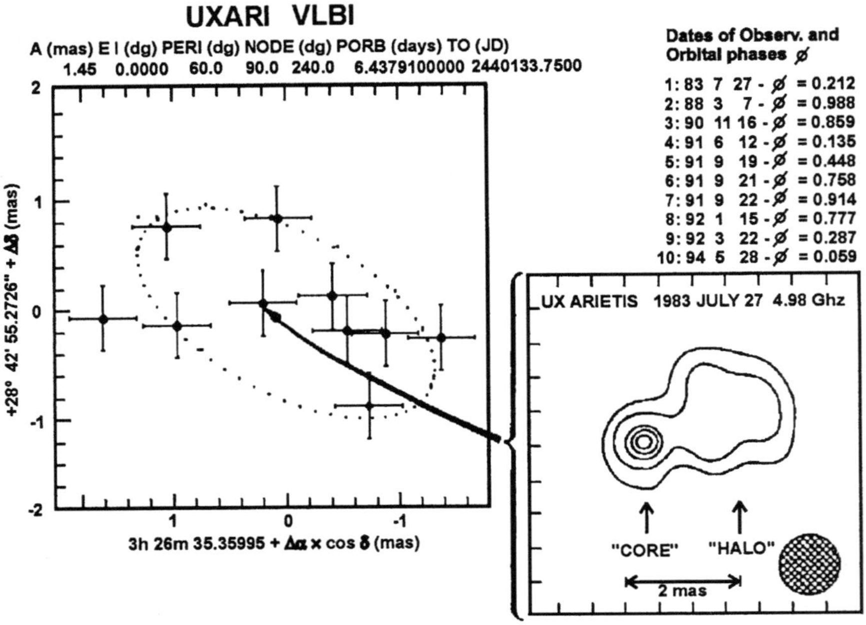

Fig. 2.7. Sky distribution of the VLBI astrometric measurements (left panel; circles with error bars) and the projected *spectroscopic* orbit of the RS CVn binary UX Ari (dotted ellipse). The right panel is the VLBI intensity map. The uncertainties of the astrometric VLBI data are only 0.3 milliarcseconds. The radius of the primary subgiant component is 0.25 milliarcseconds. The astrometric measurements suggest that the center of the radio emission wanders within the intra-system region. (After Lestrade [67]).

around both stars or from their rotation poles (the polar caps remain visible during the partial eclipse). Of course, this reenders the mapping an ill-posed process with remaining uncertainties of unknown extend but is further evidence of the existence of inter-binary magnetic fields.

Yet more evidence comes from the detection of a X-ray "superflare" on the RS CVn-binary HR 5110 (F2IV+K2IV)(Graffagnino et al. [31]). The flare emitted $4{\times}10^{36}$ erg within a three-day event. Such energies are hard to explain with a conventional two-ribbon flare model and the authors suggested that it comes from a common corona in-between the two active stars. All types of flares on the Sun, mostly compact flares and two-ribbon flares, are related to magnetic fields in the one or another way and we may presume that stellar flares are as well. The observation of a flare from the inter-binary region would thus be evidence for the existence of a strong joint magnetic field between the stars.

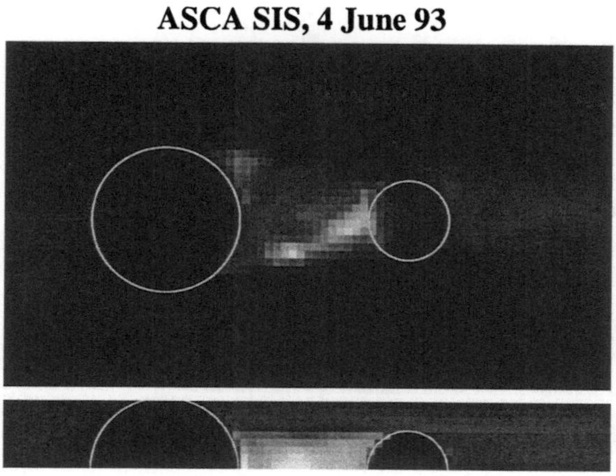

Fig. 2.8. The inter-binary X-ray emission of AR Lac. The observations were made with the Solid-State Imaging Spectrograph (SIS) onboard of ASCA in the energy range 0.4–1.5 keV. The top panel shows the orbital plane, the lower panel shows (half of) the side view. (After Siarkowski et al. [105]).

2.4.3 The Angular Momentum Distribution on the Upper Main Sequence

Now that we have identified likely mechanisms for the braking of the Sun and solar-type stars, we want to put that information into a larger picture and see what stars of higher mass tell us. This can be done by looking at a single diagram (Fig. 2.9) that plots (pseudo) angular momentum versus stellar mass as was done first by McNally [73].

It is obvious from this diagram that the more massive stars behave qualitatively differently than the lower-mass stars. The change takes place at a spectral type of ≈F2–F5. Stars cooler than this or, equivalently, stars less massive than this, must have lost a significant amount of their initial angular momentum. A spectral type of F2–5 on the main sequence coincides with the state where stars start to develop a convective envelope; and thus also a dynamo-generated magnetic field that, in turn, switches on a magnetic-braking mechanism. The earlier, more massive, stars lack this braking and can maintain their initial angular momentum gained during the contraction of the pre-stellar cloud. It is easy to show that this will lead to the $L/L_{\odot} = 100\,M^{5/3}$ dependence shown in Fig. 2.9. In his original paper, Kraft [60] used a simple mass-luminosity relation together with moments of inertia from stellar models to derive this relationship. The exponent of the power-law fit was later revised by Kawaler [52] (but see also Dicke [21] and Gray [33]). But what about binaries?

2.4.4 Rotational Velocities of Close Binaries

It was realized from the beginning that widely separated binaries (with this usually the Visual Binaries are meant) behave much like single stars since their gravitational interaction is too weak. We thus reformulate our question and ask "What about *close* binaries?", and mean binaries with orbital periods so short that tidal interaction plays a significant role in the evolution of its components. We explicitely exclude mass-exchanging systems like the Algols, β Lyraes, and

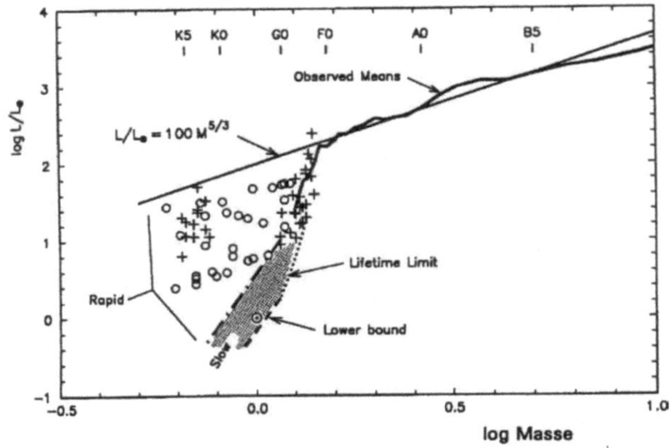

Fig. 2.9. Rotational angular momentum in solar units versus stellar mass. Both axes are logarithmic. The thick line is the observed mean for stars hotter than the Sun and follows approximately the relation $L/L_{\odot} = 100\,M^{5/3}$ (straight line). The rapid-braking and the slow-braking domains are indicated. Plusses are Pleiades stars, open circles are α Per-cluster stars. The Sun is also indicated. (After Gray [35]).

W UMas, because they exhibit a completely different mechanism to exchange angular momentum.

Let's plot the same graph as, e.g. in Fig. 2.5 for the three open cluster, but now only for components in close binaries from the field. We borrow this diagram from the book by David Gray [35] and show it in Fig. 2.10. Immediately, we see that stars in close binaries behave differently to their single-star counterparts. Binary stars hotter than ≈F5 usually rotate *slower* while binary stars cooler than that rotate *faster* than their single counterparts. The explanation was already suggested in the introduction in Sect. 2.2.1 where we found that the angular momentum stored in the orbital motion of a binary can be about a million times higher than that of the stellar rotation. Consequently, the stellar rotation will synchronize to the orbital motion (and not vice versa). For an average (close) binary system with an orbital period of, say, 10 days, this means that a rapidly-rotating, more massive star with an initial rotation period of a few days will be slowed down to the orbital value, while a slowly-rotating, less-massive star with an initial rotation period of several tens of days will be speeded up. It is this increased rotational velocity that runs, on the average, a more effective dynamo in late-type, close-binary components than in single stars.

However, simple scaling between single and binary stars proportional to rotational velocity or rotational period has been abandoned meanwhile. The feedback onto the stellar interior when a star rotates more rapidly than it should, turned out to be too complicated and needs refined mechanisms to deplete angular momentum within the stellar interior (i.e. it needs an answer to the question where the field lines are anchored in the star). For binaries, we will encounter two of

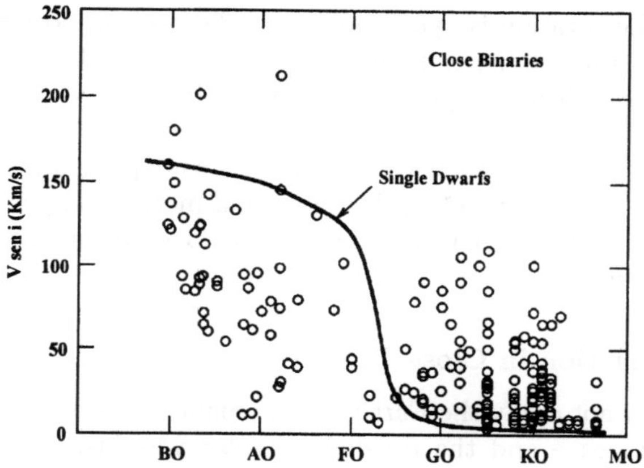

Fig. 2.10. The distribution of observed rotational velocities, $v \sin i$, for components of close binary stars (open circles). The line shows the mean for single stars. We see that stars hotter than ≈F2–5 rotate on average *slower* than single stars while stars cooler than ≈F2–5 rotate *faster* than typical single stars (After Gray [35]).

the most promising theories later in the next section. First, let us consider tidal coupling in more detail and see what happens if a binary orbit is eccentric.

2.5 Synchronisation of Stellar Rotation

2.5.1 The Earth-Moon Binary System

The term "tidal coupling" indeed stems from the observation of tides of our oceans. Tides are nothing else than the effect of differential gravitation in the sense that regions on Earth that are farther away from the moon will be less attracted to the moon's gravitation than closer ones (following Newton's $1/r^2$ dependence). A fluid is more prone to this effect because it always tends to enlarge its surface and the effect of tides is so much easier to see then (the main land also experience tides). Anyway, we may take the Earth-Moon system as an example of a binary with an orbital period of one month (actually it is a triple system because of the Sun) and see what the principle effects of tidal coupling will be.

First of all, we pretent that the Earth's water envelope (2/3 of the Earth's surface are oceans) plays the same role as the plasma of the convective envelope of a solar-type star. Since the rotation of the Earth is much faster than the orbital revolution of the Moon, i.e. one day compared to one month, the Earth will rotate several times under the tidal deformations of its envelope. This causes significant friction between the ocean and the ocean floor, i.e. between the envelope and the core. As we know, friction causes a depletion of heat which in turn is taken from the rotational energy of the Earth. Thus, it is tidal friction that slows down the rotation of the Earth. The total angular momentum in the Earth-Moon system must be conserved though and, as the consequence, the distance to the Moon increases until the rotation is synchronized to the orbital motion of the Moon. The equilibrium state at the time when the orbital and the rotational periods are equal will be reached at a period of roughly 50 days – which will happen in about a billion years from now.

For convective binary stars, two mechanisms to deplete angular momentum were proposed in the literature: tidal friction and meridional plasma motions. It is not completely decided yet which one is to be favored but it seems that they are likely complementary.

2.5.2 Tidal Friction in Close Binaries

Of course, we can not apply the Earth-Moon scenario to stars – with or without convective envelopes – and the previous chapter was just meant to illustrate the effect of tides in general. Nevertheless, the basic principle is the same but the detailed physical mechanism that depletes rotational energy is different. For stars with convective envelopes, J.-P. Zahn [139], [140] suggested that it is the change of the *viscosity* of turbulent convection cells that effectively produces enhanced friction throughout the stellar convection zone and thus converts angular momentum.

Zahn finds a surprisingly short time scale of

$$t_{\text{friction}} \approx \left(\frac{MR^2}{L}\right)^{\frac{1}{3}} \approx 1 \text{ Year} , \qquad (2.18)$$

where M is the stellar mass, R the stellar radius, and L the stellar luminosity. It suggests that any change in the turbulent convection of a star with a convective envelope – like the Sun – has a profound effect on its rotation.

If we consider a very young binary system whose components still rotate asynchronuously to the orbital motion, the tidal bulges will not immediately align along the apsidal line of the system. Instead, they will be misaligned by the difference of the angular velocities of the rotation and the orbit $(\omega_{\text{rot}} - \omega_{\text{orb}})$. A fact that has been known for over hundred years (Darwin [19])[2]. The time it takes to reach complete synchronisation can then be computed via the change of kinetic energy that is needed by the orbit to align the bulges or, equivalently, by the viscous dissipation of kinetic energy of the stellar rotation. Zahn's theory of viscous tidal friction predicts the following synchronization time scale in years

$$t_{\text{sync}} \propto f(k,q)\, t_{\text{friction}} \frac{I}{MR^2} \left(\frac{a}{R}\right)^6 \approx 10^4 \left(\frac{1+q}{2q}\right)^2 P^4 , \qquad (2.19)$$

where $f(k,q)$ is a function that includes the gyration radius k (an equivalent of the mass distribution inside the star) and the mass ratio, $q = M_2/M_1$ of the stellar components. I is the total moment of inertia and a/R the distance between the components in units of the stellar radius. P is the orbital period. Note that mass and radius in the left part of (2.19) refer always to only one of the two stellar components.

[2] G.H. Darwin; not related with Charles D., the evolution guru.

2.5.3 Meridional Motions

A purely hydrodynamical mechanism was suggested by J.-L. Tassoul & M. Tassoul [128], [129], [130]. For a star with a convective envelope it is based on the triggering of a meridional flow in the outer layers due to the non-spherical shape of one of the binary components (or both). The rotational angular momentum is thereby converted into a poleward meridional motion. Tassoul & Tassoul find a synchronisation time scale in years of

$$t_{\text{sync}} = 34 \left(\frac{L_\odot}{L}\right)^{\frac{1}{4}} \left(\frac{M}{M_\odot}\right)^{\frac{5}{4}} \left(\frac{R_\odot}{R}\right)^{3} P^{\frac{11}{4}} , \tag{2.20}$$

where L is again stellar luminosity and P the orbital period. A comparison with the tidal-friction synchronisation time scale shows that the Tassoul mechanism is significantly more effective (on average several orders of magnitude).

Table 2.1. Active Binaries with Synchronized Rotation

Star	Spectral-type	P_{orb} (days)	P_{rot} (days)
EI Eri	G5IV	1.947227	1.945
V711 Tau	G5IV+K1IV	2.83774	2.841
RS CVn	F4IV+G9IV	4.79785	4.791
II Peg	K2-3V-IV	6.724183	6.718
HU Vir	K0IV-III	10.3876	10.28
σ Gem	K1III	19.60447	19.41
V792 Her	F2IV+K0III	27.5368	27.07
HR 4665	K1III+K1III	64.44	63.75

Table 2.1 is a list of several examples of active, synchronuously rotating binaries. P_{orb} is the orbital period from radial velocities, and P_{rot} is the stellar rotation period from photometric observations.

2.5.4 Pseudosynchronuous Rotation

In a binary with an eccentric orbit, both stars will experience yet another component of differential gravitation due to the varying distance between the two stars during one orbital revolution. Based on this effect, the dutch astronomer P. Hut [49], [50] showed that a rotating star in equilibrium will eventually take on the orbital angular velocity at periastron, and not an average value along the orbit. According to Kepler's second law, this means that the synchronized rotation period will be the value corresponding to the angular velocity at periastron, which is, of course, shorter than the orbital period. The relation between this pseudo-synchronuous rotation period and the orbital period as a function of the orbital eccentricity, e, is given as a truncated series by Hut to

$$f(e) = \frac{P_{\rm pseudo}}{P_{\rm orb}} = \frac{(1+3e^2+\frac{3}{8}e^4)(1-e^2)^{\frac{3}{2}}}{1+\frac{15}{2}e^2+\frac{45}{8}e^4+\frac{5}{16}e^6}. \quad (2.21)$$

Figure 2.11 shows $f(e)$ in graphical form. For a given eccentricity one reads off the function $f(e)$ and multiplies it with the orbital period to find the pseudo-synchronous period.

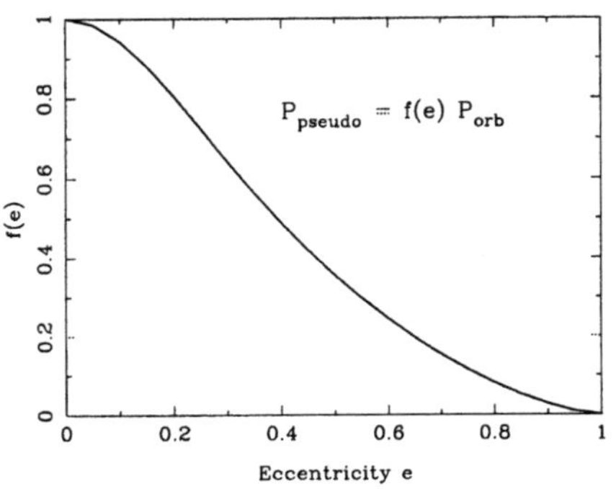

Fig. 2.11. The dependence of the synchronisation period on the orbital eccentricity. $f(e) = \frac{P_{\rm pseudo}}{P_{\rm orb}}$ as given in Eq. (2.21).

2.5.5 Asynchronous Rotation

Table 2.2 lists several examples of active binaries with asynchronous rotation. Some of them rotate faster than the synchronuous or the pseudo-synchronuous period, some of them slower. Naturally, one would now think that these systems had not had enough time to synchronize their rotation, but will do so in the future. However, all these systems contain either evolved stars like class-III giants and class-IV subgiants or are old low-mass, main-sequence stars that had already enough time to synchronize according to any of the above theories (see also the comments in Habets & Zwaan [38]). Another argument, namely that some of these systems are overflowing their Roche lobes and transfer mass from one component to the other, remains at best speculative because none of the systems in Table 2.2 show convincing evidence of mass transfer.

The puzzle has not been solved so far. Two polish astronomers, however, suggested an interesting way out of the dilemma (Glebocki & Stawikowski [29]). They suggested that the stellar rotation axis/axes in these systems is/are not perpendicular to the orbital plane and offered a statistical study based on photometric period determinations and spectroscopic $v \sin i$ measurements to pin down the inclination of the rotation axes with respect to the orbital plane. In my opinion the errors from $v \sin i$ measurements and from radii estimates are too large, especially for giant stars of low-to-moderate gravity and thus having strong surface velocity fields, as that the difference between the photometrically determined period and that from the rotational velocity can be used to obtain

Table 2.2. Active Binaries with Asynchronuous Rotation

Star	Spectral-type	e	P_{orb} (days)	P_{pseudo} (days)	P_{rot} (days)
AY Cet	G5III	0.00	56.824	...	77.22
α Aur	G0III/G9III	0.00	104.01	...	8.0
VY Pyx	K0III	0.00	45.13	...	19.34
DQ Leo	G5III-IV/A7V	0.00	71.69	...	55.0
λ And	G8IV-III	0.04	20.521	20.0	53.95
HD 181809	K1III	0.05	13.048	12.8	60.23
LU Hya	K1IV	0.13	16.54	14.9	21.0
V1285 Aql	M3.5V/M3.5V	0.20	10.319	8.3	2.9
HD 155989	G5III	0.32	122.56	73.5	30.0
TY Pic	G8-K0III	0.32	106.74	64.0	43.76
12 Cam	K0III	0.35	80.174	45.2	80.94
RZ Eri	K0IV	0.35	39.28	21.6	31.4
AZ Psc	K0III	0.50	47.12	16.5	91.2
LS TrA	K2IV/K2IV	0.52	49.431	16.3	46.19
BN Mic	K1III	0.52	63.09	20.8	61.73
HD 118234	K0.5III	0.59	59.05	14.7	64.0
HR 7578	K2-3V/K2-3V	0.69	46.817	7.9	16.5

$\sin i$ with reasonable error bars; which just means that we need better spectroscopy. As we will see later, Doppler imaging of components with large $v \sin i$ offers another – much more reliable – way to determine $\sin i$.

2.5.6 A Comparison with Observations

When we now make the step to compare theoretical predictions with actual observations, it is very import to realize the meaning of time scales. Most important, we need to realize that, if we insert today's observed stellar parameters into (2.19) and (2.20), we obtain the time scale of synchronisation of a to-be-synchronized binary but with today's parameters. For evolved stars, this can lead to incorrect comparisons. Actually, we would rather need to insert the stellar parameters before the synchronization takes place, which then requires exact knowledge of the current evolutionary status of the two binary stars *and* its history.

It is obvious that a comparison can only be made for binary systems where we have very exact knowledge of the relevant stellar parameters, e.g. masses good to within a few percent, just to name an example. This was done by Andersen et al. [2] for two binary systems each containing *two* giant stars; TZ For, an inactive eclipsing binary and Capella (α Aur), an active non-eclipsing binary. Since the present paper deals with active binaries, let us summarize some of the stellar properties of Capella and compare them to the inactive TZ For system. This is done in Table 2.3 but I refer to the paper of Andersen et al. [2] for further discussions.

Table 2.3. A comparison between TZ For and Capella

	Hot Star		Cool Star	
Parameter	Capella	TZ For	Capella	TZ For
Spectral type	G1III	F7III	G9III	G8III
Mass [$M_\odot$]	2.5±0.2	1.95±0.03	2.6±0.4	2.05±0.06
Radius [$R_\odot$]	8.3±0.5	3.96±0.09	14.1±4.1	8.32±0.12
Temperature T_{eff} [K]	5800±300	6350±100	5000±200	5000±100
Luminosity [$L_\odot$]	66	23	72	39
Gravity $\log g$ [–]	3.0±0.1	3.53±0.02	2.75±0.11	2.910±0.017
Density [$\mathrm{g\,cm^{-3}}$]	0.0067	0.044	0.0025	0.0050
Rotation period P_{rot} [days]	8.5	4.7	104:	105
$v \sin i$ [$\mathrm{km\,s^{-1}}$]	36±3	42±2	5±2	4±1
$v \sin i_{syn}$ [$\mathrm{km\,s^{-1}}$]	2.7	2.6	3.8	5.5
Orbital period P_{orb} [days]	104.02	75.67		
Eccentricity [–]	0.000	0.000		
Inclination i [°]	47	85.64±0.05		

Astrophysical parameters of single giant stars are uncertain because there are just a few such stars that can be studied in spectroscopic binary systems, and even less that are also eclipsing binaries. Moreover, Roche lobe overflow and mass exchange set a limit to the stellar radius in a close binary. Non-interacting binaries with giant components have therefore mostly very long orbital periods and are difficult to observe fully within an astronomer's lifetime. There are, however, a few systems that have sufficiently short orbital periods (see Andersen [1]). Capella is such a system and consists of a G1III and a G8-K0III component (e.g. Strassmeier & Fekel [118]) in a 104-day orbit.

The Capella binary has been classified as a RS CVn system and is included in the catalog of chromospherically active binary stars (Strassmeier et al. [121]). It is untypical in that the hotter of the two components, the G1III giant, is the active star and not the late-type G8-K0 component. Therefore, the G1 component could be covered with starspots that modulate the photospheric light with the stellar rotation period. Detecting the accompanying light variability could lead to a precise stellar rotation period for this Hertzsprung-gap giant.

Capella is usually much too bright for standard high-precision broad-band photoelectric photometry and is a challenge for modern light-sensitive instrumentation. Moreover, its long orbital period of 104 days makes it a further challenge for obtaining full phase coverage and the fact that there are two stars that make up for the observed light diminishes the observable amplitude. The only reliable (and published) photometry that we are aware of is that by Krisciunas & Guinan [61], who observed the star independently between 1981 and 1990. They found a possible 0.04 mag peak-to-peak variability in V. Their period analysis showed peaks in the power spectrum corresponding to periods of 82 days and 67 days. None of them seemed significant.

Fig. 2.12. The Wolfgang-Amadeus APT at Fairborn Observatory in southern Arizona. Left: Wolfgang, blue optimized. Middle: Amadeus, red optimized. Right: Lou Boyd, Director, fully optimized.

Therefore, we put Capella on Amadeus, one of the two Vienna "Wolfgang-Amadeus" APTs (Fig. 2.12) for routine monitoring (which is still underway). The Amadeus APT is optimized for red wavelengths with an EMI-9828 photomultiplier tube and use a narrow, 31-Å FWHM, Hα filter. In addition, all measurements of Capella were made with a 5^{m} neutral density filter in front of the spectral filter. Along with a 10-sec integration time, we were able to keep the count rates below 1 Mill. c/s.

The data from the first two observing seasons are shown in Fig. 2.13. Clearly, Capella is a variable star with an amplitude of 0.04 mag as suggested by Krisciunas & Guinan [61]. A multifrequency analysis suggests two significant (but still preliminary) photometric periods of 104±13 and 8.50±0.24 days that we interpret to be the rotation periods of the cool and the hot component of Capella, respectively. These periods, if correct, would confirm that the hotter component of Capella rotates asynchronuously while the cooler component appears to be synchronized to the binary motion.

The present position of the G1III component in the Hertzsprung gap, where it is approaching the base of the giant branch, indicates that violent changes are taking place in its internal structure; the mass of the convection zone increases rapidly as does also the total stellar moment of inertia (Sweigart et al. [127], Rutten & Pylyser [97]). We may expect that this has a profound impact on the visible surface rotation of the two Capella giants. Precise rotation periods of more stars in this evolutionary stage may thus help to further understand the angular-momentum loss in late-type stars.

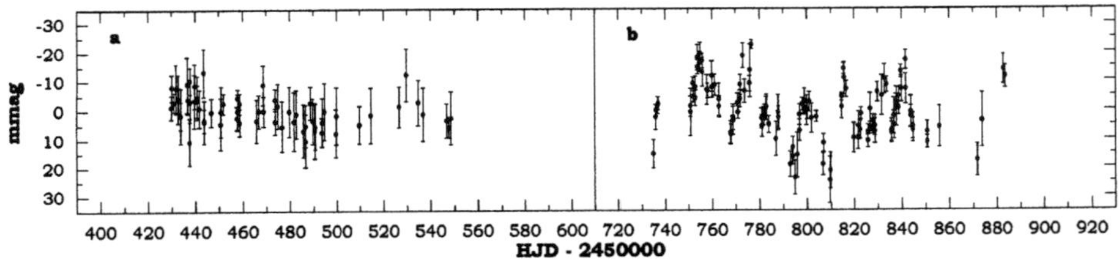

Fig. 2.13. The seasonal Hα-light curves of Capella for 1996/97 (panel **a**) and for 1997/98 (panel **b**). The error bars refer to rms magnitudes obtained by averaging count rates throughout an observing sequence. For each of the two data sets, the average magnitude is set to zero.

2.6 Stellar Rotation and Orbital-Period Variations

Orbital-period variations of binary stars are usually attributed to mass exchange from the Roche-lobe filling star to the other component. However, mass exchange can not account for alternating period changes, as observed in some Algols and in most active RS CVn binaries, because the direction of the mass exchange is predetermined and does not reverse within observationally accessible time scales (but see also, e.g. Biermann & Hall [9]). Back in 1992, Applegate [3] suggested a qualitative model in which the orbital period is modulated by the stellar activity. His model was recently quantified and expanded by Lanza et al. [65] and we will discuss it in more detail in this chapter.

2.6.1 First, Precise Stellar Rotation Periods Are Needed

In the previous as well as in the forthcoming chapters, we find the stellar rotation *period* to be an utterly important parameter. However, it is not an easy task to measure rotation periods. Formally, one could deduce a rotation period off the line broadening from a single high-resolution spectrum together with an assumption of a stellar radius, but this is notoriously uncertain due the unknown inclination of the rotation axis. Periods determined from photometry are usually a factor of 10–100 more accurate than from rotational velocities from line broadening and are an excellent tool for probing even differential rotation or specific surface phenomena like starspots, plages, and their evolutionary changes. The trivial relation between surface velocity and rotation period is

$$P_{\rm rot} = 50.6\,\frac{R\sin i}{v\sin i}, \tag{2.22}$$

with the rotation period in days, $R\sin i$ in solar radii, and $v\sin i$ in $\mathrm{km\,s^{-1}}$.

Sunspots are the most important tracers of solar rotation (cf. Bray & Loughhead [12], Zirin [141]). In as early as 1611, G. Galilei, J. Goldsmid, and Ch. Scheiner traced the positions of sunspots to actually discover solar rotation, and the same principle is nowadays also used for more distant stars (Kron [62], Hall [39] a.o.). Fig. 2.14 demonstrates its validity from disk-resolved solar observations. Note in this connection that starspots that are detected from photometric observations usually are a factor 10–1000 larger than Sunspots. The "missing-flux" problem, i.e. the question where the energy goes that is blocked by the spots becomes even worse then. This energy already amounts to 10^{36} erg in case of the Sun, which is equivalent to the energy budget of the solar chromosphere and corona combined! (For more discussions on that topic, start with Spruit [106]).

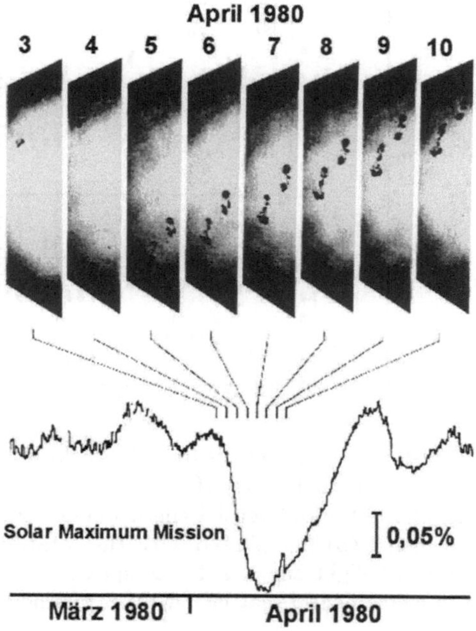

Fig. 2.14. Variations of the total solar flux (the solar "constant") due to the apparent motion of a sunspot group across the solar disk. The bar in the lower right corner represents a change of 0.05%. (Data from Hugh Hudson, after Friedman [54]).

With the advent of fully robotic telescopes, the search for rotationally modulated light variations became a relatively straightforward task. There are currently at least three groups that employ automatic photoelectric telescopes (APTs) for this reason: Tennessee State University joined together with the Mt. Wilson H&K program and operates several APTs at Fairborn observatory in southern Arizona (Henry [47]), the Vienna University Observatory also operates two APTs at the same site (Strassmeier et al. [117]), while the Catania Observatory operates an APT at their observatory on Mt. Etna in Sicily (Rodonó & Cutispoto [95]). Repeated observations are also made at several manual observatories, most noticably at ESO with the now decommissioned 50-cm telescope (e.g. Cutispoto [17]). The results of all these photometric observations are series of light curves for well over hundred spotted stars with time coverages just interrupted by the annual seasons, the day-and-night cycle and, of course, unfavorable weather conditions and – rare – telescope failures. As an example, Fig. 2.15 shows the long-term light variations of the spotted, close binary XX Tri (HD 12545).

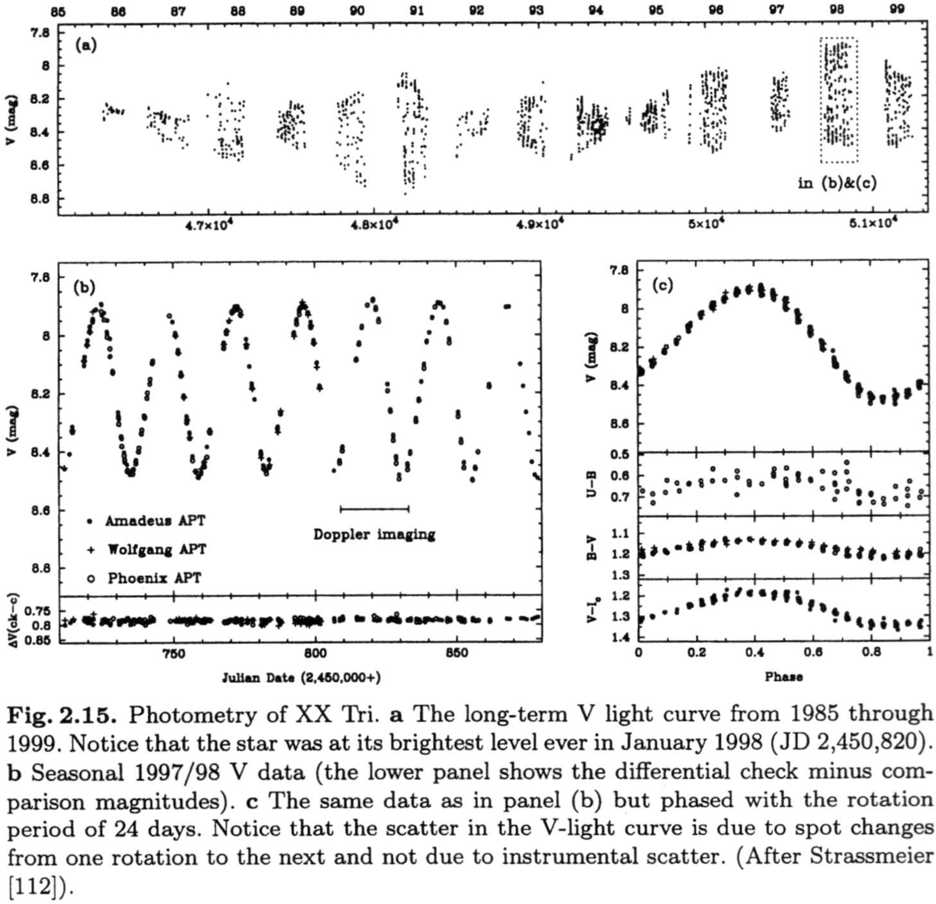

Fig. 2.15. Photometry of XX Tri. **a** The long-term V light curve from 1985 through 1999. Notice that the star was at its brightest level ever in January 1998 (JD 2,450,820). **b** Seasonal 1997/98 V data (the lower panel shows the differential check minus comparison magnitudes). **c** The same data as in panel (b) but phased with the rotation period of 24 days. Notice that the scatter in the V-light curve is due to spot changes from one rotation to the next and not due to instrumental scatter. (After Strassmeier [112]).

2.6.2 Differential Rotation

With the availability of long-term APT photometry of stars with large starspots, such as for the tidally-locked close binaries of the RS CVn or BY Dra type, it is possible to search for differential rotation in a large sample of stars. The applicability was amply demonstrated by various groups and I refer to the papers of Baliunas et al. [4], Strassmeier & Bopp [116], Lanza et al. [66], and the review by Hall [43]. The simplest technique is to determine photometric periods from consecutive parts of a photometric time-series, usually from consecutive observing seasons if the period is of the order of weeks or longer, and relate the differences, ΔP, to the average period ($\Delta P/ < P >$). In case the star is in a binary and is synchronized to the orbital motion, the orbital period can be used instead of the average period. However, there is some theoretical support for the view that the convective motion in the outer layers of close late-type binary components is governed by the orbital angular momentum (Schrijver & Zwaan [101]) and care must be exercised when data from long intervals in time are phased together.

Most eclipsing RS CVn binaries show indeed relatively complex orbital-period variations. The prototype RS CVn, for example, exhibits orbital observed-minus-computed (O-C) variations of up to 0.2 days in about 40 years (Rodonó et al. [96]; Fig. 2.16).

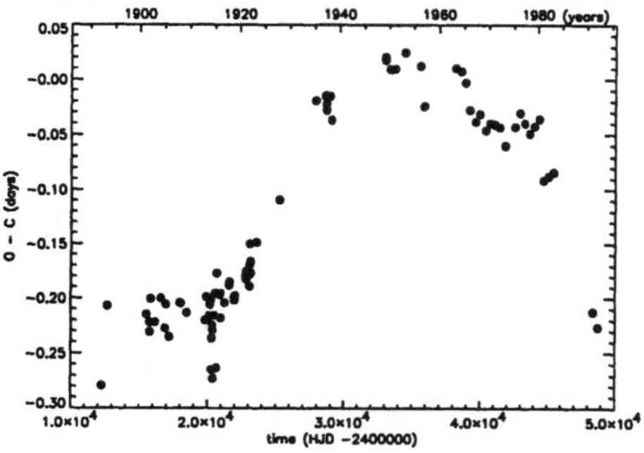

Fig. 2.16. Orbital period variations of RS CVn. The $O-C$ residuals were computed using a period of 4.797817 days. Note the long-term variation with a possible period of 100 years and smaller variations on a shorter time scale. See text. (After Rodonó et al. [96]).

Hall [42] used O-C based photometric periods of 85 active stars to infer their differential surface rotation. The range of observed rotation periods should be a measure of the strength of differential rotation and could be converted to a differential-rotation parameter, $k \equiv \Delta\Omega/\Omega$, if one assumes a restricted latitude range for the spots. After correcting the sample for their various Roche-lobe fillings by introducing a dimensionless Roche-lobe filling factor F $(0 < F < 1)$, Hall [42] found the following relation

$$\log k = -1.96(\pm 0.12) \ + \ 0.71(\pm 0.07) \ \log P_{\rm rot} \ - \ 0.38(\pm 0.16) \ F \ . \qquad (2.23)$$

In words, differential rotation becomes *much* smaller as rotation becomes faster. Most stars in Hall's sample had coefficients 10 to 1000 times smaller than the Sun ($k_\odot \approx 0.2$), i.e. they are close to being rigid rotators. Differential rotation is more suppressed the faster a star rotates. This is what non-linear dynamo theories predict. One word of caution though: Hall used the observed timings of light curve minima, i.e. the times when a spot is facing earth, to deduce the changes of the photometric period. If there is more than one spot and if they are not strictly separated by 180°, the actual light-curve minimum traces a timing that corresponds to the disk averaged maximum spot coverage and not to the longitude of a particular spot. Recent measurements of differential rotation by means of actual light-curve spot modelling (e.g. for HR7275; Strassmeier et al. [122]) and even in combination with Doppler imaging (e.g. for HU Vir; Hatzes [44], Strassmeier [109]) gave coefficients even smaller by a factor 2–4 than those predicted from Hall's relation. However, taking into account the large scatter in the relationship above, the values observed individually are still within the uncertainties of Hall's general relationship.

Unfortunately, photometric time-series modelling can not provide the sign of the differential rotation coefficient, i.e. decide whether the equatorial regions rotate faster than the polar regions or vice versa. It would require to actually resolve the stellar disk and to determine the latitudinal location of the spots that give rise for the light variations observed. This has been done with the Doppler-imaging technique and many new results regarding differential surface rotation (for binaries and single stars) become now available. Instead of reviewing the individual results I refer to the (upcoming) proceeding of the 11th "Cool Stars, Stellar Systems, and the Sun" meeting held shortly after this summer school in Puerto de la Cruz, Tenerife in early October 1999.

2.6.3 Applegate's Scenario

Photometric observations of eclipsing binaries showed that most Algols, RS CVn- W UMa-, and CV-systems[3] exhibit orbital-period variations with amplitudes of around $\Delta P/P \approx 10^{-5}$ to 10^{-6} and with periods of decades to centuries. Figure 2.16 shows a plot of the $O-C$s for the prototype RS CVn binary during the past 100 years (cf. Rodonó et al. [96]). From a study of a large sample of Algol binaries, Hall [40] found that only Algols with spectral types later than G0, i.e. stars with a sufficiently deep convective envelope, had also alternate period variations as observed in many active binaries. An example of an alternate period change, i.e. where the period increases at one time and decreases at another, is again shown in Fig. 2.16 for the prototype active binary RS Canum Venaticorum. These observations point to a connection between the orbital-period variations and stellar magnetic activity because the presence of a deep convection zone and rapid rotation almost guarantees the functioning of a stellar dynamo.

Applegate [3] postulated that changes in the gravitational quadrupole moment of the active component in a binary system can alter the orbital period while keeping the orbital angular momentum constant. If an active star's quadrupole moment increases – due to a yet unknown reason –, the gravitational pull onto the other component increases as well, and an additional centripedal acceleration takes place that moves the two stars closer together and thus increases the orbital velocity. On the contrary, if the quadrupole moment decreases, the separation increases and the orbital velocity decreases. Therefore, any mechanism that can alter the gravitational quadrupole moment of a star can also alter the orbital period; and will do so on time scales even shorter than tidal friction (cf. Lanza et al. [65]) (see also Sect. 2.5.2).

Two mechanism are discussed that can change the graviational quadrupole moment of a star, which amounts to a deviation of its spherical shape together with a change of its gravitational potential energy. One mechanism is to somehow redistribute the internal angular momentum and the other is a change of the azimuthal magnetic field. Both result in an oblateness of the star. Applegate [3] suggested that there is a certain amount of angular momentum that is periodically exchanged between the inner and the outer convection zone as a function of the magnetic activity cycle. At times when high angular-momentum material comes to the outer convective layers, the surface will spin up and become slightly oblate which, in turn, increases the gravitational quadrupole moment.

[3] Cataclysmic Variables.

The opposite happens when this material sinks to the bottom of the convection zone. This model would also require a change in luminosity of order 0.1 mag which was found to be in agreement with some observations (Hall [41]; but see also Baliunas & Soon [5] for solar observations). Photometric cycles for a set of ten active stars were recently investigated by Oláh et al. [78] who found multiperiodic periods for eigth of them with cycle periods as long as over 60 years and as short as 2.4 years (e.g. for V833 Tau; K5V+?).

On the one hand, a magnetic field of several kiloGauss would be sufficient to exert the torque through its Lorentz force that is required to also account for such an angular-momentum exhange. The underlying mechanism is a periodic change of the azimuthal field component which results in a change of the effective centrifugal acceleration within the stellar interior and thus also changes the gravitational quadrupole moment. Recently, Lanza et al. [65] linked such an angular-momentum exchange to the framework of different types of dynamo theories that were proposed to explain solar and stellar activity. Based on the fact that the orbital-period variations change sign and that the driving force for the quadrupole moment must do so as well, Lanza et al. [65] concluded that it is more likely that active stars harbour a $\alpha^2\Omega$-type[4] dynamo rather than the solar-like $\alpha\Omega$ type (for a review of dynamo theories see, e.g., the proceedings of IAU Colloq. 130, edited by Tuominen, Moss, and Rüdiger 1991).

An important ingredient of most dynamo theories is differential rotation. It is thought to be the agent to convert the originally weak poloidal field into a strong azimuthal field. This includes a conversion of kinetic rotational energy into magnetic energy. Following Lanza et al. [65] and references therein, the quadrupole moment of a rotating star is

$$Q = k\,\frac{\Omega^2\,R^5}{G} \qquad (2.24)$$

where k is the apsidal-motion constant that is tabulated for stellar models of various age, e.g., by Claret [40]. Ω is the angular rotational velocity, R the stellar radius, and G the gravitational constant. The change of Q is then given by a change in the angular velocity $\Delta\Omega/\Omega$

$$\Delta Q = Q\left(\frac{\Delta\Omega}{\Omega}\right) . \qquad (2.25)$$

The orbital period variation that corresponds to the change of Q is given by

$$\frac{\Delta P}{P} \approx \frac{\Delta Q}{M\,a^2}\,, \qquad (2.26)$$

where M is the mass of the magnetically active component, and a is the semi-major axis of the orbit. For a typical RS CVn system, k is ≈ 0.15, $R \approx 2-7\ \mathrm{R}_\odot$, and $\Omega \approx 10^{-5}$ to 10^{-4} s^{-1}. According to Eqs. (2.24–2.26) this implies that a

[4] In the absence of differential rotation or any other shear layer, Coriolis forces and turbulent convection throughout the convection zone can take over ($=\alpha^2$ dynamo).

relative change of angular velocity of $\Delta\Omega/\Omega \approx 10^{-3}$ to 10^{-2} is needed to explain the observed orbital-period variations in Algols and RS CVns (cf. Applegate [3]). For RS CVn itself, an angular velocity variation of 3.6% is required. But after including the effect of the Lorentz force, and assuming a mass of 50% of the total mass for the convective envelope, the required change is just 0.4%. This value is then in agreement with the observations in Fig. 2.16.

Lanza et al. [65] also considered the resulting balance between the kinetic energy and the magnetic energy and found that, if 1% of the kinetic energy of the convective envelope of a subgiant is converted into magnetic energy during the activity cycle, the mean field change is around 3×10^4 Gauss for the azimuthal field. The underlying poloidal field strength must be of order 100 G to drive the torsional oscillation in the first place. Recent Zeeman-Doppler images of the active binary HR1099 (=V711 Tau) by Donati [23] recovered all three components of the magnetic field on the surface of this active K subgiant. Figure 2.17 shows the separate Doppler images for the three magnetic-field components with azimuthal fields of up to ±1 kG. Donati [23] also detected orbital-period variations of HR1099 with a possible period of 18±2 years. It provides evidence that the average azimuthal field in the convection zone is of the right order, i.e. 6 kG as suggested from an upper limit of the changes in angular rotation velocity of HR1099 (Donati [23]). Because there will be some radial gradient of the field strength within the convective zone, the expected field at the surface is likely 1 kG and thus in agreement with the observations.

2.7 Rotation and Dynamo

2.7.1 The Rossby Dynamo Criterion

In the earlier sections, we saw that magnetic activity, i.e. dynamo action in solar-type stars, is always accompanied by magnetic braking. Ironically, the magnetic field generated or strengthened by the dynamo is destroying itself via the loss of stellar angular momentum and thus by braking of the stellar surface (which in turn lowers the effective dynamo action and thus the generation of the magnetic field). Durney & Latour [26] suggested that this can not go on forever, e.g. until the stellar rotation comes to a total halt. They showed that a classical $\alpha\Omega$-dynamo will switch off long before because the α-effect will cease to function. (This is the name of the effect in which a toroidal field will be twisted due to differential rotation so that at least some of it becomes poloidal. The amount of twist is called the helicity; see Fig. 2.1.)

Durney & Latour realized that enhanced turbulent convection will eventually destroy the differential rotation and, if a solar-type dynamo should work successfully, the poloidal field must be wound up quickly enough not to be destroyed. Therefore, the stellar rotation period must be shorter than the typical convective turn-over timescale:

$$P_{\mathrm{rot}} < \tau_{\mathrm{conv}} = \frac{\ell}{v_{\mathrm{conv}}}. \tag{2.27}$$

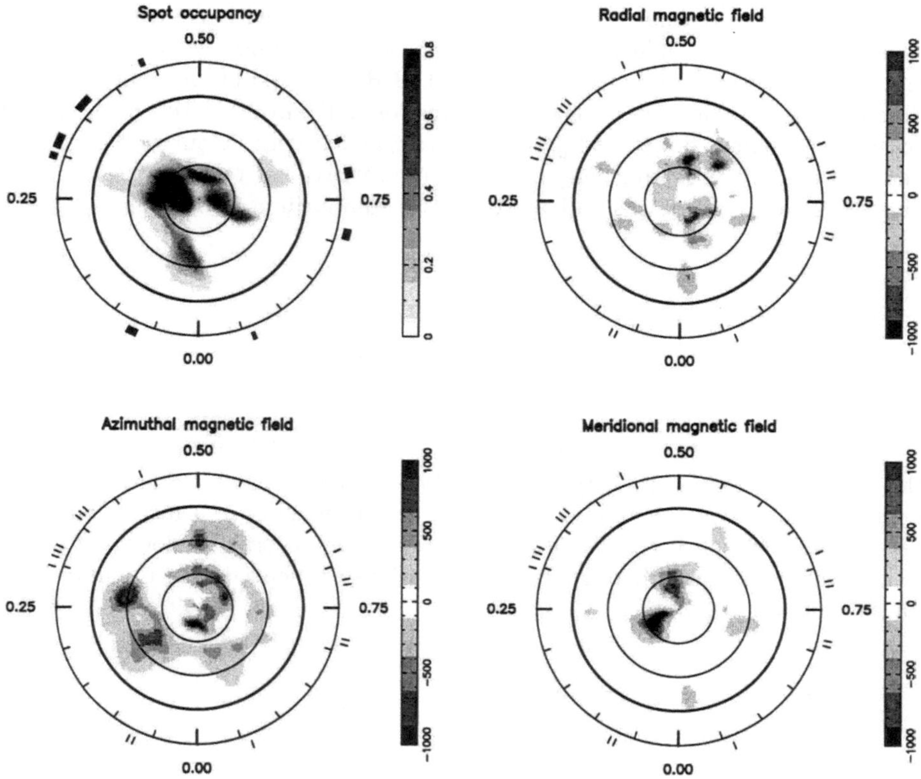

Fig. 2.17. Zeeman-Doppler image of the active K component of the close binary system HR1099 (=V711 Tau). All three field components were resolved and show field strengths of ±1 kG in 1995.94. The upper left map is the photospheric temperature map. Notice the strong azimuthal field at or near the stellar rotation pole. (After Donati [23]).

This is the classical dynamo criterium (see Rodonó [93] for a more complete summary). Stars with periods longer than this critical period should not be active. ℓ is a characteristic convective length and v_{conv} a characteristic convective velocity. The ratio of P over τ is called the Rossby-number and must be less than one for a star to be active (see Fig. 2.18):

$$Ro \equiv \frac{P_{\mathrm{rot}}}{\tau_{\mathrm{conv}}} < 1. \tag{2.28}$$

With $P_{\mathrm{rot}} = 2\pi R_\star / v_{\mathrm{rot}}$, we may rewrite (2.28) to

$$Ro = \frac{2\pi R_\star}{v_{\mathrm{rot}}} \frac{v_{\mathrm{conv}}}{\ell}, \tag{2.29}$$

and predict the (observable) minimum rotational surface velocity for active stars

$$v_{\mathrm{rot}} > 2\pi v_{\mathrm{conv}} \frac{R_\star}{\ell}. \tag{2.30}$$

While we can estimate the stellar radius, we are left with the two unknowns ℓ and $v_{\rm conv}$. What could a "characteristic" length or velocity be? Maybe the thickness of the entire convection zone (0.28 $R_\odot$ in case of the Sun)? Or the diameter of a typical convective eddy (maybe 100 km)? Or the path that a convective element flows until it is completely mixed with its surrounding? The latter is called the mixing length, ℓ, and is proportional to the pressure scale height, $H_{\rm p}$ of a perfect gas under hydrostatic equilibrium,

$$\ell = \alpha\, H_{\rm p}, \tag{2.31}$$

where α is a dimensionless parameter of the order of one and

$$H_{\rm p} \equiv -\frac{dr}{d\log p} = \frac{kT}{\mu\, m_{\rm H}\, g}\,. \tag{2.32}$$

If we compare the extremes, say the pressure scale height at the bottom of the convection zone (with a typical velocity for the overshoot region) to the full depth of the convection zone, we get alone an uncertainty of a factor four. Not yet counting the uncertainties of the adopted "characteristic" velocity. It seems to me that significantly better *stellar* observations are needed to pinpoint criteria for dynamo operation.

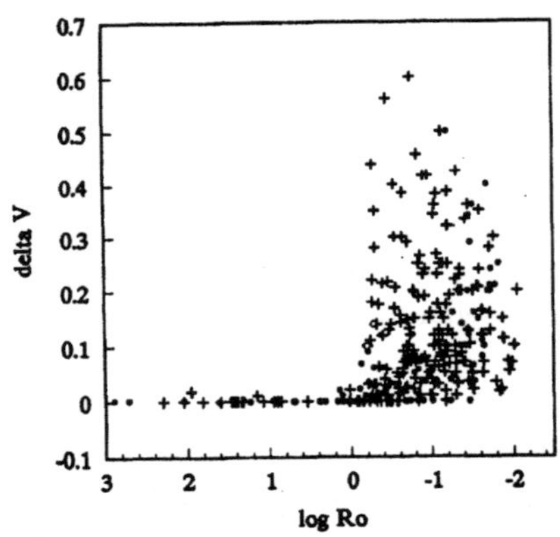

Fig. 2.18. Photospheric activity as a function of the Rossby number. "Activity" is represented by the photometric V-band amplitude in magnitudes (delta V). This sample of 359 stars includes stars throughout the HRD. However, only stars with *Ro* less than ≈ 0.65 ($\log Ro \approx -0.2$) appear to be photospherically active. (After Hall [43]).

2.7.2 Rotation-Activity Relations

All stellar dynamo theories predict increased efficiency with increased rotation (for a review on dynamo action and rotation see Rodonó [93]). This is also what has been observed from disk-averaged activity indicators since the days of O. C. Wilson's seminal paper – actually a series of papers, e.g. Wilson [138] – and its quantification for single dwarf stars by Noyes et al. [77]. These studies, and many other, adopted the strength of the Ca II H&K-core emission lines near 395 nm as the indicator of magnetic activity. Atmospheric activity is most easily measured

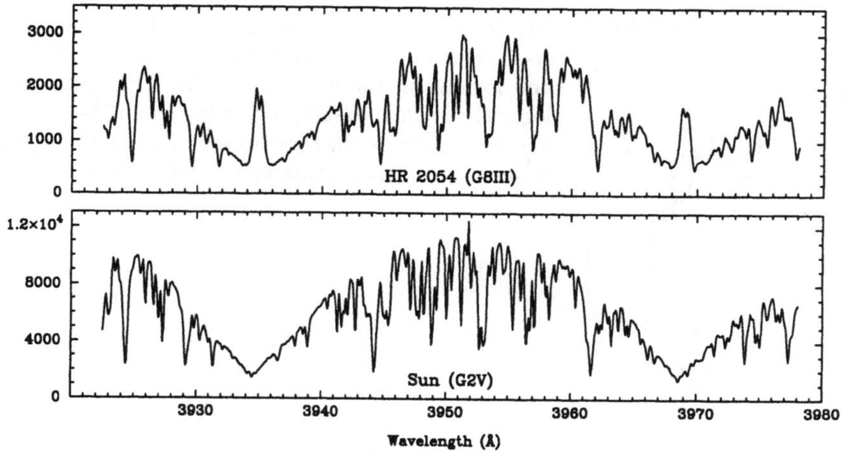

Fig. 2.19. A comparison of a Ca II spectrum of the Sun (lower panel) and that of an active binary (top panel). Notice the very strong H and K emission lines in the active-binary spectrum. Both spectra were taken with the 0.9-m coudé feed telescope at KPNO.

by the radiative losses from the layers above the photosphere since they depend on the magnetic-flux density in the respective lower layers (most dominantly from the flux density in the photosphere). The solar H&K-line cores come from a region 2000 km above the $\tau = 1$ photosphere where the local temperature has risen up to $\approx$7000 K. The ions and electrons are accordingly faster compared to the underlying photosphere and collisions and scattering between calcium atoms and fast electrons are more frequent. Figure 2.19 compares the Ca II H and K emissions from the Sun to that of an active binary.

Synoptic solar H&K observations verified the expected relation to the photospheric magnetic-field strength throughout the solar cycle (see, e.g. Schrijver [100]) and represent the basis for stellar applications. It is important to note that the disk-averaged observations of the Sun correspond rather to an "average" magnetic feature than to an averaged network. Thus, our stellar observations of surface features from disk-integrated data are on a sound basis. But again, what about binaries?

Figure 2.20 shows the dependence of activity on surface rotation. The two panels *a* and *b* plot rotation as velocity and period, respectively, and explicitely compare single stars (plusses) with binary stars (dots). We see that, on average, a comparable star in a binary rotates faster than its single counterpart and is also more active. Note that the sample in Fig. 2.20 includes only stars that are significantly more active than the Sun. Despite that there is a clear tendency of decreasing chromospheric flux with decreasing rotational velocity or period, the scatter for a given rotation is about 1 dex. This can be only partially due to observational errors in the rotation rates and strengthens the view that it is not solely rotation that scales the emerging magnetic flux (Noyes et al. [77]).

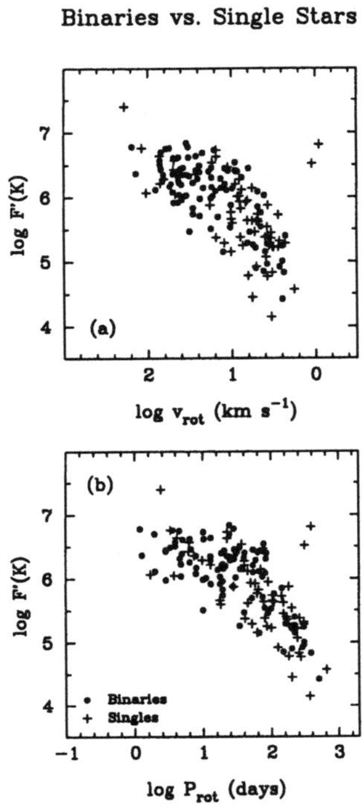

Fig. 2.20. The rotation-activity relation for evolved stars. The abscissas show $\log v_{\rm rot}$ in the upper panel (a) and $\log P_{\rm rot}$ in the lower panel (b). The ordinates are logarithmic Ca II K surface flux in erg cm^{-2}s^{-1}. Plusses and dots compare the results for single and binary stars, respectively. (After Strassmeier et al. [123]).

2.7.3 Distributed Dynamo versus Boundary-Layer Dynamo: The Red-Dwarf Puzzle

The puzzle is a simple one (if it is one after all?): if there were no radiative core in the stellar interior, and thus no overshoot region that would enable the generation and storage of magnetic fields, we would expect such a star to be magnetically inactive. This is the old controversy of a boundary-layer dynamo below the convection zone against a distributed dynamo throughout the convection zone or even just below the stellar surface (see the book by Stix [108] for further details). Canonical stellar evolution theory predicts that stars below $\approx$0.3 $M_\odot$ are fully convective and thus should not possess a boundary layer.

As was demonstrated by SOHO, local velocity fields can be used to indicate the position of the dynamo. Figure 2.21 shows in the bottom panel the distribution of the velocity of sound as a function of depth in the solar interior (the upper panel shows the rotation rates for the equatorial regions; see Fig. 2.2 for other latitudes as well). The increase of detected frequencies between 0.6–0.7 $R_\odot$ (a depth of approximately 190,000 to 270,000 km) is due to a local increase of the speed of sound, maybe acompanied with a chemical inhomogeneity, and coincides with the location of the overshoot region. That, in turn, suggests it to be a rather turbulent place, exactly what a dynamo needs.

However, HST observations of the very low-mass star Van Biesbroeck 10 (= Gliese 752B; M8Ve) in 1994 uncovered a strong UV flare and provided indirect evidence for the existence of surface magnetic fields on a fully-convective star.

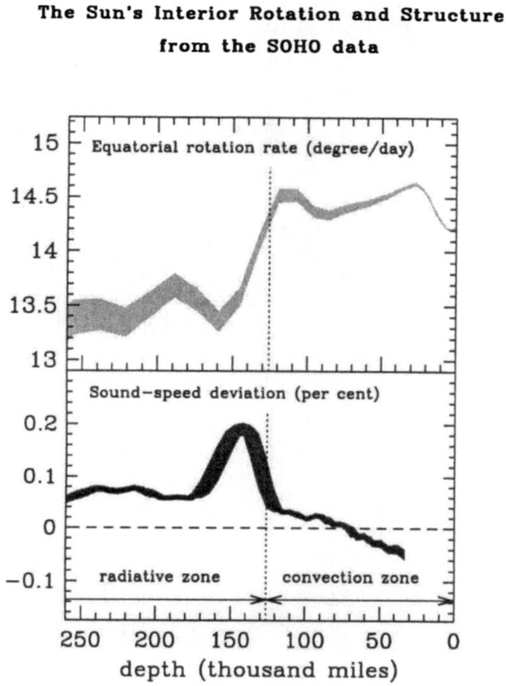

Fig. 2.21. Upper panel: A plot of the solar equatorial rotation rate as a function of depth. These data show a sharp shear layer just beneath the solar convection zone (its lower boundary is indicated by the vertical dotted line). Lower panel: Also, they indicate that the sound waves speed up in this region and provide a more turbulent regime that is necessary for a dynamo process. (After Kosovichev et al. [57]).

Actually, VB 10 (Gliese 752B) is the secondary component of a binary system just about 6 pc away. The primary is a red dwarf of just one third of the mass of the Sun, while the secondary – the star that showed the flare – is near the threshold of stellar hydrogen burning of just 0.08 $M_{\odot}$, which is certainly not massive enough to have a radiative core.

Stellar flares are caused by intense, twisted magnetic fields that accelerate electrons and even protons to much above the local velocity of sound. The rapid energy release heats the surrounding stellar atmosphere up to several 100,000 K. The most extreme flares on the Sun can even cause brief nuclear reactions on the surface, as inferred from the appearance of 2.2 MeV γ-ray emission. The outburst temperature on VB 10 was indirectly measured at 270,000 K and Linsky [68] attributed it to the presence of an intense, but unstable, magnetic field. After all, this suggests a dynamo process that operates without the presence of a boundary layer.

X-ray emission from VB 10, VB 8 (M7Ve) and Gliese 406 (M6Ve) has been detected with ROSAT, and VB 8 and VB 10 show even evidence of flaring in X-rays (c.f. Linsky et al. [69]). VB 8 was detected already with the high-resolution imager onboard the *Einstein* satellite and in the extreme ultraviolet by EUVE (see Drake et al. [25]). Most recently, Neuhäuser & Comerón [75] announced the detection of X-ray emission from a very young brown dwarf in the Chamaleon I dark cloud and Neuhäuser et al. [76] possibly found another young, bona-fide brown dwarf in the ρ-Ophiuchi cloud. Stellar X-ray emission is usually either due

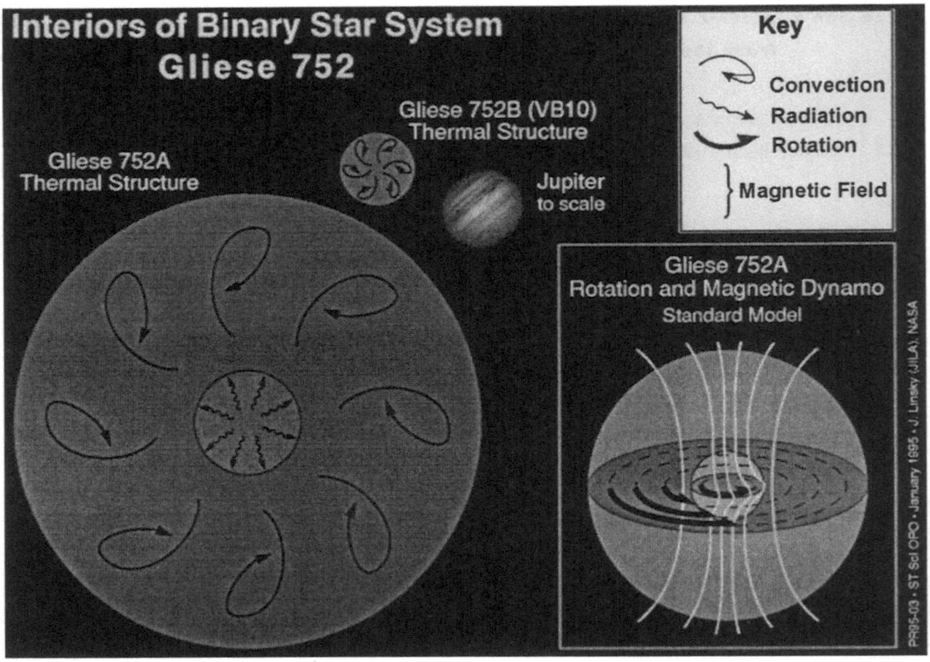

Fig. 2.22. Artistic conception of the interiors of the two components of Gliese 752A and B (the latter is also called VB 10). While the primary still has a small radiative core, the very-low mass secondary (VB 10; 0.08 $M_{\odot}$) must be fully convective. However, it was the secondary that showed a strong UV flare and thus must harbor a strong magnetic field. The insert shows a somewhat simplified standard model for the primary and emphasises the anchoring of magnetic field lines in the boundary between radiative core and convective envelope. (After STScI press release PR95-03, J. Linsky, JILA).

to an active corona or due to shocks in a massive stellar wind. A very cool, low-mass, star like VB 10, or even a brown dwarf, can not have a massive stellar wind like an O-star and the X-ray emission must be due to stellar activity and thus a magnetic field. Further observational evidence for magnetic activity on fully convective stars comes from recent radio observations. Benz et al. [8] detected a large coronal radio source above the rotation poles of the fully-convective, young flare star UV Cet B which indicated a poloidal magnetic field of strength 15–130 G.

Therefore, we face the possibility that all late-type stars – including our Sun – operate a mixture of dynamo processes. That is, a boundary-layer dynamo in the overshoot region below the convection zone as well as a distributed dynamo throughout the convection zone. At this point we enter the regime of speculation since no computations have been done with both types of dynamos simultaneously. However, there is growing evidence for a mixed dynamo. The boundary-layer dynamo is successful in explaining the various sunspot phenomena including their latitudinal distribution, Joy's law (tilt dependence with latitude), and its proper motions (Caligari et al. [13]) and also provides the correct timescales to store magnetic flux for as long as the sunspot cycle. The place of the dynamo must be sufficiently "stable" and, according to the Schwarzschild-criterium, means that the local $d\ln T/d\ln p$-gradient must be smaller than the adiabatic temperaturegradient (T temperature, p pressure)

$$\nabla \equiv \frac{d\ln T}{d\ln p} < \nabla_{\mathrm{ad}}. \tag{2.33}$$

Spruit & van Ballegooijen [107] identified the overshoot region to be such a sub-adiabatic region[5] and suggested it to be the location of the solar dynamo (see also Stix [108]). The overshoot-layer dynamo also explains the existence of high-latitude and even polar spots on very rapidly rotating stars (Schüssler & Solanki et al. [103]) but currently fails to explain the equatorial and low-latitude spots on these stars (see also Belvedére et al. [7]). On the other hand, recent Zeeman-Doppler images of the active stars LQ Hya and HR 1099 by Donati [23] show a dominant azimuthal field component (i.e. a field parallel to the surface) and Donati [23] claimed that this is a sign of a distributed dynamo operating in the convection zone just beneath the stellar surface. Since LQ Hya is a single star and HR 1099 a RS CVn-type close binary, we presume that binarity *per se* does not influence the quality of the dynamo. Furthermore, Zeeman splitting of lines in the very active M4.5Ve-dwarfs Gliese 729 and EV Lac revealed very high surface field strengths of 2–4 kG (Johns-Krull & Valenti [51]). Since late M dwarfs must have a very deep convective envelope due to their low temperatures, or are even fully convective, the observations of strong fields on these stars also favors the existence of a distributed dynamo rather than a boundary-layer dynamo.

Finally, the relation of velocity fields induced by non-radial oscillations and the appearance of magnetic surface fields must be incorporated into the next-generation dynamo models – solar and stellar ones – if real observations are to be modeled (not to mention the inner primordial field as discussed in Sect. 2.3.2). Reasoning to do so comes from the fact that the average oscillation frequency of the Sun changes along with the 11-year sunspot cycle. Pallé et al. [79] first demonstrated this from disk-integrated resonance-spectrometer measurements of frequency shifts obtained at the Observatorio del Teide. Figure 2.23 shows the relatively close relation between frequency shift and sunspot number for the previous cycle. It suggests that the Sun changes some interior (dynamic?) parameters during its sunspot or magnetic cycle just as Applegate's scenario in Sect. 2.6.3 requires. Whether it is of right order is currently unknown.

But how can we "observe" the dynamo? Well, of course, we can't do so directly, but the Sun demonstrates that we could do it indirectly; by using magnetic surface phenomena like spots as surface tracers of the dynamo action. The appropriate observation technique is called Doppler imaging, or Doppler tomography, and we will discuss it in more detail in the following Sect. 2.8.

[5] A region with $\delta \equiv \nabla - \nabla_{\mathrm{ad}} < 0$ is called subadiabatic and is stable, a region with $\delta > 0$ is called superadiabatic and is unstable.

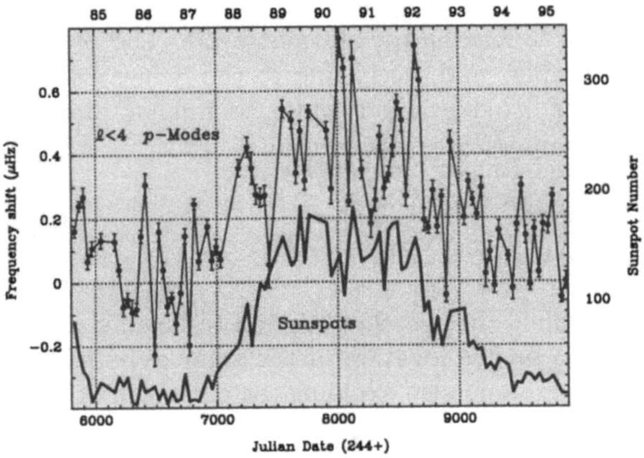

Fig. 2.23. The change of oscillation frequencies of $\ell < 4$ of the Sun as a function of time (upper curve). The lower curve plots the relative sunspot number. Time is indicated in years on the upper axis and in Julian dates on the lower axis. (Data from Pere L. Pallé, Instituto de Astrofisica de Canarias).

2.8 Doppler Imaging of Stellar Surface Structure

This section outlines the principles of the Doppler-imaging technique, summarizes the observational requirements, and presents some relevant observations and results of stars in close binaries.

2.8.1 Principle

Doppler imaging is an elaborate computational technique similar to medical tomography and inverts a series of high-resolution spectral line profiles into an "image" of the stellar surface (Deutsch [20], Vogt et al. [135], Rice et al. [92], Collier Cameron [16], Piskunov & Rice [84]). Cool starspots produce distortions in the spectral line profiles that systematically change during a star's rotation (Fig. 2.24). It is the way how these distortions change with time which allows to reconstruct the stellar surface temperature distribution. The technique is very similar to that described in the chapter by Marsh in these proceedings except that the velocity range in the line profile is restricted to the (rotating) stellar surface and the intensity range is given by the temperature contrast between spotted and unspotted photosphere.

A rotationally broadened spectral line profile, $R(\lambda)$, as a function of rotational phase, φ, can be written as

$$R_{\rm obs}(\lambda,\varphi) = \frac{\iint I_{\rm c}(M,X(M))R_{\rm loc}(M,X(M),\lambda+\Delta\lambda_{\rm D}(M,\varphi))\cos\theta\, dM}{\iint I_{\rm c}(M,X(M))\cos\theta\, dM} \tag{2.34}$$

where $I_{\rm c}$ is the continuum intensity, $R_{\rm loc}$ the local line profile, M the position on the stellar surface, $\Delta\lambda_{\rm D}$ the Doppler shift of the surface element dM at a particular rotation phase, and $X(M)$ the unknown surface parameter that is solved for. In our case $X(M) \equiv T_{\rm eff}(M)$, for Ap-stars it may be chemical abundance (or actually equivalent width).

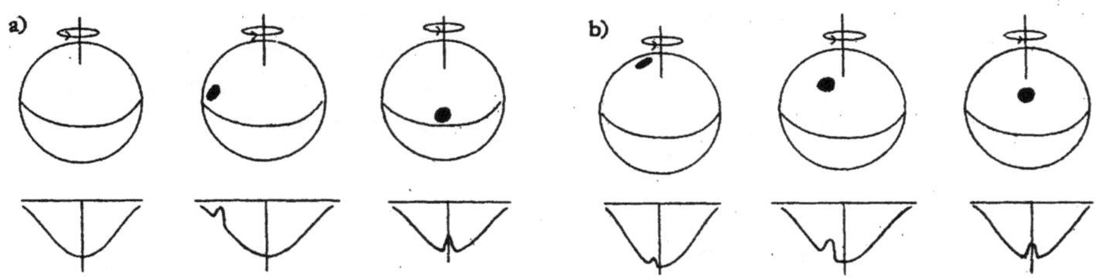

Fig. 2.24. The relation between the position of a dark spot on the stellar surface and its appearance as a bump in the broadened spectral line profile. The left three rotational sequences (a) are for a low-latitude spot and the right sequences (b) are for a high-latitude spot. (After Rice [89]).

The inverse problem amounts to finding the surface parameter $X(M)$ from the observed profiles such that

$$D_\lambda(\varphi) = \frac{1}{n_\varphi n_\lambda} \sum_{\varphi=1}^{n_\varphi} \sum_{\lambda=1}^{n_\lambda} g(\varphi, \lambda)[R_{\rm obs}(\varphi, \lambda) - R_{\rm th}(\varphi, \lambda)]^2 \quad \leq \quad \sigma^2. \tag{2.35}$$

Since this is an ill-posed problem, one introduces an additional criterion – a so-called regularisation functional $r(X(M))$ – and minimizes

$$E(X(M)) = D(X(M)) + \Lambda\, r(X(M)) \tag{2.36}$$

where Λ is the Lagrange multiplier. The choice of $r(X(M))$ is related to the choice of X, and two regularisation functionals are usually used for temperature mapping,

$$\text{Maximum Entropy}: \quad r(X(M)) = \int\!\!\int_{\rm M} X(M) \log(X(M)) dM \tag{2.37}$$

and

$$\text{Tikhonov}: \quad r(X(M)) = \int\!\!\int_{\rm M} \mid \nabla X(M) \mid^2 dM. \tag{2.38}$$

Note that the "physics" go into the computation of the local line profiles and the continuum intensities as well as the model atmospheres from which the profiles are computed via a LTE solution of the equation of radiative transfer (see, e.g., Piskunov & Rice [84], Strassmeier et al. [120]).

Continuum variations are used as additional constraint in the line-profile inversion because stars with cool spots show light and color variations that are modulated with the stellar rotation period (e.g. Rodonó & Cutispoto [94], Strassmeier et al. [114]). Simultaneous photometry in several bandpasses is required and small, but fully automatic, telescopes do a good job (see Strassmeier et al. [117]). Even amateur astronomers regularely contribute to this kind of photometric monitoring.

From solar analogy we know that starspots are regions of strong magnetic fields and cause different levels of polarization for different parts on the solar/stellar surface. Extracting spatial information from all four Stokes parameters, however, requires extremely high quality spectra (Donati [22], Semel et al. [104]).

Example: A Four-Pixel Star. Let's assume that the surface of the star that we want to Doppler image consists of just four pixels (usually one works with approximately 3000 pixels). This is convinient because then we can do the inversion simply by hand. The following example is described in more detail by Gull & Skilling [37] and also in Piskunov et al. [85].

Our "four-pixel" star looks like the following: The true pixel intensities are

I_1	I_4
I_2	I_3

known and are $I_1 = 4, I_2 = I_4 = 8$ und $I_3 = 16$. However, our "observations" can not resolve the star and gave the following results (these are synonym for the spectral line profiles observed)

$$I_1 + I_2 + I_3 + I_4 = 36, \quad I_1 + I_2 = 12, \quad I_1 + I_4 = 12. \tag{2.39}$$

Obviously, from these data we can not determine the individual intensities because we can not solve for four unknowns with just three equations, i.e. observations. This is called an ill-posed problem because there are many possible solutions that would fit the observations. Now we introduce the regularisation in form of a maximum entropy or Tikhonov functional and write out the terms in the integrals in (2.37) and (2.38):

$$\begin{aligned} r_{\mathrm{MaxEnt}} &= I_1 \log I_1 + I_2 \log I_2 + I_3 \log I_3 + I_4 \log I_4 \\ r_{\mathrm{Tikhonov}} &= (I_1 - I_2)^2 + (I_2 - I_3)^2 + (I_3 - I_4)^2 + (I_4 - I_1)^2 . \end{aligned}$$

Each of these two equations is the sought for fourth equation to solve the problem. We now substitute into (2.39) and get

$$\begin{aligned} &\text{with} \quad r_{\mathrm{MaxEnt}} : \quad I_1 = 4, \quad I_2 = I_4 = 8, \quad I_3 = 16 \\ &\text{with} \quad r_{\mathrm{Tikhonov}} : \quad I_1 = 3, \quad I_2 = I_4 = 9, \quad I_3 = 15. \end{aligned}$$

Maximum entropy provided the correct solution of the problem and Gull & Skilling [37] compared this result with the answer to a more every-day question:

"How many kangaroos are blue-eyed and left-handed if 1/3 of all kangaroos have blue eyes and 1/3 are left handed, assuming that these properties are uncorrelated". Note though that the Tikhonov regularisation provided the same perfect fit to the observations. Since the four pixels had well defined intensities, and the Tikhonov regularisation searches for the smoothest possible solution, it will smooth the mean surface gradient to an extent that is still able to reproduce the observations. The maximum entropy regularisation is designed for cases where no correlation exists between neighbouring pixels, which is the case for our four-pixel star, and thus provided the correct answer. Actual applications, however, have to deal with noise in the data (our four-pixel star data were implicitely of infinite signal) and then a Tikhonov regularisation may be as appropriate as the maximum entropy.

2.8.2 Observational Requirements

Two stellar parameters – brightness and rotational broadening – dictate the instrumental requirements for Doppler imaging. The former sets a limit for the achievable S/N ratio and the latter determines the size of a resolution element on the stellar surface. In detail, there are five observational criteria to be considered (see also Collier-Cameron [16] and Strassmeier [110]):

Ratio of Local Line Broadening to Rotational Broadening. This ratio is fixed for a given star and should be as large as possible. It limits the resolution on the stellar surface even at infinite instrumental resolution,

$$\frac{r_{\mathrm{spot}}}{R_{\mathrm{star}}} \geq \frac{FWHM_{\mathrm{star}}^{\mathrm{local\ line}}}{FWHM_{\mathrm{star}}^{\mathrm{Doppler}}} \tag{2.40}$$

where r is the spot radius (for an assumed circular feature) in units of the stellar disk radius R. The smallest $v \sin i$ of a star that was successfully Doppler imaged is currently 17.5 km s^{-1} (HD 129333; Strassmeier & Rice [124]). The upper limit of $v \sin i$ is basically determined by the achievable S/N ratio of the spectra because the lines get very shallow at the high rotation rates observed, e.g., in young open-cluster stars. The current record holder is nevertheless the single field G-giant FK Comae with a $v \sin i$ of 160 km s^{-1} (Doppler imaged by Korhonen et al. [56]).

Remark 1. Only stars with $v \sin i \geq 17$ km s^{-1} are currently useable for Doppler imaging !

Noise and Photon Statistics. Because the spot shape on the stellar surface is determined by the shape of the bump in the line profile, we aim for the highest possible S/N. Observations showed the existence of line-profile variability on the sub-percent level with respect to the continuum and if we would like to detect a

1% spot-bump amplitude ($R_{\rm bump}$) at a level that is n times the uncertainty of a spectral-line profile point ($\sigma_{\rm profile}$), we require

$$R_{\rm bump} \geq n\, \sigma_{\rm profile} \text{ with} \tag{2.41}$$

$$\sigma_{\rm profile} = f\,(\Delta t, \dot{N}_\lambda, N_{\rm RON}, \nu_{\rm sampling}, 1 - R_{\rm line}, \text{ff}, \text{sky}\dots) \tag{2.42}$$

where Δt is the integration time, $\dot{N}_\lambda$ the stellar photon flux, $N_{\rm RON}$ the read-out-noise of the CCD, $\nu_{\rm sampling}$ the sampling frequency of pixels through the line width, $1 - R_{\rm line}$ the depth of the spectral line, *ff* the "stability" of the flat field of the CCD, and *sky* the sky background contribution. Obviously we need a signal-to-noise ratio of ≈300 for a 3σ detection. Order extraction from large format echelle spectra must be done extremely carefully (see, e.g., Donati [22]) to avoid systematic line profile errors due to, e.g., a slight misalignment between dispersion direction and CCD pixel rows. Much higher S/N can be achieved by a least-squares deconvolution of individual line profiles in an echelle spectrum and consequent coadding (Donati [22], but see also e.g. Kennelly et al. [53], [54]).

The series of maps in Fig. 2.25 compares some simulations for various S/N ratios. A S/N of 150:1 represents a typical observation of a V=10th-magnitude star with a 3–4m class telescope, a spectral resolution of R=120,000 and a 50-min integration time (e.g. as achieved in paper IX for the pre-main-sequence star HDE283572 with the 3.6-m CFHT; Strassmeier & Rice [125]). The maps in the right column in Fig. 2.25 are difference maps between the original input map and the recovery and emphasize the surface regions of increased sensitivity. Overal, we conclude that S/N ratio *alone* does not significantly improve the recovery once S/N≈300:1 is surpassed.

Remark 2. The desired S/N ratio per wavelength bin is at least 200–300:1 !

Instrumental Resolution. The best resolution on the stellar surface is achieved when

$$FWHM_{\rm instr} \leq FWHM_{\rm line} \tag{2.43}$$

where $FWHM_{\rm line}$ is the width of the intrinsic (local) line profile. Of course, even for stars that would not rotate, the observed line widths are significantly broadened. For example, a solar-type star with $T_{\rm eff}$=5000 K has a thermal line width of 1.2 km s^{-1}, a microturbulence of ≈ 2 km s^{-1}, and a macroturbulence of ≈ 4 km s^{-1}. While the latter two mechanism can be deconvolved from the observations the thermal width remains as the intrinsic barrier. If we now consider the instrumental resolutions usually applied for Doppler-imaging work, i.e. $\lambda/\Delta\lambda \approx 35\,000 - 120\,000$ or ≈ 8–3 km s^{-1}, we see that, even when $\lambda/\Delta\lambda$ is 100 000, one gives away resolution on the stellar surface by approximately a factor of two. Nevertheless, simulations showed that reliable images can be reconstructed even when only five spectral resolution elements are available across the line (Piskunov & Wehlau [86], Strassmeier & Rice [124]).

Remark 3. One should aim at $\lambda/\Delta\lambda \approx 250\,000$ for stars with small $v \sin i$ and 100 000 for stars with moderate to large $v \sin i$!

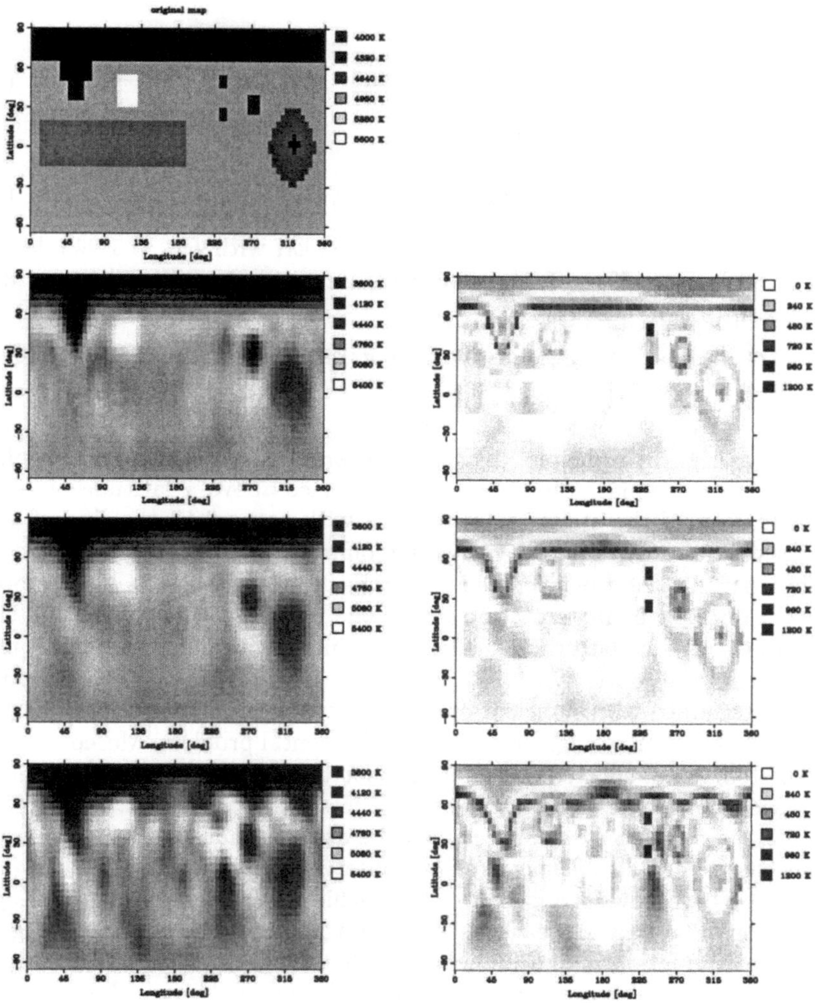

Fig. 2.25. Examples of recovery using a Tikhonov penalty function at various S/N ratios. The top image is the (artificial) input image. The others are, from top to bottom, recovered with S/N=3000:1, 300:1, and 75:1. The images in the right column are the difference maps between input and output. (The grey scale indicates the temperature difference in Kelvin. White regions indicate no difference, and black regions indicate maximum difference.) A reconstruction with S/N=900 (not shown) is practically identical to the S/N=3000 case. (After Rice & Strassmeier [91]).

"Bump" Smearing. During a CCD integration the star rotates and the spots keep moving through the profile and consequently a spot bump will be eventually smeared out. We can easily show that the smearing width, δ, in per cent of the full width of the line profile is four times the integration time divided by the stellar rotation period. With

$$\left(\frac{d\lambda}{dt}\right)_{\rm spot} = \frac{2\pi}{P_{\rm rot}}\frac{\lambda_0 v \sin i}{c} \quad \text{and} \quad \Delta\lambda_{\rm rot} \simeq \frac{\pi}{2}\frac{\lambda_0 v \sin i}{c} \tag{2.44}$$

we get

$$\delta(bump) = \frac{\dot{\lambda}\Delta t}{\Delta\lambda_{\rm rot}} \simeq 4\,\frac{\Delta t}{P_{\rm rot}}. \tag{2.45}$$

As an example consider the RS CVn-binary EI Eri with $P_{\rm rot} = 1.945$ days and $v \sin i = 50\ {\rm km\,s^{-1}}$. If the integration time $\Delta t = 1^h$ then $\delta(bump)$ is $\approx 8\%$ or $\approx 10°$ on the stellar surface.

Remark 4. Integration times must be as short as possible. Aim for $\Delta t \leq 0.01\,P_{\rm rot}$.

Data Phasing. In principle, just two line profiles, taken at opposite phases, will allow to see almost the entire stellar surface above a latitude of $-i$ (i is the inclination of the stellar rotation axis with respect to the line of sight). In practice, however, Doppler imaging requires six to eight line profiles well distributed over a rotation cycle to recover the full stellar surface. If these profiles are taken during different rotation cycles one phases the data with the stellar rotation period. The latter is usually approximated by the photometric period or, when the rotation is synchronized, by the orbital period. Both periods are usually determined with sufficient precision but aliasing in case no orbital period is known, or the star is single, remains a fundamental problem. Moreover, broad-band lightcurves of spotted stars tell us that there can be changes of the spot distribution with time scales as short as a few stellar rotations. Although this might not be true for every star it must be assured that the light curve shape did not change during the spectroscopic observations!

Perfect phase coverage is only rarely achieved in real observations, given interruptions due to bad weather, telescope-time limitations etc.. In Fig. 2.26 we show a simulation of the influence of phase gaps in the data. Firstly, two successive observations are removed from the whole evenly-spaced data (original data spacing is 20°) according to a phase gap of 0.167 or 60° in stellar longitude. Secondly, we remove three successive phases (a gap of 0.222 or 80°) and then four successive phases (a gap of 0.278 or 100°). Note that Fig. 2.26 compares the obtainable image quality for input data that have very high S/N and no external errors in order to separate the phase-gap influence from other effects. For this case, we find that our Doppler-imaging technique is not severely prone to phase gaps.

Remark 5. Ideally, the observing interval should not exceed the length of one stellar rotation period and should cover as many phases as possible!

2.8.3 Some Applications to Close Binary Stars

In the following, I will discuss some recent results on a few selected binary systems.

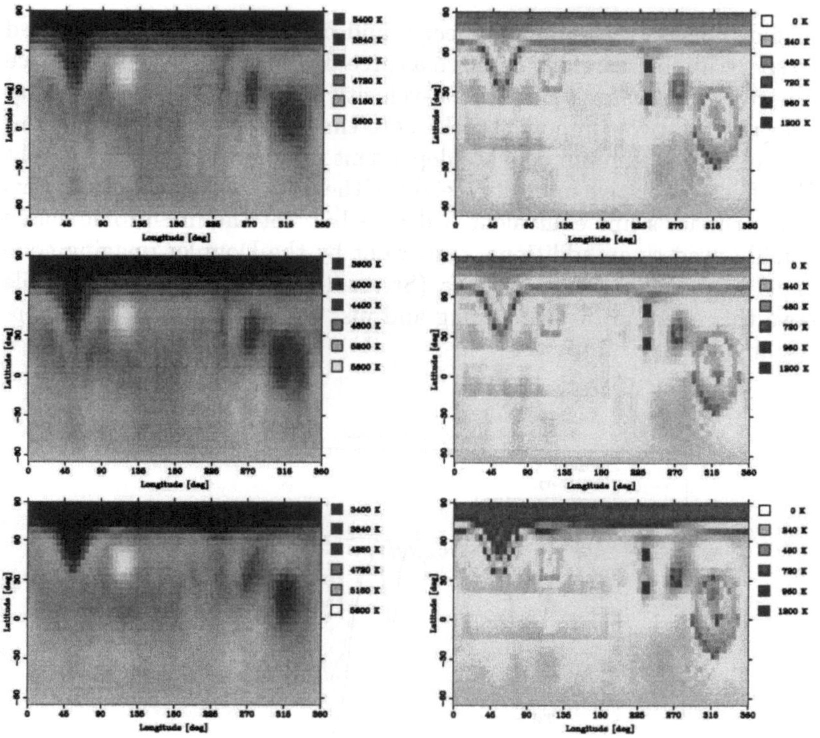

Fig. 2.26. Examples of recoveries with increasing phase gap (from top to bottom). Maps in the left column are recovered maps from the input map in Fig. 2.25, in the right column are again the difference maps. From top to bottom: phase gaps of 60° (phase 0.167) from 305° to 5°, 80° (phase 0.222) from 285° to 5°, and 100° (phase 0.278) from 265° to 5°. S/N of the data was 3000:1 in order to exclude all other effects. (After Rice & Strassmeier [91]).

A Post-main-sequence Binary: V711 Tau = HR 1099. Vogt & Penrod [134] were the first to obtain a Doppler image of V711 Tau from data in late 1981 and discovered its prominent polar spot. Because the Sun does not show spots in excess of ±40° , this discovery spurred significant interest in these types of stars. V711 Tau is a triple system and consists of a close double-lined spectroscopic pair with a spotted and rapidly rotating K1IV primary and an inactive and slowly rotating G5V secondary in a 2.8-day orbit. The tertiary is a fainter K3V star 6 arcsec away.

Subtracting the secondary spectrum from the composite spectrum is mandatory in the case of V711 Tau because, firstly, the non-negligible amount of continuum dilution due to the secondary and, secondly, the blending of the primary lines with the secondary lines near conjunction. There are several ways to remove the secondary spectrum but one has to decide already at the time of taking the observations if a particular technique will be applied. We make use of a computer program that combines two template spectra and compares it to the observed (composite) spectrum by searching through a three-dimensional parameter space to find the best estimates for $v \sin i$, radial-velocity shift, and relative intensity weight for each of the two components. We note that the primary's actual contribution to the joint continuum is phase dependent due to the star being a light variable. The effective relative intensity weight of the primary then changes. This will affect the primary's line equivalent widths – but not the line-profile shape – and is actually used as an additional constraint by the Doppler-imaging code and must not be removed from the data (Strassmeier & Bartus [113]), while others remove this effect prior to mapping and just invert the line-profile *shape* (e.g. Vogt et al. [133]).

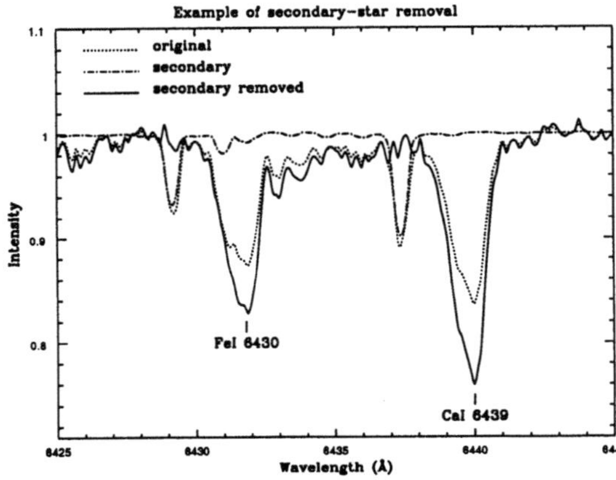

Fig. 2.27. An example of the secondary-star removal from a composite spectrum. The dotted line is the observed (composite) spectrum of V711 Tau and the dash-dotted line shows the contribution of the secondary star as a result of a two-component fit to the composite spectrum. Note that the relative continuum level of the secondary was shifted to unity for better visualization. The spectrum of the primary (full line) is then obtained by removing the secondary and re-normalizing it to the continuum level of the primary. (After Strassmeier & Bartus [113]).

Photoelectric photometry of V711 Tau through a, say, 30-arcsec diaphragm not only includes the variable primary component but also both non-variable companions. The additional light from the two stars considerably dilutes the amplitude of the photometric variations of the primary. We remove this dilution effect by applying a simple magnitude-scaling equation to the observed differential magnitudes. Additionally, the primary's light curve is folded with the periodic variations from a small ellipticity effect due to the fact that the primary is nearly filling its Roche lobe (Fekel [27]). We correct for this by removing a $\cos 2\phi$ term (ϕ being the phase) with a full amplitude of $0^{m}.026$ in V. Also plotted in Fig. 2.28 is a spot-model fit from a least-squares solution of the seasonal V and I_C light curves.

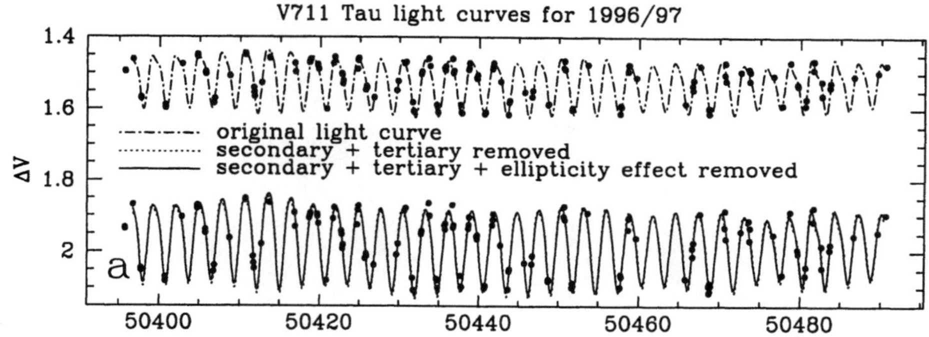

Fig. 2.28. Differential V-photometry for V711 Tau and the fit with a time-dependent spot model. The upper light curve (dots) is the original light curve prior to any preparations, while the lower light curve (dots) is after removal of the dilution due to the secondary and tertiary star and the ellipticity effect of the primary. The various line styles indicate a spot-model fit with the individual effects removed. (After Strassmeier & Bartus [113]).

At this stage, everything is ready for Doppler imaging. We first perform a full LTE spectrum synthesis in each of the wavelength regions observed by using a grid of ten model atmospheres from ATLAS-9 (Kurucz [64]) as well as a pre-calculated table of elemental abundances. We then alter the abundances – if necessary – by fitting the observed spectrum and then proceed with the altered abundance table. For each iteration a discrepancy integral between model and data is computed and minimized with the help of conjugate gradients. Simultaneous inversions of up to twenty lines with either a maximum-entropy or a Tikhonov regularisation is possible with TempMap (Rice [89], Rice & Strassmeier [90]) but we usually include no more than 10 blends per main-mapping line to keep the CPU time down. Note that TempMap inverts the V and I_C light curves simultaneously and thus further constrains the Doppler image.

Our images from late 1996 show five distinct spots (see Fig. 2.29). A large and asymmetric polar spot with an average temperature contrast of 1100 K seems to be the most prominent feature (referred to as spot A) and the map agrees well with independent results from other researchers (e.g. Donati [23]). Such a comparison is very important because, even if one has high-S/N observations, Doppler imaging remains an ill-posed problem. Together with the collection of maps by the group of S. Vogt and A. Hatzes (Vogt et al. [133]), there are now consistent maps for V711 Tau from three independent groups. All of them recover a polar spot which, in our opinion, is further evidence for its reality. We also note that the combined positions of spots A plus B, as well as C plus D agree reasonably well with the positions from our photometric two-spot time-series modelling in Fig. 2.28.

For further analysis one may compute cross-correlation functions (ccf) from two consecutive Doppler maps (Fig. 2.30). This is done by cross-correlating strips of constant latitudes as described, e.g., in Donati & Collier Cameron [24] and

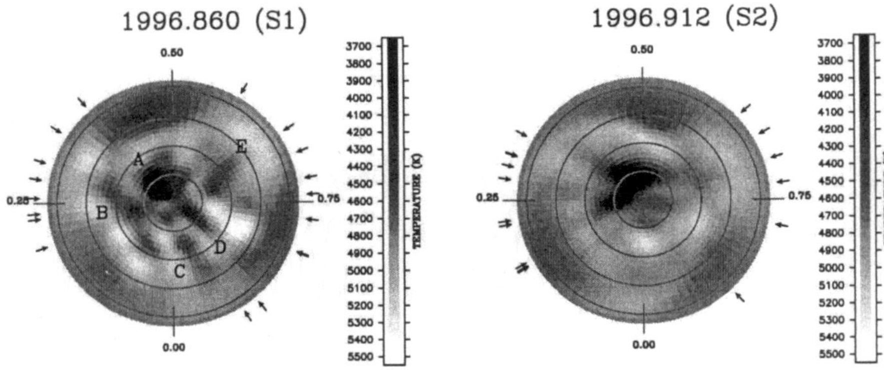

Fig. 2.29. Ca I 6439-Å images for the epochs 1996.860 and 1996.912. The maps are presented in a flattened pole-on view from latitude −40° to 90°. Latitudinal circles are drawn at separations of 30°. The phases of the spectroscopic observations are marked around the maps by arrows, latitudes are drawn in steps of 30°. Several surface features are marked A–E. (After Strassmeier & Bartus [113]).

Strassmeier et al. [115]. Such cross-correlation maps can be used to search for a consistent longitudinal migration pattern as a function of latitude, e.g. caused by differential rotation, and indeed may show a latitude-dependent phase migration for V711 Tau at one stage (as indicated by the <-shaped ccf above the equator and at a phase shift of 0.6). Since this pattern did not repeat in other ccf maps it suggests that intrinsic spot changes are common and that the surface differential rotation is likely masked by short-term changes, e.g. by a short lifetime of small spots due to local magnetic-field reconnections. High time-resolution Zeeman-Doppler images will eventually help to sort the phenomenology in the magnetic zoo.

Let us now turn to another interesting star, and to an even more complex application of the Doppler-imaging technique.

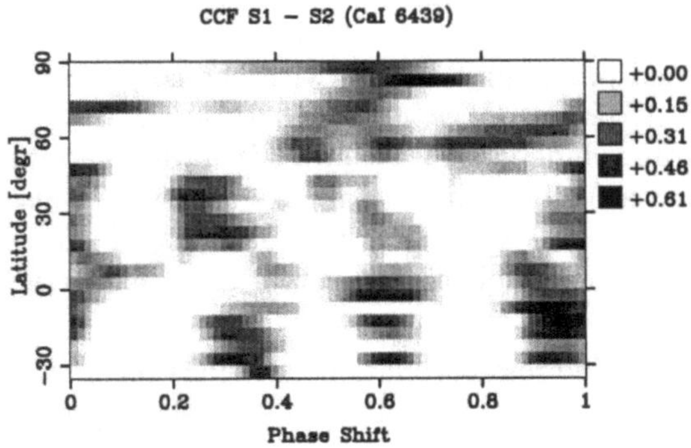

Fig. 2.30. Cross-correlation function between the two maps in Fig. 2.29. The projection style is now Mercator. The x,y-axes plot stellar latitude versus the phase shift required to optimally correlate the two maps.

A Main-Sequence Eclipsing Binary: ER Vul = HD 200391. ER Vulpeculae is a close binary with two synchronized G dwarfs (G0V+G5V) in a 0.698-day orbit. Its components are on the main sequence and both stars show large Ca II H&K emission-line fluxes and rotationally broadened line profiles (e.g. Barden [6]). No normal G dwarf would rotate that fast unless it is a young ZAMS star at the age of the α-Per cluster. ER Vul represents a laboratory to study two evolved main-sequence stars with the angular momentum of a single ZAMS star. The only other close eclipsing binaries that were Doppler imaged are YY Gem, consisting of two active dMe stars (Hatzes [45]), and V471 Tau, consisting of a K dwarf and a white dwarf (Ramseyer et al. [88]). The imaging of both of these binaries exhibit some "abnormalities" that are difficult to deal with, i.e. the rich molecular spectrum in case of YY Gem, and the wind impact from the white dwarf in case of V471 Tau, and thus renders these maps uncertain and preliminary.

The orbit of ER Vul is very close to circular with a radius of just 4 $R_\odot$ and its inclination of 67° causes partial eclipses. It is the combination of the fast rotation of two solar-type stars, and thus the large line broadening, with the occurence of eclipses that makes this system special. The eclipses will even help to further constrain the Doppler imaging because the transiting star will act like a mask when it moves across the eclipsed component.

Vincent et al. [132] presented a version of the Doppler-imaging technique that includes this additional information (see also Fig. 2.31).

In principle, the technique is the same as for a single star but that the Doppler shift in the line profiles now also includes the orbital motion of both binary components. The elements of the orbital motion are assumed to be known to

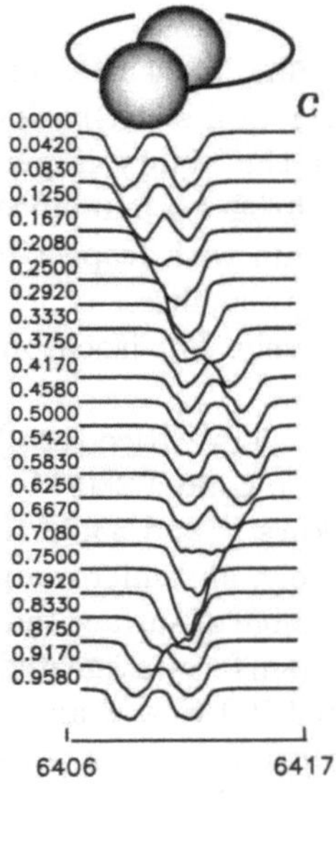

Fig. 2.31. Artificial data set for an eclipsing binary star. Phases 0.25 and 0.75 mark the conjunctions. In such a case, the Doppler shift of a particular surface feature includes the orbital motion of the respective component. The eclipsing star acts like a mask and limits the visibility time of low-latitude features on the eclipsed component (in case of a partial eclipse as is shown in this figure). Note that polar features would not be eclipsed. (After Vincent et al. [132]).

arbitrary precision, the stars are spherical, and the spectrum synthesis is precalculated, tabulated, and then interpolated. The explicit solution of the radiative transfer for both atmospheres (if not identical) allows to includes line blends and preserves line equivalent width. This is important because the relative line depths need to be modelled along with continuum variations due to spots. Furthermore, short-period binaries have very large radial-velocity variations and thus large orbital Doppler shifts. Therefore, large strips of spectrum need to be synthesized in order to reproduce the combined spectrum, and this is CPU intensive.

Following the numerical mapping experiments of Vincent et al. [132] and their principle verification, Piskunov [83] presented Doppler images for both components of ER Vul (Fig. 2.32). The temperature distribution obtained is shown in Fig. 2.32. Hot spots are recovered with $\Delta T \approx 1200$ K near the substellar points and are presumably due to the reflection effect. Cool regions were also detected on both components but seem to be unrelated to the relative positions of the two stars. A large feature on the cooler secondary star extends almost accross the entire disk, best seen at phase 0.42 in Fig. 2.32. Such enormous cool spots are usually only seen on rapidly-rotating giants. The "record" holder in this context is the K0 giant in the single-lined spectroscopic binary XX Tri (HD 12545). A single spot sixty times larger than the largest sunspot group (or equivalently twenty times the projected solar disk) was detected by Strassmeier [112]. However, despite that the eclipse in the ER Vul system helps to constrain the solution, the mapping of ER Vul outside of eclipse still suffers from a north-south ambiguity. The fact that the enormous cool feature on the secondary appears near quadrature calls for a confirmation of this, otherwise impressing, result. Future observations, maybe from someone at this summer school, will clearify the situation.

A Pre-main-sequence Binary: V824 Ara = HD 155555. V824 Ara consists of a G5 IV primary star and a K0 V-IV secondary star in a short-period orbit of $P = 1.68$ days. Pasquini et al. [81] suggested that the binary is part of the young disk population in agreement with its high Li I 6708-Å abundance. The position of the two stellar components in the H-R diagram with respect to the pre-main-sequence tracks of D'Antona & Mazzitelli [18] gives masses of 1.12 and 0.99 solar masses for the primary and secondary, respectively. The isochrone that fits both stars best suggests an age of ≈18 Myr.

Fig. 2.33 shows our Doppler images of V824 Ara (Strassmeier & Rice [126]). These maps were derived with the maximum-entropy inversion code TEMPMAP and presumed an inclination of 50° and a $v \sin i$ of 36.8±1 km s^{-1} and 33.7±1.5 km s^{-1} for the primary and secondary component, respectively, and micro- and macroturbulences of 2.0 km s^{-1} with solar abundances. The range of surface temperatures on the primary is 3300–5400 K with an average (surface-integrated) value of 4900 K while the secondary shows a range of 3400–5200 K and an average of 4550 K. The most significant feature on the primary is a complex equatorial region covering the longitudes between $\ell \approx 220°$ through about $\ell = 330°$.

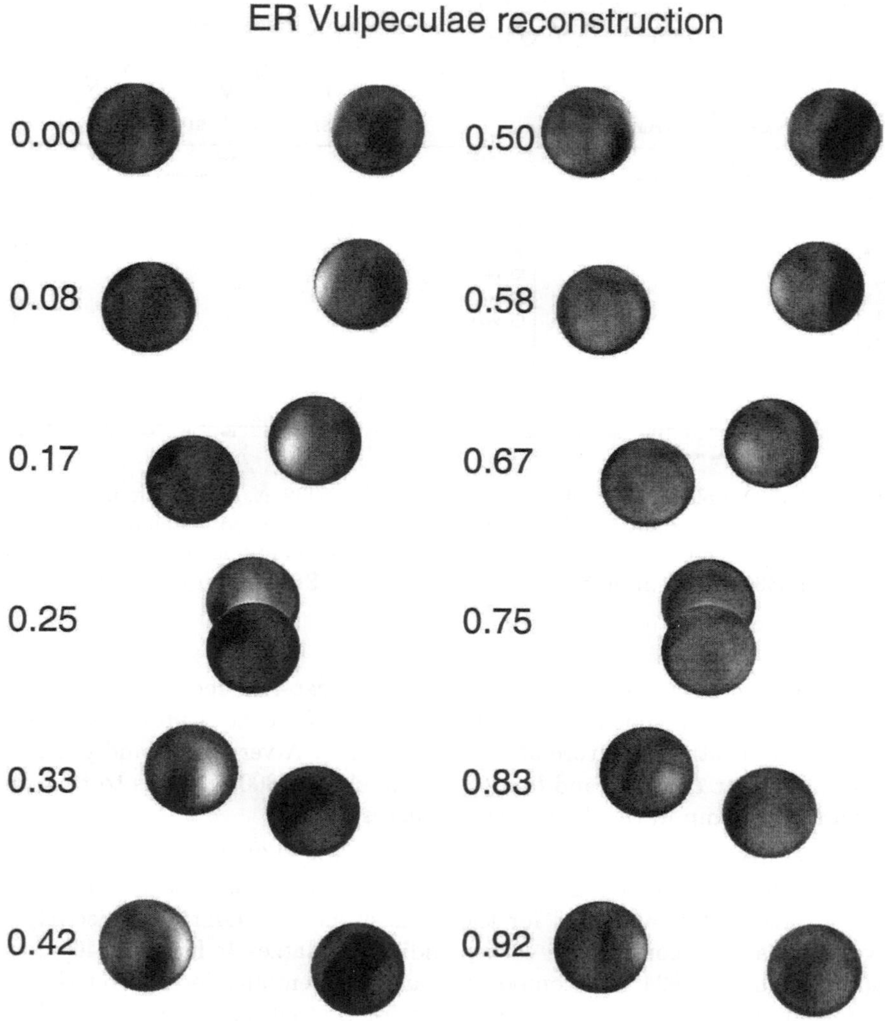

Fig. 2.32. Doppler images of ER Vul. The images are shown at 12 phases (indicated by the number) throughout the entire orbit. The relative separation of the two components is on the same scale as the diameter of the images. The G0 primary star is on the right at phase 0.00. (After Piskunov [83]).

Its coolest part consists of a large, somewhat elongated spot tilted against the stellar equator and located at $\ell \approx 300°$. Our primary-star maps also recover a cool and slightly decentered polar spot with a temperature difference of approximately 1700 K relative to the adopted "unspotted" photosphere of 5400 K. An appendage at $\ell \approx 180°$, that reaches down to approximately $+60°$ in latitude, is also evident.

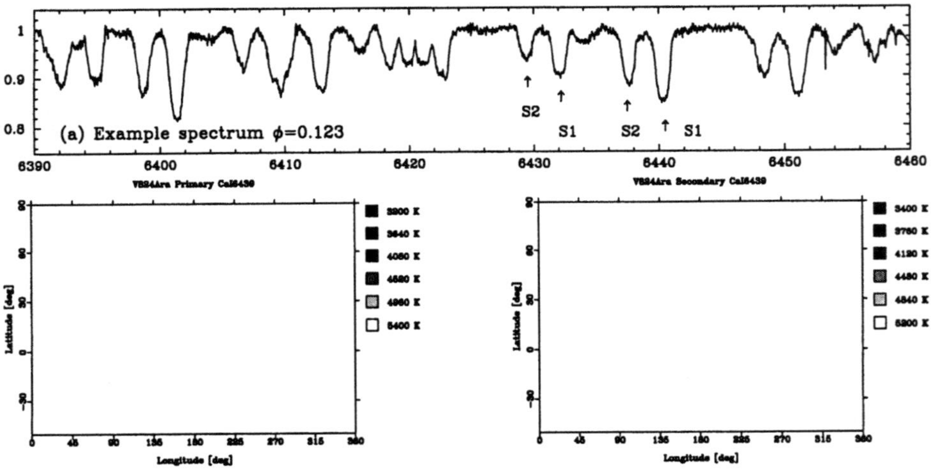

Fig. 2.33. Doppler images of V824 Ara from Ca I 6439 Å. Lower left panel: primary component. Lower right panel: secondary component. The top panel, **a**, shows a representative spectrum of the combined V824 Ara spectrum and identifies the Ca I 6439 Å and Fe I 6430 Å lines from both components. (After Strassmeier & Rice [126]).

The secondary star does not show a polar cap-like spot but has instead a cool high-latitude spot at $\ell \approx 270°$ and $b \approx +70°$ that is probably connected with a lower latitude feature at around $\ell = 225°$. A very small and particularly cool feature at $\ell = 130°$ and $b \approx +30°$ with $\Delta T \approx 1800$ K seems to be required by a sharp bump in the data at these phases.

A comparison of the two Doppler images in Fig. 2.33 does not show an immediately obvious spot concentration at the facing hemisphere's central meridians (longitudes of 90° and 270° for the primary and secondary, respectively) and we thus can not confirm the earlier finding by Hatzes & Kürster [46] that the spots on both V824 Ara components are preferentially located on the facing hemispheres. Nevertheless, our maps suggest that either component is likely influenced by the activity of the other component.

V824 Ara is the only known (close) pre-main-sequence binary where both components can be Doppler imaged independently. It thus allows – at least in principle – the determination of the inclination axes for both components separately. From numerical inversions of artificial test data it was amply demonstrated that the misfit between the data and the model (χ^2) was minimized when the correct inclination is adopted (e.g. Kürster et al. [63], Strassmeier [112]). The run of the χ^2 distribution for both components of V824 Ara for a large range of inclinations is shown in Fig. 2.34. It has a flat minimum of width ≈20° at about the same inclination range for both stars (40°–62° for the primary and 48°–67° for the secondary, respectively). The quality and sampling of the present data limits the precision at which the inclination can be determined. We do not regard the difference of 6° as significant. Therefore, we conclude that the rotation axes of the two components of V824 Ara have equal inclination with respect to the orbital plane.

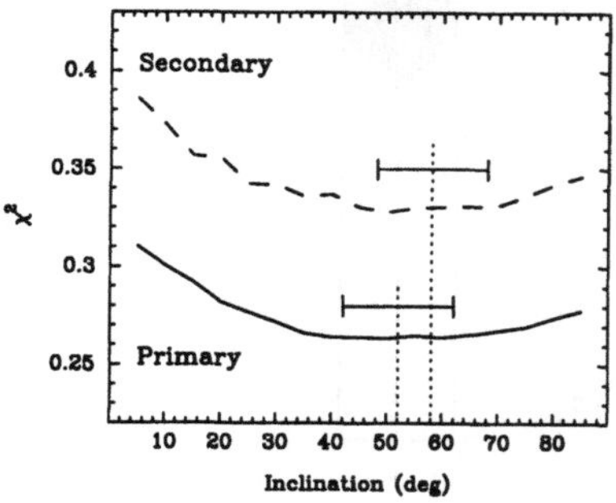

Fig. 2.34. The misfit between the data and the predicted profiles (χ^2) of V824 Ara as a function of stellar inclination. The full line is for the primary (G5), the dashed line for the secondary (K0). The most likely values are indicated with a vertical dotted line, and the range of equally likely values is shown as horizontal bars. Note that both ranges significantly overlap, which suggests that the inclinations are very likely identical for both components. (After Strassmeier & Rice [126]).

The existence of a polar or very high-latitude spot along with large low-latitude features may indicate that a mixture of solar and non-solar magnetic flux behavior exists. Firstly, the Sun does not show a polar spot at all and, secondly, does not exhibit spots larger than a fraction of a percent of the visible hemisphere and their occurence is limited to just a narrow equatorial band. Such a bimodality of the spot distribution on pre-main-sequence stars was recently predicted in the flux-tube modelling by Granzer et al. [32]. To make a more detailed comparison for V824 Ara, we adopted the MHD code used by Granzer et al., originally designed by Caligari et al. [13] and Schüssler et al. [102], in combination with a model of the solar convection zone and spun it up to the rotation period of V824 Ara ($\Omega = 16 \times \Omega_{\odot}$). Two underlying stellar pre-main-sequence models are computed matching the masses of the two components of V824 Ara. Hydrogen burning was assumed to contribute 0.01% of the total luminosity (according to a nominal age of $\approx$18 Myr).

The results are shown in Fig. 2.35 in form of a stellar crossection and a surface spot-probability function for each stellar component. The crossections show the path of the crests of the emerging flux tubes plotted as lines. The spot-probability pattern is shown as a function of latitude. Larger width of the pattern corresponds to higher probability of magnetic flux emergence. The two main conclusions from this figure are, firstly, we would expect to see magnetic flux only above a latitude of approximately 35–40° and no spots below that latitude, i.e., near the equator (for both stars). Secondly, if there is indeed a polar spot on the primary and a high-latitude feature on the secondary they must have formed after the flux-tube emergence because the current model suggests no flux tubes at latitudes above 70°. A possible mechanism to transport flux to the pole is meridional circulation. Recent time-series Doppler images of the RS CVn binary HR 1099 suggest such a poleward migration scenario (Vogt et al. [133], Strassmeier & Bartus [113]) and would be a possible explanation for the polar activity on V824 Ara.

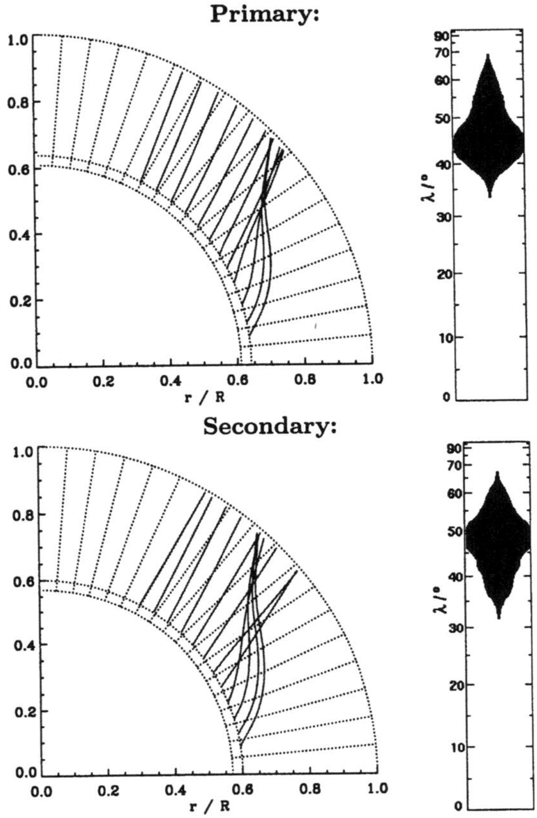

Fig. 2.35. The predicted latitude distribution of emerging flux tubes for the two stellar components of V824 Ara. The stellar models were computed with an updated version of the Kippenhahn-code for pre-main-sequence models matching V824 Ara. A radial cross section (left) is shown with the trajectories of the summit of rising flux loops. The small sector below the convection zone at r/R=0.6 indicates the width of the overshoot region. The right panels show the spot-probability function on the stellar surface as a function of latitude λ. (After Strassmeier & Rice [126] and Granzer et al. [32]).

Two Contact Binaries: AE Phe and YY Eri. Webbink [193] suggested a scenario in which W UMa-type contact binaries with common envelopes evolve into rapidly rotating single stars. The progenitors of W UMa binaries are likely to be very active short-period RS CVn systems while the single, rapidly-rotating, after-W UMa mergers could be identified with the class of the overactive FK Comae giants (for a review see Guinan & Giménez [36]). If so, one must assume that the magnetic flux observed in the RS CVn stars is conserved throughout the W UMa stage (because the FK Comae stars are magnetically very active and otherwise a mechanism must be invented that dilutes or even switches the dynamo off and on again). In any case, cool W UMa systems should be magnetically active and possibly show spots on their surfaces. In the meantime, this has been amply demonstrated from light-curve modeling of W UMa's.

However, there are two major problems with Doppler imaging of W UMa stars. Firstly, their high rotational velocities of up to 200 km s^{-1} and thus their shallow lines and, secondly, the non-spherical shape of the two stellar components. While the radiative transfer in a W UMa-star atmosphere, and the computation of the specific intensities over the stellar disk, remains a source of uncertainty, its relative impact is likely to be weak due to the enormous line broadening. Nevertheless, it seems mandatory to perform test reconstructions before applying the mapping technique to real data. This was done by Piskunov [83] and for this lecture I follow his paper.

Figure 2.36 shows such a test reconstruction. The stellar parameters of the F8-type contact-system Y Sextantis with a mass ratio of 1:2 and an orbital period of 0.42 days were adopted. The projected rotational velocity is 180 km s^{-1}. The gravitational darkening coefficient, usually denoted β, was set to the value of 0.08 suggested by Lucy [104] for stars with convective envelopes. A single spot of rectangular shape and 500 K cooler was placed on the equator of the primary component. Solving the forward problem for such a systemic configuration in a spectral region containing a strong Fe I line (5506.78 Å in this case) generates a set of simulated observations. These artificial data, degraded to a S/N ratio of 100:1 by salting in some random noise, are then inverted to obtain the Doppler image (in the right column in Fig. 2.36). The reconstruction of the spot is reasonably good, just the sharp edges are lost due to the low S/N of the data, and further improvements could be made by including simultaneous photometry.

Maceroni et al. [72], [71] presented the, so far, only maps of contact binaries. They presented two maps of the southern W UMa systems AE Phe and YY Eri. The YY Eri map is reprinted in Fig. 2.37. Both stars have quite short periods of around 0.3 days but are relatively bright (8-th magnitude). Nevertheless, the stringent S/N requirement and that of preventing phase smearing (see Sect. 2.8.2) limits one to the use of strong lines like Hα. The disadvantage of such strong lines is that they are contaminated by chromospheric emission and velocity fields. They are still sensitive to cool photospheric spots but they do not (opposite to the optically-thin lines) map a particular layer in the stellar atmosphere. Therefore, Maceroni et al. [72] did not attempt to map the surface temperature but instead mapped equivalent width relative to the solar value, i.e. $W = S * W_{\odot}$, where $W_{\odot}$ is the equivalent width of the solar line and S is a geometrical function only dependent on the position of the surface (it actually contains the Roche geometry as well). The maps in Fig. 2.37 show the resulting S distribution with different grey scales. $S = 1.83$ corresponds to approximately 350 K above the average value and $S = -0.5$ to 400 K below the mean. The maps are not yet conclusive but indicate the expected irradiance patterns and show the stellar poles to be hotter than the backs and the equator. Whether, and how, this is related to the ongoing mass exchange is presently not understood but newer and better Doppler maps of contact binaries may allow a better insight in the origin of mass transfer.

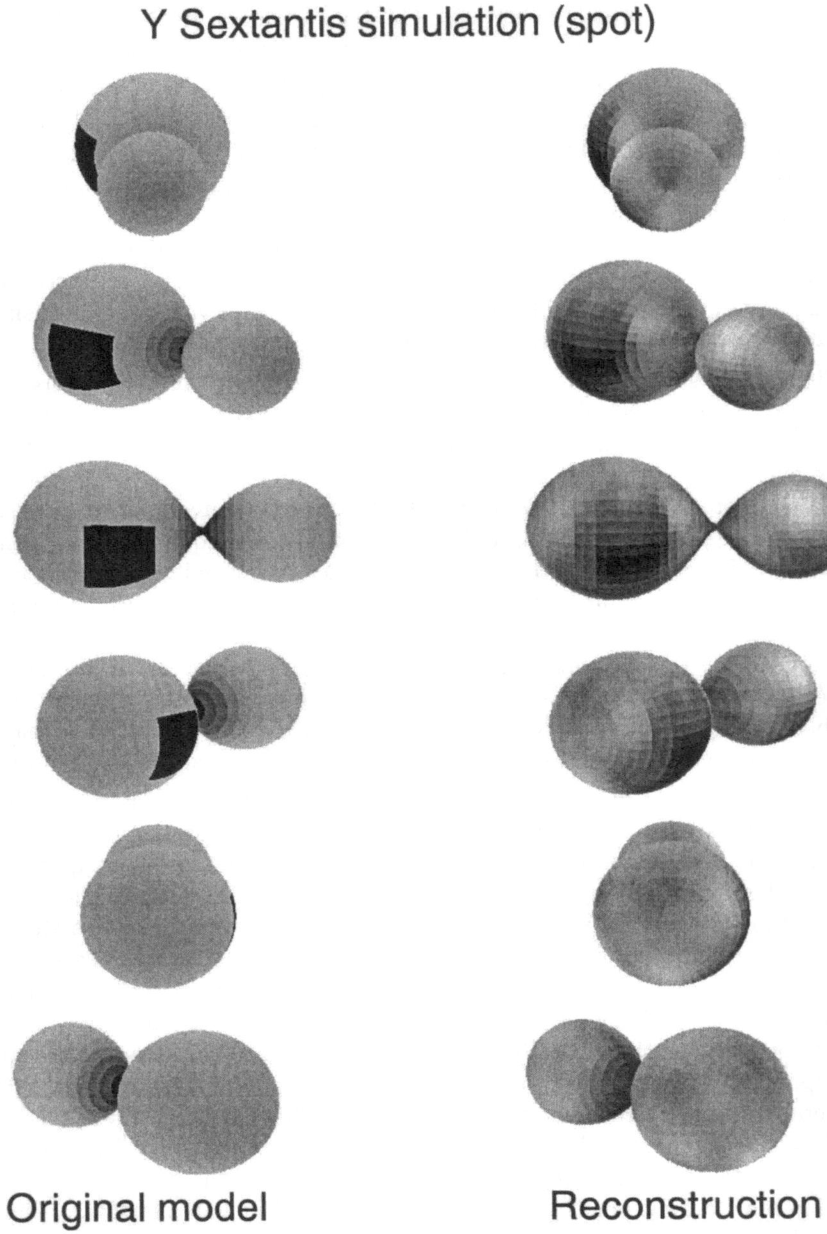

Fig. 2.36. Test reconstruction of a simulated W UMa-type contact binary. The system parameters were adopted to simulate the F8 contact binary Y Sextantis. A single, rectangular cool spot is placed on the primary star. Left: input model. Right: reconstruction. (After Piskunov [83]).

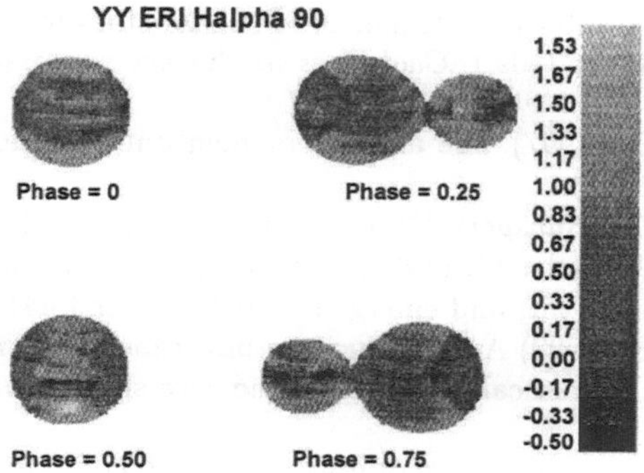

Fig. 2.37. Doppler images of the contact binary YY Eri. The maps were obtained from high-resolution Hα line profiles and plot the Hα equivalent width S relative to the solar value (see text). Approximately, $S = 1.83$ corresponds to temperatures 350 K above the mean value and $S = -0.5$ to 400 K below the mean. (After Maceroni et al. [72]).

Acknowledgements

The author wishes to thank John Rice, Artie Hatzes and Yvonne Unruh-Gazey for many discussions on the Doppler-imaging technique, Hans Zinnecker for the idea to check on the alignment of rotation axes in binary stars, and Th. Granzer and M. Schüssler on the many aspects of flux-tube simulations. Scans of figures and/or eps files were promptly provided by Carla Maceroni as well as Nuccio Lanza and J.-F. Donati. Yvonne Unruh-Gazey read the entire manuscript. Of course, I also thank the EADN organizer, Carlos Lázaro, for his engagement in this summer school. This work is supported by the Austrian Science Foundation grants S7301-AST and S7302-AST.

References

1. Andersen J. (1991) Accurate masses and radii of normal stars, A&AR 3, 91
2. Andersen J., Clausen J. V., Nordström B., Tomkin J., Mayor M. (1991) TZ Fornacis: stellar and tidal evolution in a binary with a fully-fledged red giant (Absolute dimensions of eclipsing binaries. XVII), A&A 246, 99
3. Applegate J. H. (1992) A mechanism for orbital period modulation in close binaries, ApJ 385, 621
4. Baliunas S. L., Horne J. H., Porter A. et al. (1985) Time-series measurements of chromospheric Ca II H and K emission in cool stars and the search for differential rotation, ApJ 294, 310
5. Baliunas S. L., Soon W. H. (1995) Are variations in the length of the activity cycle related to changes in brightness in solar-type stars?, ApJ 450, 896
6. Barden S. C. (1985) A study of short-period RS Canum Venaticorum and W Ursae Majoris binary systems: the global nature of Hα, ApJ 295, 162
7. Belvedére G., Lanza A. F., Sokoloff D. (1998) A dynamo scenario for polar and whole-surface starspot generation, Solar Physics 183, 435
8. Benz A. O., Conway J., Güdel M. (1998) First VLBI images of a main-sequence star, A&A 331, 596
9. Biermann P., Hall D. S. (1973) A model to explain alternate period changes in Algol-like binaries, A&A 27, 249

10. Bouvier J. (1997) The surface and internal rotation of low-mass stars in young clusters, in Pallavicini R. & G. Micela (eds.), *Cool stars in clusters and associations*, Mem.S.A.It. 68, p. 881
11. Bouvier J., Forestini M., Allain S. (1997) The angular momentum evolution of low-mass stars, A&A 326, 1023
12. Bray R. J., Loughhead R. E. (1979) *Sunspots*, Dover Publ. Inc., New York
13. Caligari P., Moreno-Insertis F., Schüssler M. (1995) Emerging flux tubes in the solar convection zone. I. Asymmetry, tilt, and emergence latitude, ApJ 441, 886
14. Charbonneau P., MacGregor K. B. (1992) Angular momentum transport in magnetized stellar radiative zones. I. Numerical solutions to the core spin-up model problem, ApJ 387, 639
15. Claret A. (1995) Stellar models for a wide range of initial chemical compositions until helium burning. I. From X=0.60 to X=0.80 for Z=0.02, A&AS 109, 441
16. Collier Cameron A. (1992) Modelling stellar photospheric spots using spectroscopy, in Byrne P. B. & D. J. Mullan (eds.), *Surface Inhomogeneities on Late-Type Stars*, Lecture Notes in Physics, Vol. 397, Springer-Verlag, Berlin, p. 33
17. Cutispoto G. (1996) Long-term photometric monitoring of active stars. V. $UBV(RI)_c$ photometry collected in Feb.–Mar. 1990, A&AS 119, 281
18. D'Antona F., Mazzitelli I. (1997) Evolution of low-mass stars, in Pallavicini R. & G. Micela (eds.), *Cool stars in clusters and associations*, Mem.S.A.It. 68, p. 807
19. Darwin G. H. (1879) Phil. Trans. Royal Soc. 170, p. 1
20. Deutsch A. (1958) Harmonic analysis of the periodic spectrum variables, in Lehnert B. (ed.), *Electromagnetic Phenomena in Cosmological Physics*, IAU Symp. 6, Cambridge Univ. Press, Cambridge, p. 209
21. Dicke R. H. (1970) The rotation of the Sun, in A. Slettebak (ed.), IAU Colloq. 4, *Stellar Rotation*, Reidel, Dordrecht, p. 289
22. Donati J.-F., Semel M., Carter B. D., Rees D. E., Cameron A. C. (1997) Spectropolarimetric observations of active stars, MNRAS 291, 658
23. Donati J.-F. (1999) Magnetic cycles of HR 1099 and LQ Hydrae, MNRAS 302, 457
24. Donati J.-F., Collier Cameron A. (1997) Differential rotation and magnetic polarity patterns on AB Doradus, MNRAS 291, 1
25. Drake J. J., Stern R. A., Stringfellow G., Mathioudakis M., Laming J. M., Lambert D. L. (1996) Detection of quiescent EUV emission from the vey low-mass dwarf Van Biesbroeck 8: evidence for a turbulent field dynamo, ApJ 469, 828
26. Durney B. R., Latour J. (1978) Geophys. Ap. Fluid Dyn. 9, 241
27. Fekel F. C. (1983) Spectroscopy of V711 Tau (=HR 1099): fundamental properties and evidence for starspots, ApJ 268, 274
28. Friedman H. (1987) *Die Sonne*, Spektrum Akademischer Verlag, Heidelberg
29. Glebocki R., Stawikowski A. (1997) Alignment of rotational axes in asynchronous late type binaries, A&A 328, 579
30. Gough D. O. (1991) Internal solar rotation, in Catalano S. & J. Stauffer (eds), *Angular Momentum Evolution in Young Stars*, Kluwer, Dordrecht, p. 271
31. Graffagnino V. G., Wonnacott D, Schaeidt S. (1995) HR 5110 superflare: an interbinary flare identified?, MNRAS 275, 129
32. Granzer Th., Schüssler M., Caligari P., Strassmeier K. G. (1999) Distribution of starspots on cool stars. II. Pre-main-sequence and ZAMS stars between 0.4 $M_\odot$ and 1.7 $M_\odot$, A&A, in press
33. Gray D. F. (1988) *Lectures on spectral-line analysis: F, G, and K stars*, The Publisher, Ontario, Canada

34. Gray D. F. (1991) Rotation of evolved stars, in Catalano S. & J. Stauffer (eds.), *Angular Momentum Evolution of Young Stars*, Kluwer, Dordrecht, p. 183
35. Gray D. F. (1992) *The observation and analysis of stellar photospheres*, Cambridge University Press, Cambridge
36. Guinan E. F., Giménez A. (1993) Magnetic activity in close binaries, in J. Sahade et al. (eds.), *The Realm of Interacting Binary Stars*, Kluwer, Dordrecht, p. 51
37. Gull S. F., Skilling J. (1983), in Roberts J. A. (ed.), *Indirect Imaging*, Proc. Symp. Sydney, Cambridge Univ. Press, Cambridge
38. Habets G. M. H. J., Zwaan C. (1989) Asynchronuous rotation in close binary systems with circular orbits, A&A 211, 56
39. Hall D. S. (1972) A T Tauri-like star in the eclipsing binary RS Canum Venaticorum, PASP 84, 323
40. Hall D. S. (1989) The relation between RS CVn and Algols, Space Sci. Rev. 50, 219
41. Hall D. S. (1991) A connection between long-term luminosity variations and orbital period changes in chromospherically active binaries, ApJ 380, L85
42. Hall D. S. (1991) Learning about stellar dynamos from long-term photometry of starspots, in Tuominen I., Moss D. & Rüdiger G. (eds.), IAU Colloq. 130, *The Sun and Cool Stars: activity, magnetism, dynamos*, Springer, Berlin-Heidelberg, p. 353
43. Hall D. S. (1994) The active dynamo stars RS CVn, BY Dra, FK Com, Algol, W UMa, und T Tau, Mem.S.A.It. 65, 73
44. Hatzes A. P. (1998) Spot activity and the differential rotation on HD 106225 derived from Doppler tomography, A&A 330, 541
45. Hatzes A. P. (1995) Doppler images of the dMe binary YY Gem, in Strassmeier K. G. (ed.), Poster Proceedings: IAU Symp. 176 *Stellar Surface Structure*, Univ. Vienna, p. 90
46. Hatzes A. P., Kürster M. (1999) Doppler images of the pre-main sequence binary V824 Ara, A&A 346, 432
47. Henry G. W. (1995) The development of precision robotic photometry, in Henry G.W. & J.A. Eaton (eds), *Robotic telescopes: current capabilities, present developments, and future prospects for automated astronomy*, PASPC 79, p. 44
48. Hünsch M., Schmitt J. H. M. M., Schröder K. P., Reimers D. (1996) ROSAT X-ray observations of a complete, volume-limited sample of late-type giants, A&A 310, 801
49. Hut P. (1981) Tidal evolution in close binary systems, A&A 99, 126
50. Hut P. (1982) Tidal evolution in close binary systems for high eccentricity, A&A 110, 37
51. Johns-Krull C. M., Valenti J. A. (1996) Detection of strong magnetic fields on M dwarfs, ApJ 459, L95
52. Kawaler S. D. (1987) Angular momentum in stars: The Kraft curve revisited, PASP 99, 1322
53. Kennelly E. J., Walker G. A. H., Merryfield W. J. (1992) ApJ 400, L71
54. Kennelly E. J. et al. (1998) ApJ 495, 440
55. Kippenhahn R., Möllenhoff C. (1975) *Elementare Plasmaphysik*, BI Wissenschaftsverlag, Mannheim/Wien/Zürich
56. Korhonen H., Berdyugina S. V., Hackman T., Duemmler R., Ilyin I. V., Tuominen I. (1999) Study of FK Coma Berenices. I. Surface images for 1994 and 1995, A&A 346, 101
57. Kosovichev A. G., Schou J., Scherrer P. H. et al. (1999) Internal structure and rotation of the Sun: first results from the MDI data, IAU Symp. 181, in press

58. Kraft R. P. (1965) Studies of stellar rotation. I. Comparison of rotational velocities in the Hyades and Coma clusters, ApJ 142, 681
59. Kraft R. P. (1967) Studies of stellar rotation. V. The dependence of rotation on age among solar-type stars, ApJ 150, 551
60. Kraft R. P. (1970) in Herbig G. H. (ed.) *Spectroscopic Astrophysics*, Berkeley, Univ. of Calif. Press, p. 385
61. Krisciunas K., Guinan E. (1990) Photometry of Capella (Dec 1981 to Apr 1990), IBVS 3548
62. Kron G. E. (1947) The probable detection of patches of varying brightness on AR Lacertae B, PASP 59, 261
63. Kürster M., Schmitt J. H. M. M., Cutispoto G. (1994) Doppler imaging with a CLEAN-like approach. II. A photospheric image of AB Doradus (=HD 36705), A&A 289, 899
64. Kurucz R. L. (1993) ATLAS-9, CD-ROM No. 13, CfA-Harvard
65. Lanza A. F., Rodonó M., Rosner R. (1998) Orbital period modulation and magnetic cycles in close binaries, MNRAS 296, 893
66. Lanza A. F., Rodonó M., Zappalá R. A. (1994) Fourier analysis of spotted star light curves as a tool to detect stellar differential rotation. II. Spots' evolution and binarity, A&A 290, 861
67. Lestrade J.-F. (1996) VLBI for probing large-scale magnetic structures of stars, in Strassmeier K. G. & J. L. Linsky (eds.), *Stellar Surface Structure*, IAU Symp. 176, Kluwer, Dordrecht, p. 173
68. Linsky J. L. (1995) STScI Press Release PR95-03
69. Linsky J. L., Wood B. E., Brown A., Giampapa M. S., Ambruster C. (1995) Stellar activity at the end of the main sequence: GHRS observations of the M8Ve star VB 10, ApJ 455, 670
70. Lucy L. B. (1967) Gravity darkening for stars with convective envelopes, Zf.f.Ap. 65, 89
71. Maceroni C., Vilhu O., van't Veer F., van Hamme W. (1994) Surface imaging of late-type contact binaries. I. AE Phoenicis and YY Eridani, A&A 288, 529
72. Maceroni C., van't Veer F., Vilhu O. (1991) Surface imaging of W Ursae Majoris contact binaries, ESO Messenger 66, 47
73. McNally D. (1966) The distribution of angular momentum among main-sequence stars, The Observatory 85, 166
74. Mestel L., Weiss N. O. (1987) Magnetic fields and non-uniform rotation in stellar radiative zones, MNRAS 226, 123
75. Neuhäuser R., Comerón F. (1999) Science 282, 93
76. Neuhäuser R., Briceno C., Comerón F. et al. (1999) Search for X-ray emission from bona-fide and candidate brown dwarfs, A&A 343, 883
77. Noyes R. W., Hartmann L. W., Baliunas S. L., Duncan D. K., Vaughan A. H. (1984), Rotation, convection, and magnetic activity in lower main-sequence stars, ApJ 279, 763
78. Oláh K., Kollath Z., Strassmeier K. G. (1999) A&A, submitted
79. Pallé P. L., Régulo C., Roca Cortés T. (1989) Solar cycle induced variations of the low ℓ solar acoustic spectrum, A&A 224, 253
80. Parker E. N. (1955) Hydromagnetic dynamo models, ApJ 122, 293
81. Pasquini L., Cutispoto G., Gratton R., Mayor M. (1991) HD 155555: a nearby pre-main sequence binary?, A&A 248, 72
82. Pinnsonneault M. H., Kawaler S. D., Sofia S., Demarque P. (1989) Evolutionary models of the rotating Sun, ApJ 338, 424

83. Piskunov N. E. (1996) Doppler imaging of eclipsing binaries, in Strassmeier K. G. & J. L. Linsky (eds.), IAU Symp. 176 *Stellar Surface Structure*, Kluwer, Dordrecht, p. 45
84. Piskunov N. E., Rice J. B. (1993) Techniques for Surface Imaging of Stars, PASP 105, 1415
85. Piskunov N. E., Tuominen I., Vilhu O. (1990) Surface imaging of late-type stars, A&A 230, 363
86. Piskunov N. E., Wehlau W. H. (1990) Mapping stellar surfaces from spectra of medium resolution, A&A 233, 497
87. Pizzo V., Schwenn R., Marsch E., Rosenbauer H., Mühlhäuser K.-H., Neubauer F. M. (1983) Determination of the solar wind angular momentum flux from the *Helios* data—an observational test of the Weber and Davis theory, ApJ 271, 335
88. Ramseyer T. F., Hatzes A. P., Jablonski F. (1995) Doppler imaging of V471 Tau, AJ 110, 1364
89. Rice J. B. (1996) Doppler imaging of stellar surfaces, in Strassmeier K. G. & J. L. Linsky (eds.), IAU Symp. 176 *Stellar Surface Structure*, Kluwer, Dordrecht, p. 19
90. Rice J. B., Strassmeier K. G. (1998) Doppler imaging of stellar surface structure. VII. The very young, single K2-dwarf LQ Hydrae, A&A 336, 972
91. Rice J. B., Strassmeier K. G. (1999) Doppler imaging from artificial data, A&A, in preparation
92. Rice J. B., Wehlau W. H., Khokhlova V. L. (1989) Mapping stellar surfaces by Doppler imaging: technique and application, A&A 208, 179
93. Rodonó M. (1991) Magnetic activity and rotation, in Catalano S. & J. Stauffer (eds), *Angular Momentum Evolution in Young Stars*, Kluwer, Dordrecht, p. 207
94. Rodonó M., Cutispoto G. (1992) Long-term photometric monitoring of active stars. I. 1988-89 UBV photometry with the Phoenix APT, A&AS 95, 55
95. Rodonó M., Cutispoto G. (1994) A long-term program of monitoring active close binaries with the Catania automated photometric telescope, in D'Antona F. et al. (eds), *Evolutionary Links in the Zoo of Interactive Binaries*, Mem.S.A.It. 65, p. 83
96. Rodonó M., Lanza A. F., Catalano S. (1995) Starspot evolution, activity cycle and orbital period variation of the prototype active binary RS Canum Venaticorum, A&A 301, 75
97. Rutten R. G. M., Pylyser F. (1988) Magnetic structure in cool stars. XV. The evolution of rotation rates and chromospheric activity of giants, A&A 191, 227
98. Schatzman E. (1954) Remarques sur l'evolution des étoiles. III. Interprétation des rotations axiales, Ann. d'Astrophys. 17, 300
99. Schmitt J. H. M. M. (1997) Stellar Coronae, in De Greve J.P., Blomme R. & Hensberge H. (eds.), IX EADN school, Lecture Notes in Physics Vol. 497, Springer-Verlag, Berlin Heidelberg, p. 277
100. Schrijver C. J. (1996) The magnetic field of the nearest star: a paradigm for stellar activity, in Strassmeier K. G. & J. L. Linsky (eds.), *Stellar Surface Structure*, IAU Symp. 176, Kluwer, Dordrecht, p. 1
101. Schrijver C. J., Zwaan C. (1991) Activity in tidally interacting binaries, A&A 251, 183
102. Schüssler M., Caligari P., Ferriz-Mas A., Solanki S. K., Stix M. (1996) Distribution of starspots on cool stars. I. Young and main-sequence stars of 1 $M_{\odot}$, A&A 314, 503
103. Schüssler M., Solanki S. K. (1992) Why rapid rotators have polar spots, A&A 264, L13

104. Semel M., Donati J.-F., Rees D. E. (1993) Zeeman-Doppler imaging of active stars. III. Instrumental and technical considerations, A&A 278, 231
105. Siarkowski M., Preś P., Drake S. A., White N. E., Singh K. P. (1996) Corona(e) of AR Lacertae. II. The spatial structure, ApJ 473, 470
106. Spruit H. C. (1991) Influence of starspots on internal stellar structure, in Byrne P. B. & Mullan D. J. (eds.), *Surface inhomogeneities on late-type stars*, Springer Verlag, Berlin-Heidelberg, p. 78
107. Spruit H. C., van Ballegooijen A. A. (1982) Stability of toroidal flux tubes in stars, A&A 106, 58
108. Stix M. (1989) *The Sun*, Springer Verlag, Berlin-Heidelberg.
109. Strassmeier K. G. (1994) Rotational-modulation mapping of the active atmosphere of the RS Canum Venaticorum binary HD 106225, A&A 281, 395
110. Strassmeier K. G. (1995) Stellar surface structure: Doppler imaging with the VLT UVES, in Walsh J. & Danziger I. J. (eds.), *Science with the VLT*, Springer, Berlin, p. 87
111. Strassmeier K. G. (1997) *Aktive Sterne: Laboratorien der solaren Astrophysik*, Springer Verlag, Wien-New York.
112. Strassmeier K. G. (1999) Doppler imaging of stellar surface structure. XI. The super starspots on the K0 giant HD 12545: larger than the entire Sun, A&A 347, 225
113. Strassmeier K. G., Bartus J. (1999) Doppler imaging of stellar surface structure. XII. Rapid spot changes on the RS CVn binary V711 Tau, A&A, submitted
114. Strassmeier K. G., Bartus J., Cutispoto G., Rodonó M. (1997a) Starspot photometry with robotic telescopes. Continuous UBV and VRI photometry of 23 stars in 1991–1996, A&AS 125, 11
115. Strassmeier K. G., Bartus J., Kővári Zs., Weber A., Washuettl A. (1998) Doppler imaging of stellar surface structure. VIII. The effectively single and rapidly-rotating G8-giant HD 51066 = CM Cam, A&A 336, 587
116. Strassmeier K. G., Bopp B. W. (1992) Time-series photometric spot modeling. I. Parameter study and application to HD 17433 = VY Arietis, A&A 259, 183
117. Strassmeier K. G., Boyd L. J., Epand D. H., Granzer T. (1997b) Wolfgang-Amadeus: The University of Vienna twin automatic photoelectric telescope, PASP 109, 697
118. Strassmeier K. G., Fekel F. C. (1990) The spectral classification of chromospherically active binary stars with composite spectra, A&A 230, 389
119. Strassmeier K. G., Fekel F. C., Gray D. F., Hatzes A. P., Schmitt J. H. M. M., Solanki S. K. (1998) Evolved stars: what happens to activity off the main sequence, in Donahue R. A. & J. A. Bookbinder (eds.), *Cool Stars, Stellar Systems, and the Sun*, Cambridge, PASPC 154, p. 257
120. Strassmeier K. G., Lupinek S., Dempsey R. C., Rice J. B. (1999) Doppler imaging of stellar surface structure. X. The FK Comae-type star HD 199178 = V1794 Cygni, A&A 347, 212
121. Strassmeier K. G., Hall D. S., Fekel F. C., Scheck M. (1993) A catalog of chromospherically active binary stars (second edition), A&AS 100, 173
122. Strassmeier K. G., Hall D. S., Henry G. W. (1994) Time-series photometric spot modeling. II. Fifteen years of photometry of the bright RS CVn binary HR7275, A&A 282, 535
123. Strassmeier K. G., Handler G., Paunzen E., Rauth M. (1999) Chromospheric activity in G and K giants and their rotation-activity relation, A&A 281, 855

124. Strassmeier K. G., Rice J. B. (1998) Doppler imaging of stellar surface structure. VI. HD 129333 = EK Dra: a stellar analog of the active young Sun, A&A 330, 685
125. Strassmeier K. G., Rice J. B. (1998) Doppler imaging of stellar surface structure. IX. A high-resolution image of the weak-lined T Tauri star HDE 283572 = V987 Tauri, A&A 339, 497
126. Strassmeier K. G., Rice J. B. (1999) Doppler imaging of stellar surface structure. XIII. The pre-main-sequence binary V824 Ara = HD 155555, A&A, submitted
127. Sweigart A.V., Greggio L., Renzini A. (1989) The development of the red giant branch. I. Theoretical evolutionary sequences, ApJS 69, 911
128. Tassoul J.-L. (1987) On the synchronization in early-type stars, ApJ 322, 856
129. Tassoul J.-L. (1988) On orbital circularization in detached close binaries, ApJ 324, L71
130. Tassoul J.-L., Tassoul M. (1990) A time-dependent model for synchronization in close binaries, ApJ 359, 155
131. Uchida Y., Sakurai T. (1985) Magnetodynamical processes in interacting magnetospheres of RS CVn binaries, in Kundu M. R. & G. D. Holman (eds.), IAU Symp. 107 *Unstable current systems and plasma instabilities in astrophysics*, Reidel, Dordrecht, p. 281
132. Vincent A., Piskunov N. E., Tuominen I. (1993) Surface imaging of eclipsing binary stars. I. Techniques, A&A 278, 523
133. Vogt S. S., Hatzes A. P., Misch A. A., Kürster M. (1999) Doppler imagery of the spotted RS CVn star HR 1099 (=V711 Tau) from 1981 – 1992, ApJS 121, 547
134. Vogt S. S., Penrod G. D. (1983) Doppler imaging of spotted stars: application to the RS Canum Venaticorum star HR 1099, PASP 95, 565
135. Vogt S. S., Penrod G. D., Hatzes A. P. (1987) Doppler images of rotating stars using maximum entropy image reconstruction, ApJ 321, 496
136. Webbink R. F. (1976) The evolution of low-mass close binary systems. I. The evolutionary fate of contact binaries, ApJ 209, 829
137. Weiss W. W., Baglin A. (eds.) (1993) IAU Colloq. 137, *Inside the Stars*, PASPC Vol. 40.
138. Wilson O. C. (1976) Absolute magnitudes of stars from widths of chromospheric Ca II emission lines, ApJ 205, 823
139. Zahn J.-P. (1977) Tidal friction in close binary stars, A&A 57, 383
140. Zahn J.-P. (1989) Tidal evolution of close binary stars. I. Revisiting the theory of the equilibrium tide, A&A 220, 112
141. Zirin H. (1988) *Astrophysics of the Sun*, Cambridge Univ. Press, Cambridge

3 Cataclysmic Variables

Józef Smak

N. Copernicus Astronomical Center
Bartycka 18, 00-716 Warsaw, Poland
e-mail: jis@camk.edu.pl

Abstract. Topics covered by this review are: observational data, basic parameters and standard model for non-magnetic cataclysmic variables (CV's), structure of accretion disks in CV's and their thermal instability, dwarf nova outbursts, boundary layer, and origin of emission lines.

3.1 Introduction

The field of cataclysmic variables (in short: CV's) is so broad that a selection of topics to be covered by the present series of lectures had to be made. It was guided by two factors. The first was the personal interest and preference of the author. The second – was the intention to provide the students and future readers with a minimum of basic information about those observational facts (e.g. emission lines from disks in close binary systems) and theories (e.g. models of accretion disks) which are important not only in the field of cataclysmic variables but also in other areas, ranging from Algols to quasars.

Bibliography. Due to the lack of space no attempt was made in this article to present an extensive bibliography on CV's and their accretion disks. The most recent books, conference volumes, and reviews, containing a more detailed treatment of many problems discussed below and – in particular – detailed lists of references to individual papers, are: Cordova 1995, Duschl et al. 1994, Frank et al. 1992, Hack and la Dous 1993, Kato et al. 1998, King 1995, la Dous 1994, Osaki 1996, Lin and Papaloizou 1996, Papaloizou and Lin 1995, Sellwood and Goodman 1999, Sion 1999b, Wheeler 1993, Wickramasinghe et al. 1997.

Units and notation. The units used throughout this review will be those of the *cgs* system (unless specifically noted otherwise, like – for example – in the case of stellar masses and radii expressed in solar units, or the orbital period given alternatively in seconds, hours, or days). In Sect. 3.2, following the common custom in the area of binaries, capital letters will denote system parameters in absolute (i.e. *cgs*) units, while small letters – their dimensionless equivalents; for example: the radius of the disk will be either R_d or – in units of the orbital radius – $r_d = R_d/A$. In further sections, however, following the common practice in the field of accretion disks, small letters will also be used for some parameters, including those denoted earlier with capital letters; for example: r_d in Sect. 3.3-3.5 will be the same as R_d in Sect. 3.2. To avoid possible confusion another warning concerning this point will be given in Sect. 3.3.3. In the case of accretion the radial component of velocity of the accreting material v_r will be *negative.*

Consequently the accretion rate $\dot{M}$ will also be *negative* and some formulae will contain its absolute value $|\dot{M}|$. In the text, however, we shall simply write $\dot{M}$.

Terminology. Standard, self-explaining terminology will be used, with only few items requiring special comments at this point. The mass ratio will be defined as $q = M_2/M_1$, with the *primary* component being the white dwarf. The term "Keplerian", used commonly in application to an accretion disk, is imprecise and can be misleading. It actually refers to the *circular* Keplerian motion of an element of the disk around the much more massive central object. The orbital velocity, which is given in this case by (3.12), is commonly referred to as the "rotational"velocity, and this can also be misleading.

Equations. Equations will be numbered with extra letters being added in some cases to indicate their applicability: "(K)"– for Keplerian motion (in the sense defined above), and "(S)"– for steady-state accretion.

3.2 Cataclysmic Variables as Binary Systems

3.2.1 Standard Model

Cataclysmic variables, in spite of a considerable diversity of their observed characteristics and types of behavior, can be almost uniquely described by the following, relatively simple definition:

(a) they are binary systems;

(b) the primary component is a white dwarf;

(c) the secondary component is a low mass main sequence star filling its Roche lobe; and

(d) the mass transfer rate is between, roughly, 10^{-11} and $10^{-8} M_{\odot}$/yr.

Concerning point (c) we should add that in few cases (best lnown examples being T CrB, GK Per) the secondary can be an evolved star. Point (d) requires two comments. The lower limit to the mass transfer rate is a natural one, imposed by the existence of gravitational waves, which – at the shortest orbital periods – are the main factor responsible for removing the angular momentum from the system and thereby forcing the secondary to remain in contact with its Roche lobe and transfer mass at such a rate. The upper limit is connected with the structural behavior of the accreting white dwarf: at $\dot{M} > 10^{-8} M_{\odot}$/yr it becomes a giant and the binary system belongs to a different category of objects (e.g. symbiotic stars).

The *standard model* of a CV, corresponding to that definition, is shown in Fig. 3.1 In addition to the two stellar components, it includes the accretion disk around the white dwarf formed as a result of the mass transfer and the hot spot, at the edge of the disk, resulting from the collision of the stream with the outer parts of the disk.

It should be added, however, that this model does not apply to *magnetic* CV's, which contain white dwarfs with strong magnetic fields and do not possess regular accretion disks (see Marsh 1999). The present article will be devoted exclusively to *non-magnetic* CV's with accretion disks.

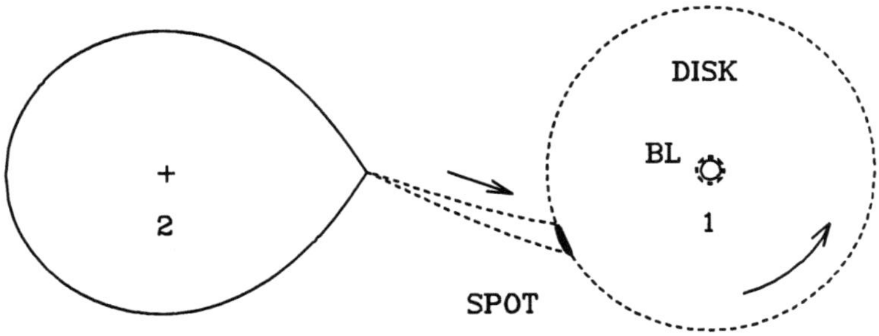

Fig. 3.1. Standard model of a cataclysmic varaible. Rotation of the system and of the disk is counterclockwise. BL denotes the boundary layer.

3.2.2 Observational Data and Basic Parameters

The bulk of data pertaining to the determination of basic system parameters comes from observations in the optical part of the spectrum, from near ultraviolet to near infrared. On the other hand, observations in the far UV and in the X-ray region are crucial primarily for studying specific physical processes.

The relative contributions from the two stellar components, from the disk, and from the hot spot, differ considerably from one system to another. This implies that for a given CV some of the observational data reviewed below might not be available, seriously limiting the possibility of determining its system parameters. Fortunately, even in such cases when only a partial set of data is available, they can be supplemented with additional relations discussed in Sect. 3.2.3 and 3.2.4, and this often permits all the parameters to be determined.

The following general rules can be mentioned at this point. In the optical part of the spectrum the contribution from the white dwarf is generally quite small, except for the quiescent dwarf novae of very short periods. It is significant or even dominant, however, in the far UV (see Sion 1999b). On the other hand the contribution from the much cooler secondary becomes large at longer wavelengths, particularly in the near IR, and at longer orbital periods.

Concerning the contribution from the disk and from the hot spot, we may first note that one of the most characteristic signatures of a CV spectrum are the strong emission lines (primarily of hydrogen and helium), formed in the disk atmosphere and chromosphere. They are usually double (see Sect. 3.2.4). Superposed on such a double profile, there is often an additional, S-wave component, produced in the vicinity of the hot spot, its radial velocity reflecting the local motions of the emitting atoms in that particular area. The relative contributions from the disk and from the hot spot to the optical continuum depend primarily on the mass transfer and accretion rates. At higher mass transfer rates, which result in steady-state or stationary accretion, with $\dot{M}_{accr} = \dot{M}_{tr}$, the disk dominates in the global flux, while the contribution from the hot spot is much smaller. The same is true in dwarf novae during their outbursts. In dwarf novae at quiescence, however, when $\dot{M}_{accr} \ll \dot{M}_{tr}$, it is the hot spot that may provide as much as 50 percent of the optical flux, while the contribution from the disk is very small.

3.2.3 Spectroscopic Observations

Representative optical spectra of CV's can be found, for example, in the survey papers by Honeycutt et al. (1987) and Williams (1983). When the absorption lines of the secondary are visible in the spectrum, they can be used to determine its spectral type and – more importantly – its radial velocity variations. Radial velocities of the white dwarf can be measured indirectly from the emission lines originating in the accretion disk. In some systems broad wings of the absorption lines, confirming the white dwarf nature of the primary, are also visible. By fitting theoretical profiles to them it is possible to determine the effective temperature and gravity of the white dwarf. As mentioned above, the white dwarf contributes significantly or even dominates in the far UV. Consequently, observations in that spectral region (in particular with the IUE and HST) provide crucial information not only about the temperature of the white dwarf, but also about the structure of its outer layers, rotation and the chemical composition (cf. Sion 1999ab).

Concerning the observed values of K_1 and K_2 one must remember that the situation is often complicated by two effects. In the case of the secondary it is the effect of its irradiation by the boundary layer: The temperature distribution on the surface of the secondary is non-uniform (simply speaking – its hemisphere facing the primary is hotter), and that makes the radial velocities measured from the resulting absorption line profiles no longer representative for the true stellar velocity. In particular, the observed value of K_2 depends on the amount of irradiation and differs systematically from the true K_2. The best documented example is SS Cyg (Hessman et al. 1984). In the case of the primary, its radial velocities being measured from the emission lines, any significant departure from the crucial assumption of disk's axial symmetry may affect the observed value of K_1 (one of the possible sources of concern is the presence of the S-wave component; see above). Obvious tests for the presence of any such spurious effects include (a) deviations of the observed radial velocity curves from the expected sinusoidal shape, (b) discordant values of the two systemic velocities, γ_1 and γ_2, and (c) discordant values of the two conjuction phases, $\varphi_{\circ,1}$ and $\varphi_{\circ,2}$; in the case of eclipsing systems they should also agree with photometric $\varphi_\circ$.

3.2.4 Photometry: Eclipse Light Curves

CV's with orbital inclinations larger than, roughly, $i \approx 65 - 75°$ are eclipsing systems. The nature of eclipses depends on the relative contributions from the two stellar components, the disk, and the hot spot.

In the case of *dwarf novae at quiescence* the eclipse is due to an occultation by the secondary of the white dwarf, of the hot spot, and of the disk. Fig. 3.2 presents a schematic light curve of such an eclipsing dwarf nova, showing (a) the so-called orbital hump, or shoulder, lasting for about one-half of the cycle; it is due to the phase dependent contribution from the hot spot, which during

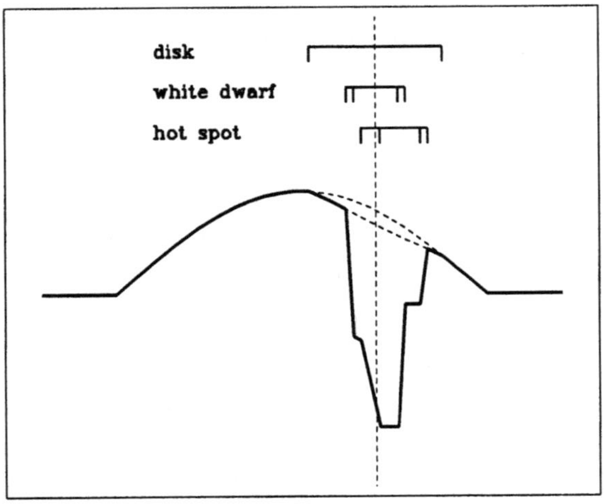

Fig. 3.2. Schematic light curve of an eclipsing dwarf nova at quiescence. Vertical dotted line marks $\varphi = 0$ – the phase of the true conjunction. The bars on top mark the durations of eclipses. Note that the eclipses of the white dwarf and of the hot spot are total, while that of the disk – only partial. The large hump, with a maximum prior to eclipse, is due to the hot spot.

the other half of the cycle is hidden behind the disk (see Fig. 3.1) and (b) the superposition of three eclipses. Their details are:

(1) The eclipse of the disk is very shallow and can usually be identified only by making a detailed analysis of the light curve. This is due to the fact that the relative contribution from the disk is small and that its occultation is only partial.

(2) The eclipse of the white dwarf is symmetric around $\varphi = 0$. In fact it is this eclipse that can best be used to define the true zero-phase.

(3) Due to an asymmetric location of the hot spot (see Figs. 3.1 and 3.7), its eclipse is shifted in phase and its ingress lasts longer than its egress.

(4) In the case of both eclipses, their partial phases are rather short, indicative of the small dimensions of the white dwarf and the hot spot. This simplifies considerably the analysis.

In general, the four phases of contacts depend on the geometrical parameters of the system. When the duration of the partial eclipse is short, we can use the mean phases of ingress and egress, i.e. $\varphi_{12} = (\varphi_1 + \varphi_2)/2$ and $\varphi_{34} = (\varphi_3 + \varphi_4)/2$, which do not depend on the dimensions of the eclipsed body. In the case of the white dwarf $\varphi_{12}(WD) \equiv \varphi_{34}(WD)$ depends only on the dimension of the secondary (which is determined by the mass ratio) and on the inclination. In the case of the hot spot $\varphi_{12}(HS)$ and $\varphi_{34}(HS)$ depend also on its location, which can be described with the spot's radial distance r_s and its position angle θ. Eliminating θ, as a function of q and r_s, using the shape of particle trajectories (see Fig. 3.7), we can write the following three relations:

$$\varphi_{12}(WD) = f_1(q, i) \tag{3.1}$$

$$\varphi_{12}(HS) = f_2(q, i, r_s) \tag{3.2}$$

$$\varphi_{34}(HS) = f_3(q, i, r_s) \tag{3.3}$$

These relations are sufficient to determine the three parameters: q, i, and r_s. Concerning r_s it should be added that that the hot spot is usually assumed to be located exactly at the edge of the disk, i.e. that $r_s \equiv r_d$ (and $R_s \equiv R_d$).

To complete this discussion we should mention that at lower inclinations, close to $i \approx 65°$, the white dwarf may not be eclipsed and, if so, only the hot spot eclipse can be analyzed. Such a situation is observed in the case of two well known dwarf novae U Gem and WZ Sge.

In the case of *CV's with steady-state accretion* and of dwarf novae during their outbursts, the eclipse is primarily due to an occultation of the disk by the secondary. Several methods have been devised to analyse such eclipses in order to determine relevant parameters, including those of the disk. Their detailed discussion is beyond the scope of the present article. Two comments, however, must be made with regard to oversimplifying assumptions of some of those methods. The first concerns the contribution from the hot spot. The second – the finite geometrical thickness of the disk.

The hot spot contributes a small, but non-negligible amount of the total flux. It is responsible for a "hump", with a 10–15 percent amplitude, just prior to the eclipse, and for a marked asymmetry of the eclipse itself. This can be taken into account by decomposing, in a relatively simple way, the observed light curve into its spot and disk components and using the *disk* light curve in further analysis. An example of such a decomposition, presented in Fig. 3.3, clearly shows the effects involved. Neglecting this step and using the observed light curve must lead to spurious results.

The disk has a finite geometrical thickness and this produces two effects. First, in addition to the flux coming from its surface, there is a non-negligible contribution from its edge, or outer belt, which – in the case of a steady-state disk – is much cooler than the central parts. Secondly, at inclinations closer to 90° the disk surface is not only viewed at a very oblique angle but becomes partly obscured by the disk's edge (see Sect. 3.2.4). Neglecting those effects inevitably leads to spurious results. For example, the widely used MEM technique (Horne 1993) produces flat distributions of the disk temperature, $T(R)$, which turn out to be simply an artifact, resulting from the assumption of a flat disk (Smak 1994b).

Finally, let us mention that when the secondary contributes significantly to the total flux, its presence can be seen in the light curve in the form of (a) the secondary eclipse, resulting from its partial occultation by the disk, and (b) the "ellipticity" observed out of eclipses. Both effects are most easily detected in the infrared light curves.

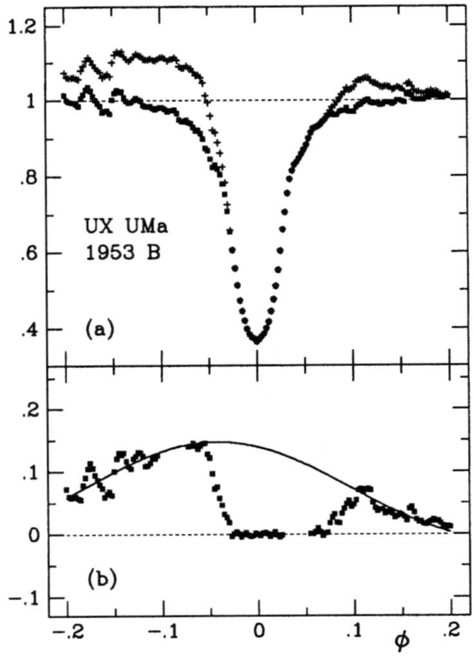

Fig. 3.3. Decomposition of the blue light curve of UX UMa into its spot and disk components (Smak 1994a). The observed light curve is shown by crosses in part (a). Part (b) is the reconstructed spot light curve (squares), clearly showing its eclipse; the continuous line is the theoretical spot light curve without eclipse effects. The reconstructed disk light curve is shown by squares in part (a).

3.2.5 The Secondary Component

We shall show below that when the secondary component fills its Roche lobe and obeys a specific mass-radius relation, then its mass uniquely determines the orbital period of the binary system. It is the very existence of this $P - M_2$ relation that often helps to determine the basic parameters of a CV's even in those cases when the observational data would normally be insufficient for this purpose.

Let us assume that the secondary is less massive than the primary. In such a case (more precisely – when $q = M_2/M_1 \leq 0.8$) its dimensionless mean radius can be approximated with (see Paczyński 1971):

$$r_2 \;=\; R_2/A \;=\; \frac{2}{3^{4/3}}\;\mu^{1/3} \tag{3.4}$$

where $\mu = M_2/(M_1 + M_2)$. Combined with the Kepler Law it gives

$$R_2 \;=\; \frac{2}{3^{4/3}}\;(GM_2)^{1/3}\;\left(\frac{P}{2\pi}\right)^{2/3} \tag{3.5}$$

This is equivalent to a general relation between the mean density of the secondary which fills its Roche lobe and the orbital period (Faulkner et al. 1972):

$$P\ \langle\rho\rangle^{1/2} = \left(\frac{3\pi}{G}\right)^{1/2} \left(\frac{3^{4/3}}{2}\right)^{3/2} = 3.78 \times 10^4 \qquad (3.6)$$

Assuming the mass-radius relation in the form

$$(R_2/R_\odot) = C_R\ (M_2/M_\odot)^\beta \qquad (3.7)$$

and combining it with (3.5), we get

$$P = 3.18 \times 10^4\ {C_R}^{3/2}\ (M_2/M_\odot)^{(3\beta-1)/2} \qquad (3.8)$$

For lower main sequence stars, with $C_R \approx 1$ and $\beta \approx 0.88$, we get

$$P \approx 3.18 \times 10^4\ (M_2/M_\odot)^{0.82}\ [\mathrm{s}] \approx 8.8\ (M_2/M_\odot)^{0.82}\ [\mathrm{hrs}] \qquad (3.9)$$

In the case of ultra-short-period CV's (see Paczyński 1967), with their secondaries being very low mass, degenerate stars, for which

$$R_2 \approx 0.0126\ (1+X)^{5/3}\ {M_2}^{-1/3} \qquad (3.10)$$

where X is the hydrogen content, we get from (3.8)

$$P \approx 45\ \frac{(1+X)^{5/2}}{M_2/M_\odot} \qquad (3.11)$$

From (3.9) and (3.11) it follows the existence of the minimum orbital period for hydrogen-rich CV's, which turns out to be slightly longer than 1 hour. This is also connected with the evolution of CV's (see Paczyński and Sienkiewicz 1983).

3.2.6 Emission Line Profiles

The shapes of the profiles of the emission lines from disks can be discussed without going into the complex details of the mechanism(s) of their origin (see Sect. 3.5.3). This is due to the fact that the Doppler broadening due to the rotation of the disk, which defines their characteristic double-peaked shape, dominates over other broadening factors. The simplest model is then the one involving a flat, Keplerian disk, with axial symmetry (see Robinson et al. 1993 and references therein).

With the Keplerian rotational velocity given by

$$V_K = \left(\frac{GM_1}{R}\right)^{1/2} \qquad (K) \qquad (3.12)$$

we obtain the lines of constant radial velocity on the surface of the disk shown in Fig. 3.4, which helps to understand the origin of the double-peaked profile:

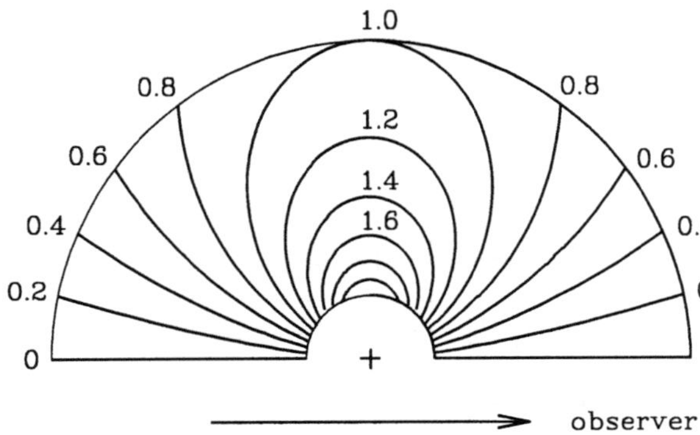

Fig. 3.4. Lines of constant radial velocity on the surface of a flat, Keplerian disk. Radial velocities are expressed in dimensionless units such that 1 corresponds to the rotational velocity at the outer radius of the disk. The other half of the disk is symmetric with respect to the one shown, with radial velocities being of opposite sign.

those parts which contribute to the profile near radial velocity equal to the outer rotational velocity of the disk have the largest area. The resulting profile is given by

$$F(U) \sim \int_{R_1}^{R_2} \frac{R^{3/2}\, f(R)\, \mathrm{d}R}{[1-(U/V)^2]^{1/2}} \qquad (K) \qquad (3.13)$$

where U is the radial velocity observed at inclination $i = 90°$, $V = V(R) \sim R^{-1/2}$ is the Keplerian velocity of rotation, R_1 and R_d are the inner and outer radii of the disk, $R_2 = min[R_d, R_d(V_d/U)^2]$, $V_d = V(R_d)$, and $f(R)$ is the emissivity or the surface density distribution of emitting atoms. For inclinations $i \neq 90°$, U should be multiplied by $\sin i$. Introducing dimensionless variables such that $r = R/R_d$, $v = V/V_d$, and $u = U/V_d$, and then substituting $x = ur^{1/2}$ we get

$$F(u) \sim \int_{r_1}^{r_2} \frac{r^{3/2}\, f(r)\, \mathrm{d}r}{(1-u^2 r)^{1/2}} \sim \frac{1}{u^5} \int_{x_1}^{x^2} \frac{2x^4 f(x^2/u^2)\, \mathrm{d}x}{(1-x^2)^{1/2}} \qquad (K) \qquad (3.14)$$

where $r_2 = min(1, u^{-2})$, $x_1 = ur_1^{1/2}$, and $x_2 = min(u, 1)$.

The radial dependence of emissivity can be determined from observations in a number a ways (see references in Robinson et al. 1993). The integral equation – (3.13) or (3.14) – can be inverted and used to reconstruct the radial dependence of emissivity $f(R)$ directly from the observed line profile. The same can be done *via* Doppler tomography (see Robinson et al. 1993, Marsh 1999). Results show that the radial dependence of emissivity can be approximated by $f(R) \sim R^{-\beta}$, with – typically – $\beta \sim 1.5 - 2.0$. Introducing such a form for $f(R)$ into (3.14) we get

$$F(u) \sim u^{2\beta-5} \int_{x_1}^{x_2} \frac{x^{4-2\beta}\, \mathrm{d}x}{(1-x^2)^{1/2}} \qquad (K) \qquad (3.15)$$

Figure 3.5 shows line profiles calculated from (3.15) for various combinations of r_1 and β (Smak 1981). The main conclusions resulting from these profiles are

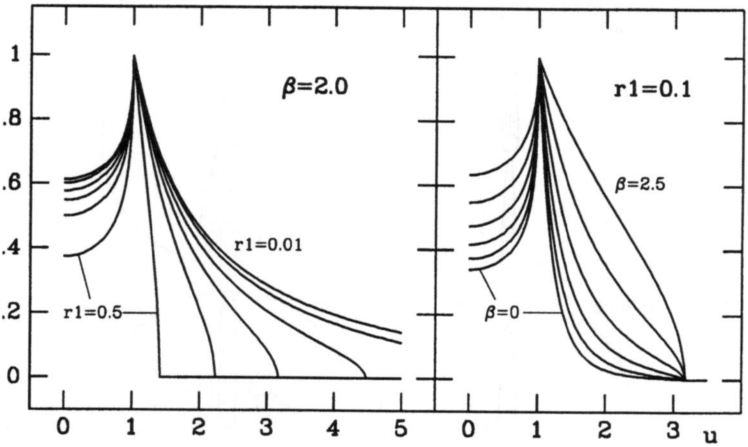

Fig. 3.5. Emission line profiles from a flat, Keplerian disk with emissivity given by $f(R) \sim R^{-\beta}$. *Left panel* shows their dependence on the radial width of the disk, $r_1 = R_1/R_d$ being the inner radius in units of the outer radius. The profiles correspond to $r_1 = 0.01$, 0.02, 0.05, 0.1, 0.2, and 0.5. *Right panel* shows the dependence on the parameter β. The profiles correspond to $\beta = 0$, 0.5, 1, 1.5, 2, and 2.5. Radial velocity u is normalized to the rotational velocity at the outer radius of the disk. (For economy of space only half-profiles are shown).

1. The line profile is double-peaked.
2. The peak velocity corresponds to the rotational, Keplerian velocity at the *outer* radius of the disk, $V_{\rm peak} = V_d \sin i$.
3. The velocity at the extreme edge of the line corresponds to the rotational, Keplerian velocity at the *inner* radius of the disk.

This model is, obviously, a major oversimplification. In a more realistic approach one should include the effects of the finite instrumental resolution, and the Doppler and Stark broadening. All of them affect the profile in a similar way, making the peaks less pronounced and the wings more extensive. In particular, the peak velocity is no longer identical with the rotational velocity at the outer radius. Approximately we have (Smak 1981) $V_{\rm peak} \approx 1.05\, V_d \sin i$.

$V_d \sin i$, as measured from the separation of the peaks in the double profile, provides an important information not only about the disk, but also about the mass ratio. Using (3.12) and the standard expression for the orbital velocity of the primary component we get

$$\frac{K_1}{V_d \sin i} = \frac{\mu}{(1-\mu)^{1/2}}\, {r_d}^{1/2} = \frac{q}{(1+q)^{1/2}}\, {r_d}^{1/2} \qquad (K) \qquad (3.16)$$

where $r_d = R_d/A$ is the dimensionless outer radius of the disk. When both K_1 and K_2 are known, (3.16) can thus be used to determine the outer radius of the disk r_d and, with A also known, R_d. When only K_1 is known, but r_d is also known (for example from eclipse analysis; see Sect. 3.2.2), then (3.16) determines the

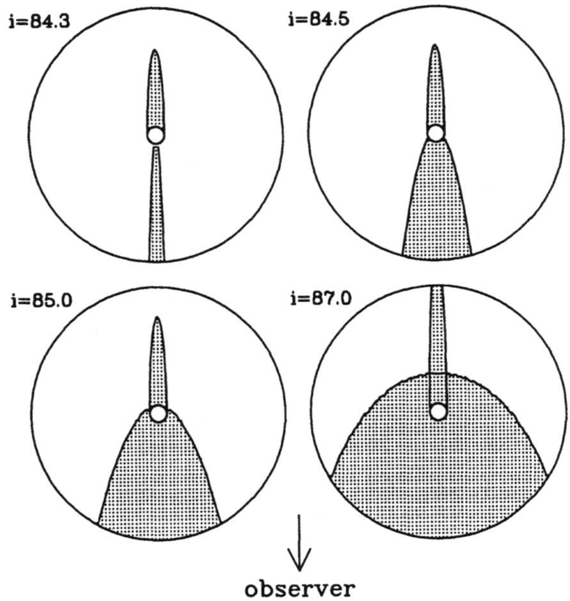

Fig. 3.6. Obscuration patterns on the surface of the disk with $z/r = 0.1$. The narrow area, behind the primary, is due to its shadow.

mass ratio and, consequently, M_1, M_2, A, and R_d. The same applies of course to the case when only K_2 is known.

The assumption of a flat disk works well only at low inclinations. At inclinations close to 90°, due to the disk's finite thickness, its surface becomes partly obscured by the disk edge. This is shown in Fig. 3.6. From a comparison with Fig. 3.4 we can conclude that, at inclinations where the obscuration is still only moderate, it will affect primarily the central part of the line profile making the depression between the two peaks deeper. At higher inclinations, however, the central part of the disk is also shadowed an then the entire profile is affected.

3.2.7 Problems

2-1. Assuming that the secondary is a spherical star with (dimensionless) radius r_2, derive exact equivalents of relations (3.1)-(3.3). Note that in this case μ will be replaced with r_2. (Derivation of (3.1) can actually be found in any standard textbook!).

2-2. Which of the parameters of an eclipsing dwarf nova at quiescence can be determined when the observational data consist of P, K_1, and $\varphi_{12}(WD)$ (defined in Sect. 3.2.1)?

2-3. Same as is 2-2, when the observational data include: P, K_1, K_2, and $\varphi_{12}(WD)$. Which of the standard assumptions described in Sect. 3.2.1 can be verified in this case?

2-4. What kind of data would be needed to verify the assumption $r_s \equiv r_d$?

2-5. Using (3.6) calculate the mean density of the secondary for orbital periods ranging from 1 to 10 hours. At what periods is this density of the order of 100 g/cm^3, indicating that the secondary must be a degenerate star?

2-6. Adopting, for simplicity, relations (3.9) and (3.11), determine the minimum value of the orbital period and the corresponding mass of the secondary.

2-7. Derive equation (3.16).

2-8. Derive the shape of obscuration areas shown in Fig. 3.6.

3.3 Accretion Disks

3.3.1 Formation of an Accretion Disk

To understand the basic physics involved in the formation of an accretion disk in a binary system we begin by considering the motion of the stream flowing from the internal Lagrangian point L_1 towards the primary. Under the assumption that the mass of a stream element is negligible compared to the masses of the two stars (restricted three-body problem), using dimensionless variables, such that $M_1 = (1-\mu)$, $M_2 = \mu$, $A = 1$, and $P = 2\pi$, this can be done relatively simply. In particular, for low velocities of escape from L_1 (in dimensionless units: $v \ll 1$), the problem has only one free parameter – the mass ratio q, or $\mu = M_2/(M_1 + M_2) = q/(1+q)$.

The top part of Fig. 3.7 shows the trajectory of a particle (or stream element) for the mass ratio $q = 0.25$ or $\mu = 0.2$. After passing near the primary the particle begins to move around it on a rosette-like trajectory or, more precisely, on an elliptical orbit undergoing rapid precession under the influence of the secondary. Its parameters are uniquely determined by the mechanical energy and the angular momentum carried by the particle. Replacing such a single particle with a stream we immediately conclude that, due to collisions resulting from intersecting trajectories, the energy of the stream elements must be dissipated. It can easily be shown that, at a given angular momentum, which is conserved, the minimum of energy corresponds to a circular Keplerian orbit, its radius being uniquely determined by that angular momentum. Denoting its dimensionless equivalent by r_h and dimensionless momentum by h we have

$$r_h = \frac{h^2}{1-\mu} \qquad (K) \tag{3.17}$$

Since h depends only on μ it follows that r_h is also a function of μ. Calculations show that r_h decreases with increasing μ (see bottom part of Fig. 3.7). The first extensive calculations of this kind, leading to the conclusions presented above, were made by Kruszewski (1966, 1967), his results being later confirmed by Lubow and Shu (1975). Consequently, "angular momentum radius" r_h is often called the "Lubow-Shu radius" (although the name "Kruszewski-radius" would appear more appropriate).

At this point – within purely mechanical approximation – we conclude that the material supplied by the stream must eventually form a Keplerian ring of

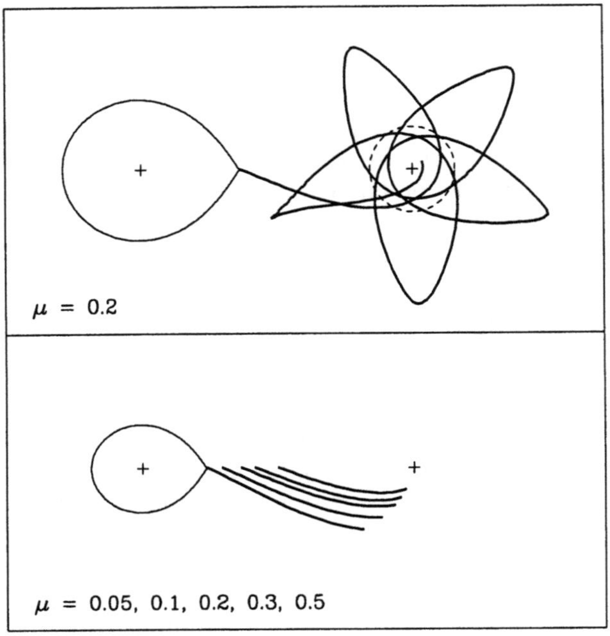

Fig. 3.7. Particle trajectories in a binary system. *Top:* Trajectory of a particle for $\mu = 0.2$. Dotted circle is the circular orbit with radius r_h defined by the particle's angular momentum. *Bottom:* Collection of trajectories showing dependence of their shapes on μ. Each trajectory begins at L_1 and ends at r_h, to show their dependence on μ. The secondary is shown only for $\mu = 0.05$.

radius r_h, rotating (or rather revolving) around the primary. As the density in the ring increases, the pressure effects become non-negligible and due to them the ring should eventually become a torus.

Including now the viscosity changes the situation in a substantial way. Viscous interactions result in an exchange of the angular momentum between the elements. Those which give away part of their angular momentum to their neighbours move – as a result – closer to the central star. The others – due to the acquired excess of the angular momentum – move outwards. The ring (or torus) spreads out into a disk.

3.3.2 General Equations

The equation of continuity and equations of motion can be, in general, written as

$$\frac{\partial \rho}{\partial t} + \text{div}\,(\rho\,\boldsymbol{v}) = 0 \tag{3.18}$$

and

$$\rho\,\frac{\mathrm{D}\boldsymbol{v}}{\mathrm{D}t} = -\,\text{grad}\,P + \text{div}\,\hat{\mathbf{s}} - \rho\,\nabla\Psi \tag{3.19}$$

where

$$\frac{\mathrm{D}\boldsymbol{v}}{\mathrm{D}t} = \frac{\partial \boldsymbol{v}}{\partial t} + (\boldsymbol{v} \cdot \nabla)\, \boldsymbol{v} \tag{3.20}$$

Ψ is the potential and $\hat{\mathbf{s}}$ is the viscosity tensor (stress tensor)

$$\hat{\mathbf{s}} = \begin{bmatrix} s_{xx} & s_{xy} & s_{xz} \\ s_{yx} & s_{yy} & s_{yz} \\ s_{zx} & s_{zy} & s_{zz} \end{bmatrix} \tag{3.21}$$

with

$$s_{xx} = 2\,\mu\, \frac{\partial v_x}{\partial x} + \lambda \operatorname{div} \boldsymbol{v} \tag{3.22}$$

$$s_{xy} = \mu \left(\frac{\partial v_x}{\partial y} + \frac{\partial v_y}{\partial x} \right) \tag{3.23}$$

and similar expressions for other elements. μ and λ are, respectively, the shear and bulk viscosity coefficients per unit volume.

3.3.3 Assumptions

The assumptions that will successively be made in the following considerations are:

A1. The disk is axially symmetric,
A2. The disk is geometrically thin,
A3. The accretion is stationary (S), and
A4. The disk is Keplerian (K).

We begin by making use of assumptions A1 and A2 which permit considerable simplification of the equations allowing, in particular, the radial and vertical structure of the disk to be treated separately. The first of them implies the use of cylindrical coordinates with all derivative $\partial/\partial\theta \equiv 0$. The second assumption will eventually allow us to "average" all radial equations over z. In preparation for that step we will neglect the z-derivatives of velocity components.

Assumptions A3 and A4 will be introduced later.

Important note: From now on, following the common custom in this area, small letters will be used to designate some of the parameters, such as the coordinates (e.g. z), velocities (e.g. v_r, v_θ), radial distances and radii (e.g. r, r_1, r_d). Note that the radial component of velocity of the accreting material v_r will be *negative.* Consequently the accretion rate $\dot{M}$ will also be *negative.*

3.3.4 The Radial Structure

We begin with the equations of the radial structure. With assumptions A1 and A2 the continuity equation becomes

$$\frac{\partial \rho}{\partial t} + \frac{1}{r}\frac{\partial}{\partial r}\,(r \rho v_r) = 0 \tag{3.24}$$

Multiplying (3.24) by $2\pi r$ and integrating it in z over the vertical thickness of the disk, i.e. from $-z_\circ$ to $z_\circ$, we get

$$\frac{\partial}{\partial t}(2\pi r\Sigma) + \frac{\partial}{\partial r}(2\pi r\Sigma\overline{v}_r) = \frac{\partial}{\partial t}(2\pi r\Sigma) + \frac{\partial \dot{M}}{\partial r} = 0 \tag{3.25}$$

where

$$\Sigma = \int_{-z_\circ}^{z_\circ} \rho\, dz \tag{3.26}$$

is the surface density, $\overline{v}_r$ is the z-averaged radial component of the velocity, and

$$\dot{M} = 2\pi r\Sigma\overline{v}_r \tag{3.27}$$

is the local accretion rate.

Applying the same same assumptions to the equations of motion we get for the tangential component of the velocity v_t

$$\rho\left(\frac{\partial v_t}{\partial t} + v_r\frac{\partial v_t}{\partial r} + \frac{v_r v_t}{r}\right) = \frac{\partial}{\partial r}\left[\mu\left(\frac{\partial v_t}{\partial r} - \frac{v_t}{r}\right)\right] + \frac{2\mu}{r}\left(\frac{\partial v_t}{\partial r} - \frac{v_t}{r}\right) \tag{3.28}$$

Introducing the angular momentum $h = v_t r$ and angular velocity $\omega = v_t/r$, after simple calculations we get

$$\rho\left(\frac{\partial h}{\partial t} + v_r\frac{\partial h}{\partial r}\right) = -\frac{1}{r}\frac{\partial}{\partial r}(r\, t_{\theta r}) \tag{3.29}$$

where

$$t_{\theta r} = -\mu\, r^2\frac{\partial \omega}{\partial r} \tag{3.30}$$

is the (θ, r) element of the shear stress tensor in cylindrical coordinates.

Combining the continuity equation (3.24) with the equation of tangential motion (3.29) we obtain the following equation for the conservation of the angular momentum

$$\frac{\partial}{\partial t}(\rho h) + \frac{1}{r}\frac{\partial}{\partial r}(r\rho h v_r) + \frac{1}{r}\frac{\partial}{\partial r}(r\, t_{\theta r}) = 0 \tag{3.31}$$

Multiplied by $2\pi r$ and integrated over z it becomes

$$\frac{\partial}{\partial t}(2\pi r\,\Sigma h) + \frac{\partial}{\partial r}(\dot{M}\, h) + \frac{\partial g}{\partial r} = 0 \tag{3.32}$$

where

$$g = \int_{-z_\circ}^{z_\circ} 2\pi r\, t_{\theta r}\, dz \tag{3.33}$$

is the momentum of viscous forces. Putting $\mu = \nu\rho$, where ν is the viscosity coefficient per unit mass, and using (3.30) we get another expression

$$g = -2\pi\, r^3\frac{\partial \omega}{\partial r}\Sigma\overline{\nu} \tag{3.34}$$

where $\bar{\nu}$ is the z-averaged value of ν.

Equation (3.32) can be simplified using second part of (3.25):

$$2\pi r \, \Sigma \, \frac{\partial h}{\partial t} \; + \; \dot{M} \, \frac{\partial h}{\partial r} \; + \; \frac{\partial g}{\partial r} \; = \; 0 \tag{3.35}$$

Getting $\dot{M}$ from this equation and inserting it into the second part of (3.25) produces

$$\frac{\partial \Sigma}{\partial t} \; = \; \frac{1}{r} \, \frac{\partial}{\partial r} \, \left\{ \left[r\Sigma \, \frac{\partial h}{\partial t} \; - \; \frac{\partial}{\partial r} \left(r^3 \frac{\partial \omega}{\partial r} \Sigma \bar{\nu} \right) \right] \; / \; \frac{\partial h}{\partial r} \right\} \tag{3.36}$$

This is the general equation describing the time-evolution of an accretion disk.

We now introduce assumptions A3 and A4 (Sect. 3.3.3). Assumption of stationary accretion implies that all time-derivatives vanish and $\dot{M}$ = const. Integrating (3.35) under this assumption gives

$$\dot{M} \; (h - h_i) \; + \; (g - g_i) \; = \; 0 \qquad (S) \tag{3.37}$$

where h and g are parameters of an element at radial distance r, while h_i and g_i refer to the inner edge of the disk at, or near, the stellar surface. Postponing a more detailed discussion of this problem till Sect. 3.5.1, we only note here that the location of the inner edge of the disk is defined by the condition that viscous interactions should no longer operate there, and that occurs when $\partial \omega / \partial r = 0$, i.e. when $\omega(r)$ reaches its maximum. If so, we can put $g_i = 0$ and this allows us to rewrite (3.37) as

$$\dot{M} \; (h - h_i) \; + \; g \; = \; 0 \qquad (S) \tag{3.38}$$

Turning to assumption A4 we note that the disk becomes Keplerian (or nearly Keplerian) when its elements move inwards along tightly wound spiral trajectories, with $v_r \ll v_t$. Neglecting v_r, using standard, Keplerian formulae for v_t, h, and ω, and assuming $r_i \approx r_1$, i.e. that the disk extends to the surface of the white dwarf (see Sect.3.5.1), we then obtain the "Keplerian"version of (3.34), (3.36), and (3.38):

$$g \; = \; 3\pi \; h \; \Sigma \bar{\nu} \qquad (K) \tag{3.39}$$

$$\frac{\partial \Sigma}{\partial t} \; = \; \frac{3}{r} \, \frac{\partial}{\partial r} \, \left[r^{1/2} \, \frac{\partial}{\partial r} \left(r^{1/2} \; \Sigma \bar{\nu} \right) \right] \qquad (K) \tag{3.40}$$

$$\dot{M} \; (GMr)^{1/2} \; \left[1 \; - \; (r_1/r)^{1/2} \right] \; + \; g \; = \; 0 \qquad (SK) \tag{3.41}$$

Combining the last of these equations with g from (3.39) gives a condition on $\Sigma \bar{\nu}$ for a stationary, Keplerian disk:

$$\Sigma \bar{\nu} \; = \; \frac{(-\dot{M})}{3\pi} \; \left[1 \; - \; (r_1/r)^{1/2} \right] \qquad (SK) \tag{3.42}$$

3.3.5 The Energy Equation

The energy equation per unit volume can be written as

$$\rho\,\frac{dE_m}{dt} + \rho\,\frac{dE_t}{dt} = \rho\,\frac{dQ}{dt} + \rho\,\frac{dW}{dt} \tag{3.43}$$

where E_m and E_t are the mechanical and thermal (internal) energy, $\rho\, dQ/dt$ is the heat supplied to the element, and $\rho\, dW/dt$ – the work performed over the element. From the equations of motion we have

$$\rho\,\frac{dE_m}{dt} = \sum_{jk} v_k\,\frac{\partial s_{jk}}{\partial x_j} \tag{3.44}$$

$$\rho\,\frac{dW}{dt} = \sum_{jk} \frac{\partial}{\partial x_j}\,(v_k\; s_{jk}) \tag{3.45}$$

where v_k, and s_{jk} are the components of the velocity vector and the viscous stress tensor in the rectangular system of coordinates, with $j, k = 1, 2, 3$. We also have

$$\rho\,\frac{dQ}{dt} = -\ \mathrm{div}\ \boldsymbol{F} \tag{3.46}$$

where $\boldsymbol{F}$ is the flux which carries away the energy dissipated by viscous interactions inside the element.

From those four equations we get for the flux

$$\mathrm{div}\ \boldsymbol{F} = -\ \rho\,\frac{dE_t}{dt} + \sum_{jk} s_{jk}\,\frac{\partial v_k}{\partial x_j} \tag{3.47}$$

which in the case of a flat, axially symmetric disk becomes

$$\mathrm{div}\ \boldsymbol{F} = -\ \rho\,\frac{dE_t}{dt} - p\ \mathrm{div}\ \boldsymbol{v} + \mu\left(r\,\frac{\partial\omega}{\partial r}\right)^2 \tag{3.48}$$

Finally, in the case of thermal equilibrium, assuming that p div $\boldsymbol{v}$ is negligible and that $|\partial F/\partial r| \ll |\partial F/\partial z|$, we obtain

$$\frac{dF}{dz} = \mu\left(r\,\frac{\partial\omega}{\partial r}\right)^2 \tag{3.49}$$

which in the Keplerian case reduces to

$$\frac{dF}{dz} = \frac{9}{4}\,\mu\,\omega^2 \qquad (K) \tag{3.50}$$

Equations (3.49) and (3.50) simply describe the intuitively obvious fact that in the case of a flat disk the energy dissipated inside the element by the viscous interactions is carried away by the flux in the vertical direction ("up" and "down", since the disk has two sides).

In the case of an axially symmetric disk in thermal equilibrium, i.e. when $dE_t/dt = 0$, equation (3.43) and its z-integrated version take the form

$$\rho \frac{\partial E_m}{\partial t} + \rho\, v_r \frac{\partial E_m}{\partial r} + \operatorname{div} \boldsymbol{F} + \frac{1}{r}\frac{\partial}{\partial r}(r t_{\theta r} \omega) = 0 \qquad (3.51)$$

$$2\pi r\, \Sigma \frac{\partial \overline{E}_m}{\partial t} + \dot{M} \frac{\partial \overline{E}_m}{\partial r} + 2\pi r\; 2F_\circ + \frac{\partial}{\partial r}(g\omega) = 0 \qquad (3.52)$$

where $F_\circ$ is the flux at the surface of the disk ($z = z_\circ$) and "2" accounts for the disk's two sides. Making use of the continuity equation (3.25) we get

$$\frac{\partial}{\partial t}(\,2\pi r\, \Sigma \overline{E}_m) + \frac{\partial}{\partial r}(\dot{M}\overline{E}_m) + 2\pi r\; 2F_\circ + \frac{\partial}{\partial r}(g\omega) = 0 \qquad (3.53)$$

In the case of stationary accretion this becomes

$$\dot{M}\left(\frac{\partial \overline{E}_m}{\partial r} - \omega \frac{\partial h}{\partial r}\right) + g \frac{\partial \omega}{\partial r} + 2\pi r\; 2F_\circ = 0 \qquad (S) \qquad (3.54)$$

which in the Keplerian case, with g from (3.39), reduces to

$$F_\circ = \frac{9}{8}\,\omega^2\, \Sigma \overline{\nu} \qquad (SK) \qquad (3.55)$$

Finally, with $\Sigma\overline{\nu}$ from (3.42), we obtain

$$F_\circ = \sigma\, T_e^{\,4} = \frac{3}{8\pi}\,|\dot{M}|\,\frac{GM_1}{r^3}\left[1 - (r_1/r)^{1/2}\right] \qquad (SK) \qquad (3.56)$$

This relation shows an interesting property of an accretion disk. If we condider a single surface density element which – somehow – dissipates its mechanical energy and gives away its angular momentum at a rate which makes it spiral in with velocity v_r, then we have

$$\Sigma \frac{d\overline{E}_m}{dt} = \Sigma \frac{d\overline{E}_m}{dr}\,|\,v_r| = 2\,F_\circ \qquad (SK) \qquad (3.57)$$

where "2" again accounts for the two sides of the disk. This gives

$$F_\circ = \frac{1}{8\pi}\,|\dot{M}|\,\frac{GM_1}{r^3} \qquad (SK) \qquad (3.58)$$

Note that the coefficient in front of (3.56) is 3 times larger, making $F_\circ$ at large distance $\sim$ 3 times higher, and that at small radial distances, due to the factor in the square bracket, $F_\circ$ resulting from that equations is much smaller. This is due to the fact that the viscous couple transports energy from the inner to the outer parts of the disk.

Let us now summarize main results. In the case of non-stationary accretion the evolution of a Keplerian disk is described by (3.40). To calculate that evolution we need an additional relation, $\Sigma\overline{\nu} = f(\Sigma)$, between the z-integrated viscosity $\Sigma\overline{\nu}$ and the surface density Σ. Such a relation can be obtained only by integrating equations of the vertical structure. This will be discussed in Sect.3.3.7.

Equation (3.56) is the basic relation in the case of stationary accretion. Together with (3.55) it provide a relation between three important parameters: the local flux, or effective temparature, the accretion rate, and the z-integrated viscosity (including their dependence on r). It can be written as

$$F_\circ = \sigma T_e^4 \sim |\dot{M}| \sim \Sigma\bar{\nu} \qquad (SK) \tag{3.59}$$

3.3.6 Radiative Properties of a Stationary Accretion Disk

Equation (3.56) can be used already at this point, *without any information concerning the viscosity*, to discuss radiative properties of stationary accretion disks.

The luminosity of an accretion disk with radius r_d can be obtained by integrating (3.56). The result is

$$L_d = L_\circ \left[1 - x_d^{-1}\left(3 - 2\, x_d^{-1/2}\right)\right] \qquad (SK) \tag{3.60}$$

where $x_d = r_d/r_1$ and

$$L_\circ = \frac{1}{2}\,\frac{GM_1}{r_1}\,|\dot{M}| \qquad (SK) \tag{3.61}$$

is the luminosity of a disk extending to infinity. Note that this formula could be obtained also from (3.58), or – simply – from the energy difference between infinity and r_1.

$L_\circ$ and L_d are the mean luminosities. Observed luminosities depend on the orbital inclination. In the simplest case of a flat disk, taking into account only the geometry and the limb darkening, we have

$$L(i) = \langle L\rangle \;(\,1 - u + u\cos i\,)\;\left(\frac{1}{2} - \frac{u}{6}\right)^{-1} \cos i \tag{3.62}$$

where u is the limb darkening coefficient. Adopting $u = 0.6$ we have

$$L(i) = \langle L\rangle \;(\,1.5\cos i + 1)\;\cos i \tag{3.63}$$

This relation is often used to correct the observed absolute magnitudes for the effect of inclination. Obviously, however, it cannot be applied to inclinations close to 90°.

In the black body approximation the spectral energy distribution of the disk is given by (see Smak 1989)

$$\nu\, F_\nu = 45\,\pi^{-4}\, L_\circ\, f\left(\frac{\nu}{\nu_*}, x_d\right) \qquad (SK) \tag{3.64}$$

where

$$\nu_* = \frac{k}{h}\left[\frac{3}{8\pi\sigma}\,\frac{GM_1}{r_1^{\,3}}\,|\dot{M}|\right]^{1/4} \tag{3.65}$$

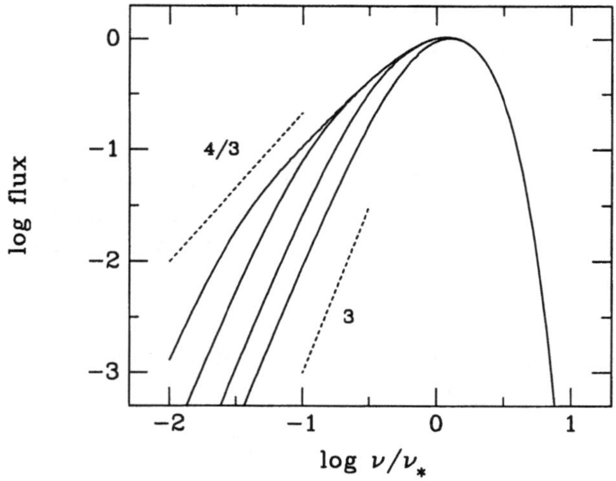

Fig. 3.8. Normalized, dimensionless spectral energy distributions of black body disks. Solid curves, from right to left, correspond to disks with $x_d = r_d/r_1 = 10$, 25, 100, and 500. Dotted lines represent $f \sim \nu^{4/3}$ and $f \sim \nu^3$.

with h being here the Planck constant, and

$$f\left(\frac{\nu}{\nu_*}, x_d\right) = \left(\frac{\nu}{\nu_*}\right)^4 \int_1^{x_d} \frac{x \, \mathrm{d}x}{\exp\left[x^{3/4}\left(1 - x^{-1/2}\right)^{-1/4} \nu/\nu_*\right] - 1} \tag{3.66}$$

Using these equations we can calculate the spectral energy distributions of black body disks as a function of only one parameter: $x_d = r_d/r_1$. They are shown in Fig. 3.8. Note that for very large disks the assymptotic solution at the low frequency end of the spectrum is $f \sim \nu^{4/3}$. For very small disks, however, we have $f \sim \nu^3$, corresponding to the Rayleigh-Jeans law. Therefore the observed shape of the spectrum cannot, in general, be used as a diagnostic tool for steady-state accretion.

3.3.7 The Vertical Structure

Equations of the vertical structure of the disk closely resemble those of the stellar structure. From (3.26) we have

$$\frac{d\Sigma_z}{dz} = \rho \tag{3.67}$$

where we use index "z"to denote $\Sigma(z)$; note that Σ, as defined by (3.26) and used elsewhere in the text, is Σ_z at $z = z_\circ$. The equation of hydrostatic equilibrium in the thin disk approximation is

$$\frac{dP}{dz} = -\frac{GM_1}{r^3} z \rho \tag{3.68}$$

In addition we have (3.50) as the equation of "energy generation" and the usual equation involving the vertical transport of energy

$$\frac{dT}{dz} = \nabla \frac{T}{P} \frac{dP}{dz} \tag{3.69}$$

where $\nabla = \min(\nabla_r, \nabla_c)$, and ∇_r, ∇_c are the radiative and convective gradients.

To integrate these equations we obviously need an expression for μ which appears in (3.50). Viscosity, however, remains unknown and therefore it appears useful to begin with the first two equations only.

Crude Approximation. Equation of hydrostatic equilibrium can be integrated (or averaged) crudely to give

$$\overline{P} \approx \frac{GM}{r^3} \frac{z_\circ \Sigma}{2} \tag{3.70}$$

We now adopt the equation of state for perfect gas, which can be written as

$$\overline{P} = \frac{k}{\mu H} \frac{\Sigma}{2z_\circ} \overline{T} \tag{3.71}$$

where μ is here the mean molecular weight, and $\Sigma/2z_\circ$ is the average density. Combining these two equations we get for the average temperature (which will be needed later in the α-disk model)

$$\overline{T} = C_T \frac{\mu H}{k} \frac{GM_1}{r^3} {z_\circ}^2 \tag{3.72}$$

with $C_T = 1$.

Polytropic Approximation. Adopting the standard polytropic relation

$$P = K \rho^{1+\frac{1}{n}} \tag{3.73}$$

and the equation of state for perfect gas we integrate the equation of hydrostatic equilibrium and get (Paczyński 1978)

$$\rho = \rho_c \left[1 - \left(\frac{z}{z_\circ} \right)^2 \right]^n \tag{3.74}$$

$$T = T_c \left[1 - \left(\frac{z}{z_\circ} \right)^2 \right] \tag{3.75}$$

where ρ_c and T_c are the "central" density and temperature at $z = 0$:

$$\rho_c = \left[\frac{1}{2K(n+1)} \frac{GM_1}{r^3} {z_\circ}^2 \right]^n \tag{3.76}$$

$$T_c = \frac{\mu H}{k} \frac{1}{2(n+1)} \frac{GM_1}{r^3} z_\circ^2 \qquad (3.77)$$

These solutions show one important property of the disk: the distribution in z of the density, temperature and pressure are rather flat. This is best demonstrated by the ratio of mean-to-central density, $\bar{\rho}/\rho_c$, which varies only from 1 for $n = 0$ to about 0.5 for $n = 3$, and the ratio of mean-to-central temperature $\bar{T}/T_c = 2/3$ (see Paczyński 1978). The mean temperature is now given by (3.72) with $C_T = 1/3(n+1)$, i.e. smaller than in the earlier case of the crude approximation; for example $C_T = 1/3$ for $n = 0$ and $C_T = 1/12$ for $n = 3$.

Geometrical Thickness: $z_\circ/r$. At this point we can already verify our assumption A2 (Sect.3.3.3). From (3.72) we have

$$\frac{z_\circ}{r} = \left[\left(\frac{k}{\mu H} \overline{T} \right) / \left(C_T \frac{GM_1}{r} \right) \right]^{1/2} \qquad (3.78)$$

With, for example, $M_1 = 1 M_\odot$, $r = 1 \times 10^{10}$, $\overline{T} = 1 \times 10^4$K, $\mu = 0.5 - 1.0$, and $C_T = 1/12 - 1$, we get $z_\circ/r = 0.01 - 0.04$. This confirms nicely our initial assumption (for further discussion of $z_\circ/r$ see Sect.3.3.9).

3.3.8 α-Disks

In spite of many efforts in this area, the nature of viscosity in the accretion disks still remains unknown and the famous α-prescription proposed by Shakura and Sunyaev (1973) remains the only way to overcome our ignorance in this respect.

With turbulence being among the possible candidates, by analogy with the turbulent viscosity which is proportional to the mean velocity and mean size of turbulent eddies, Shakura and Sunyaev assumed

$$\mu = \rho \nu = \alpha \rho v_s z_\circ \qquad (3.79)$$

where v_s is the velocity of sound and α is a dimensionless parameter; note that in the case of subsonic turbulence $\alpha < 1$.

For perfect gas the velocity of sound can be, using (3.72), and then assuming Keplerian motion, written as

$$v_s = \frac{k}{\mu H} T = C_T \frac{GM_1}{r^3} z_\circ^2 = C_T v_t^2 \left(\frac{z_\circ}{r} \right)^2 \qquad (K) \qquad (3.80)$$

With (3.72), (3.78), and (3.80) the viscosity can now be expressed as follows

$$\mu = \alpha \bar{\rho} C_T^{1/2} h \left(\frac{z_\circ}{r} \right)^2 = \alpha C_T^{-1/2} \overline{P} / \omega \qquad (K) \qquad (3.81)$$

At this point we can already verify the assumption of the Keplerian, or near-Keplerian, motion. From (3.55) and (3.56), using (3.81) (for simplitcity with $C_T = 1$) we have for the flux

$$F_\circ = \frac{9}{8} \omega^2 \alpha \Sigma h \approx \frac{3}{8\pi} (2\pi r \Sigma \mid v_r \mid) \omega^2 \qquad (K) \qquad (3.82)$$

from which it follows that

$$\frac{|v_r|}{v_t} \approx \frac{3}{2}\,\alpha \left(\frac{z_\circ}{r}\right)^2 \qquad (K) \tag{3.83}$$

As can be seen, for a flat disk $|v_r|$ is indeed much smaller than v_t.

The most important consequence of the α-disk approximation is, however, that we can now use (3.81) to complete the flux equation (3.56). In practice this requires, of course, specifying the value of C_T, which appears in (3.81) as a result of a preliminary, approximate integration of the equation of hydrostatic equilibrium. As it turns out, however, α – as *defined originally* by (3.79) – becomes also a proportionality factor in the resulting expression for the viscous stress and this opens up a possibility of an *alternative definition* of this parameter. Using this approach, which has been commonly adopted in the field of CV's (see Osaki 1966), one gets from (3.56)

$$\frac{dF}{dz} = \frac{3}{2}\,\alpha\,\omega\,P \qquad (K) \tag{3.84}$$

This, in fact, is the *practical, numerical definition* of α.

The four equations of the vertical structure can now be integrated and that, in turn, permits the construction of the time-dependent models of disks. Before presenting the results of such numerical integrations, we will pause to discuss the characteristic time scales which can already be estimated at this point.

3.3.9 The Time Scales

The four characteristic time scales are: the dynamic time scale in r:

$$\tau_d \sim \omega^{-1} \qquad (K) \tag{3.85}$$

the dynamic time scale in z:

$$\tau_z \sim \frac{z_\circ}{v_s} \sim \tau_d \qquad (K) \tag{3.86}$$

the viscous (i.e. accretion) time scale:

$$\tau_\nu \sim \frac{r}{|v_r|} \sim \tau_d \left[\alpha\left(\frac{z_\circ}{r}\right)^2\right]^{-1} \qquad (K) \tag{3.87}$$

and the thermal time scale:

$$\tau_{th} \sim \Sigma\, {v_s}^2 \,/\, F_\circ \sim \tau_d \,/\, \alpha \qquad (K) \tag{3.88}$$

Note that with $(z_\circ/r)^2 \ll 1$ and $\alpha < 1$ we have

$$\tau_d \sim \tau_z < \tau_{th} \ll \tau_\nu \tag{3.89}$$

3.3.10 The $\Sigma - T_e$ Relation, Stability, and Disk Thickness

As already mentioned, the problem of the vertical structure of the disk closely resembles that of the stellar structure. With z considered as independent variable, we have four equations – (3.67), (3.68), (3.69), and (3.84) – for the four dependent variables: Σ_z, ρ, T, and F, with the equation of state used to express P as a function of ρ and T. As in the stellar case, from the excess of two boundary conditions it follows the existence of two relationships between Σ, $F_\circ$, and $z_\circ$. Note that those parameters are equivalent to the mass, radius, and luminosity in the stellar case and the two relations are equivalent to the mass–radius–luminosity relations.

The assumptions and procedures used in the numerical integrations are similar to the stellar case (for details see references in Cannizzo 1993). There are some delicate points which are worth mentioning here. First, that (3.84) which was derived for *z-averaged* parameters is now used locally. Secondly, that there are problems in connection with the application of the mixing-length theory to the disk case, which are solved only in an approximate way. Fortunately, in the case of a geometrically thin disk this does not appear to result in major uncertainties or errors. Finally, let us emphasize that the four equations are solved for an assumed set of values of the radial distance r, the central mass M_1, and the viscosity parameter α. Of these, only r and M_1 are the obvious "free" parameters, their values to be chosen specifically to describe different parts of the disk in a given binary system. α, however, is not only an unknown parameter but, in principle, could also depend on other parameters. It is therefore a matter of an arbitrary decision that we adopt *one* value of α to apply to the entire disk.

The main result of numerical integrations is the relation between $F_\circ$ and Σ. In fact, it is a relation between four parameters: $F_\circ$, T_e, $\dot{M}$, and $\Sigma\bar{\nu}$ (see 3.59 and discussion there). For reasons, which will become immediately apparent, this relation is usually presented as the $\Sigma - T_e$ relation (Fig. 3.9). It has a characteristic S-shape which reflects the dependence on temperature of the opacity: due to the ionization of hydrogen it has a high maximum at temperatures around 1×10^4K. In the case of a thin disk this corresponds, depending on other parameters, to $T_e \sim 6000 - 8000$°K. For the same reason, as long as the disk is thin and the surface density is not very low or very high (i.e. for the range of Σ shown in Fig. 3.9), the $\Sigma - T_e$ relations calculated for different values of r are of nearly identical shape (see Fig. 3.9); they are only shifted in Σ and, slightly, in T_e. Fig. 3.9 also shows two important features. At large distances and higher temperatures the assumption of a thin disk begins to be violated (this could become a problem in the case of long-period Algols, but not in the case of CV's). The disk is optically thick, except at small distances and at low temperatures where it becomes optically thin. Dependence on α is similar (see Fig. 3.9): curves with lower α's are shifted to the right, while those with higher α's – to the left.

We now turn to the problem of stability. Starting with stability against local perturbations in Σ, let us assume for their shape

$$\epsilon \cos\left(\frac{2\pi}{\lambda} r\right) \tag{3.90}$$

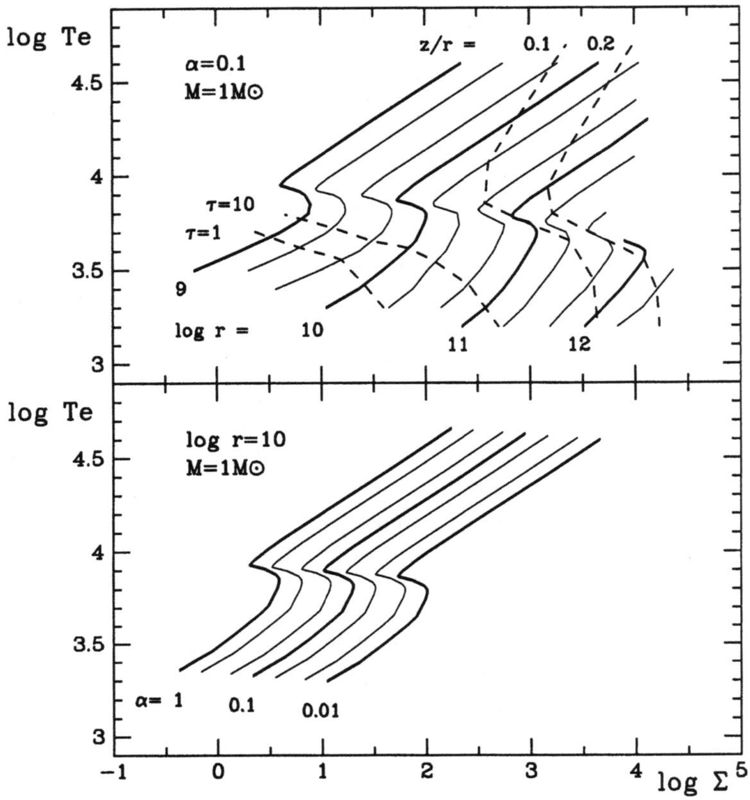

Fig. 3.9. *Top:* $\Sigma - T_e$ relations for $M_1 = 1M_\odot$, $\alpha = 0.1$ and a series of values of r. Broken lines mark *loci* of constant constant optical thickness and of constant values of $z_\circ/r$. *Bottom:* The same with $M_1 = 1M_\odot$ and $r = 1 \times 10^{10}$ showing the dependence on α.

with $\lambda \ll r$, and that – locally – we have

$$F = \sigma T_e^{\,4} \sim \Sigma^s r^q \tag{3.91}$$

From (3.48) with $\Sigma\bar{\nu}$ from (3.55), we get

$$\frac{\partial \Sigma}{\partial t} \sim - s\,\epsilon \cos\left(\frac{2\pi}{\lambda}\,r\right) \tag{3.92}$$

This implies that an instability occurs when $s < 0$. Therefore, of the three branches of the $\Sigma - T_e$ relation, it is the middle branch which is unstable.

More importantly, the middle branch of the $\Sigma - T_e$ relation is also thermally unstable. This can be shown in a qualitative way even without performing any model calculations. Let us assume that a model deviating from thermal equilibrium is located in the $\Sigma - T_e$ diagram to the *right* of the standard relation (calculated, as described above, for the case of equilibrium). Since at a given T_e

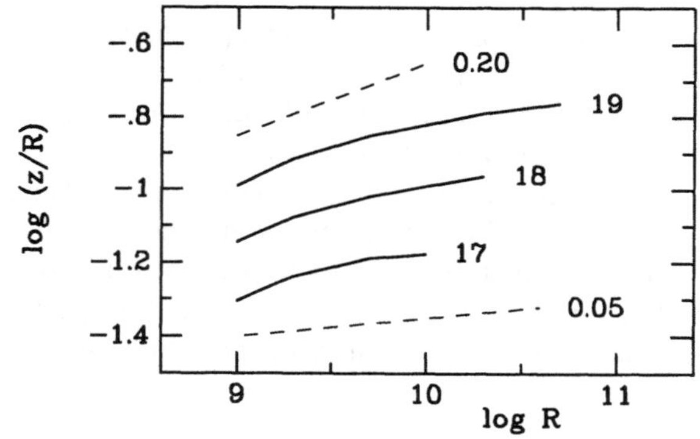

Fig. 3.10. Thickness of a steady-state disk as a function of radial distance for three accretion rates: $\log \dot{M} =$ 17, 18, and 19. Broken lines correspond to $z_\circ/r \sim r^{0.05}$ and $\sim r^{0.20}$.

the flux generated inside the disk by viscosity, F_ν, must be, roughly, proportional to Σ, while the emerging flux is $F_\circ = \sigma T_e{}^4$, it follows that the difference $F_\nu - F_\circ$ must be used for *heating* the disk material. Conversely, a model located to the *left* of the equilibrium $\Sigma - T_e$ relation must undergo *cooling*. Let us now consider a perturbation in temperature. In the case of the hot (upper) branch of the $\Sigma - T_e$ relation such a perturbation will produce a reaction of opposite sign, resulting in the return to equilibrium. The same is true for the cool (lower) branch. In the case of the middle branch, however, a positive (negative) perturbation in T will result in further heating (cooling), i.e. in an instability. This instability provides basic mechanism for dwarf nova outbursts, which will be discussed in the next Section.

The second relationship resulting from integrations involves $z_\circ$. Since we are interested primarily in the dependence of this parameter on the accretion rate, we express this relation as $z_\circ = f(F_\circ)$ and replace $F_\circ$ with $\dot{M}$ from (3.56). The results, plotted in Fig. 3.10, show that in the case of steady-state accretion the disk is slightly *concave*

$$\frac{z_\circ}{r} \sim r^\xi \tag{3.93}$$

with ξ between 0.05 and 0.20. Furthermore, due to its dependence on temperature (3.78), $z_\circ/r$ increases with $\dot{M}$. At $\dot{M} = 10^{19}$ it already exceeds $z_\circ/r = 0.1$ and it is clear that at higher accretion rates our assumption of a thin disk would begin to be violated.

3.3.11 Problems

3-1. From (3.56) find the radial distance at which the temperature reaches its maximum.

3-2. Derive (3.62).

3-3. How does the assumption of a flat disk is used in the derivation of (3.68)?

3-4. Derive an expression for $\bar{\rho}/\rho_c$ for polytropic disks.

3-5. Using (3.78) for the case of steady-state accretion, express z_o/r as a function of the accretion rate and radial distance. Assume $\bar{T} \approx T_e$ and, for simplicity, neglect the square bracket in (3.56).

3-6. Using an approximate formula $\Sigma = f(T_e, r, \alpha)$ of the form suggested by Fig. 3.9, derive an expression for the mass of a hot, steady-state disk as a function of r_d, $\dot{M}$, and α. For the mass of the central star adopt $M_1 = 1M_\odot$.

3-7. How does the vertical structure of the disk and, in particular, the shape of the $\Sigma - T_e$ relations, depend on the mass of the central star M_1? Repeat now 3-6, including M_1.

3.4 Dwarf Nova Outbursts

3.4.1 Dwarf Novae

Dwarf novae form a subclass of CV's, their characteristic photometric property being quasi-periodic outbursts with amplitudes from 2 to 6 amplitudes and recurrence periods (or cycle lengths) ranging from weeks to months or even years. Two important points have to mentioned immediately after such a short definition. First, that their behavior is far from being regular: for a given dwarf nova the durations of individual outbursts and quiescent intervals vary considerably from cycle to cycle in an irregular way. For example, in the best studied case of SS Cyg, its outburst cycles range from 15 to 95 days, while outburst durations – from 2 to 22 days. Secondly, as a group, dwarf novae are far from being homogeneous, with three main subtypes being: the U Geminorum stars, the Z Camelopardalis stars, and the SU Ursae Majoris stars. The U Gem stars represent relatively simple outburst behavior which can be explained quite well with the disk instability model presented below. In the case of the other two subtypes their behavior is more complex.

The *Z Cam stars*, in addition to longer or shorter intervals of dwarf nova activity, show longer or shorter "standstills" of constant brightness. It is commonly accepted that they represent a borderline case with the mass transfer rate almost sufficient to satisfy the stability condition (3.94). When it gets only slightly enhanced, due likely to irradiation of the secondary by the boundary layer (Meyer and Meyer-Hofmeister 1983), the accretion becomes stationary; the dwarf nova behavior returns when it falls again below the critical level.

The *SU UMa stars*, in addition to normal outbursts show less frequent *superoutbursts* with somewhat higher amplitudes and much longer durations. This group consists mostly of dwarf novae with the shortest orbital periods ($P_{orb} < 2$ hrs), with notable exceptions being TU Men ($P = 2.58$ hrs) and U Gem ($P = 4.89$ hrs), its 1985 superoutburst lasting for the record long 40 days. The characteristic feature of superoutbursts are the superhumps – light variations with a period few percent longer than the orbital period, which appear shorthly after superoutburst maximum and persist through its decline. The superoutbursts are either due to the enhanced mass transfer rate caused by irradiation by the secondary (Osaki 1985), or due to the tidal instability of the outer parts of the disk (Osaki 1996), or both.

3.4.2 Disk Instability Model for Dwarf Nova Outbursts

It has long been suspected that the dwarf nova outbursts are due to non-stationary accretion. According to an original hypothesis by Osaki (1974), the accretion in dwarf novae between outbursts is limited to the outer parts of the disk: the material supplied by the secondary collides with the disk producing the hot spot, but remains stored in the disk's outer parts. Then, due to an instability (not identified at that time), this material is suddenly accreted onto the white dwarf producing an outburst. This hypothesis proved to be correct in early 1980-ies, starting with the pioneering work by Meyer and Meyer-Hofmeister (1981), identifying the nature of the instability, and followed by several papers in which the disk instability model was developed.

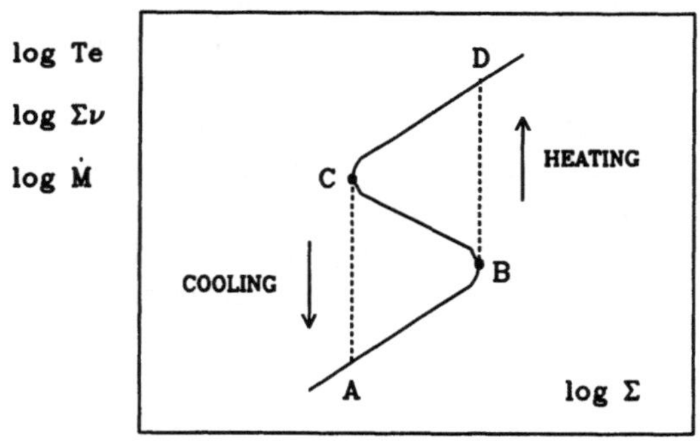

Fig. 3.11. The $\Sigma - T_e$ relation and the resulting limit cycle mechanism for the dwarf nova behavior (see text for details).

The main cause of dwarf nova outbursts is the thermal instability of their disks. If so, we can already at this point explain why such outbursts are shown only by some, but not all CV's. The thermal instability, as it follows from considerations presented in Sect.3.3.9, must occurs whenever the temperature of any part of the disk, predicted for the steady-state case, falls in the range covered by the middle, unstable branch of the $\Sigma - T_e$ relation. Under steady-state conditions the lowest temperature in the disk, as predicted by (3.56), occurs in its outermost parts, at $R = R_d$. In order to avoid instability that temperature must belong to the hot, stable branch of the $\Sigma - T_e$ relation, i.e. it must be *higher* than the temperature T_{crit} at the critical point C (see Fig. 3.11). With an approximate expression for T_e from (3.56) this *stability* condition can be written as

$$\dot{M}_{tr} \, \frac{M_1}{{r_d}^3} \; < \; \frac{8\pi}{3G} \, \sigma \, {T_{crit}}^4 \tag{3.94}$$

where the right hand side depends only on T_{crit} which, as can be seen from Fig. 3.9, is only a weak function of α, r, and also of M_1. The left hand side can be evaluated from observational data for individual CV's and then we find that, indeed, in the case of dwarf novae the stability condition (3.94) is *not* fulfilled, while in the case of other, stationary accretion CV's, their basic parameters are such that they comply with this condition. In fact, the main factor here is the mass transfer rate, which in the case of dwarf novae is systematically *lower*.

We now consider – first in a qualitative way – the consequences of the thermal instability. Within a critical range of the mass-transfer/accretion rates, corresponding to the middle branch of the $\Sigma - T_e$ relation, stationary accretion is replaced with a cyclic behavior, alternating between the two stable branches (Fig. 3.11). Let us consider, for example, the outer parts of the disk. Remembering, that the $\Sigma - T_e$ relation represents in fact a more general relationship given by (3.59), we obtain the following qualitative description of the four main phases of the resulting limit cycle. On branch AB (corresponding to quiescence) the *local* accretion rate ($\dot{M}_{accr}$) is lower than the rate at which the material is supplied from the outside ($\dot{M}_{tr}$) and, consequently, the surface density increases. When the critical density is reached (and exceeded) at point B, the thermal instability develops, resulting in a rapid (on a thermal time scale) transition to branch CD (corresponding to the outburst maximum). There the situation is reversed: with $\dot{M}_{accr} > \dot{M}_{tr}$ the surface density decreases until critical point C is reached, where the thermal instability develops again, resulting in a transition to the lower branch. Note that the same picture would be obtained for any ring inside the disk, except that in that case, instead of the mass transfer rate, we would consider the rate at which the material is supplied from the outer parts of the disk.

With those conclusions, applicable only to the *local* behavior, it remains to be seen how the limit-cycle mechanism operates on a *global* scale, i.e. in the case of the entire disk. In particular – whether an instability occuring somewhere in the disk propagates inward and outward in a coherent way, necessary to reproduce the observed dwarf nova behavior. Time dependent model calculations involve numerical integrations of the basic diffusion equation (3.40), with $\Sigma\bar{\nu}$ provided by the $\Sigma - T_e$ relation. Crucial role in those calculations is played by the outer boundary conditions which, in particular, determine the outer radius of the disk and its variations during the dwarf nova cycle. Those boundary conditions involve the the following three effects:

(1) The tidal removal of the angular momentum from the outer parts of the disk; the standard, commonly adopted value of the effective tidal radius of the disk is $R_{tid} \sim 0.9 R_{Roche}$, as resulting from analytical considerations by Papaloizou and Pringle (1977).

(2) The deposition of the stream material; note that its specific angular momentum (see (3.17) in Sect.3.3.1) is lower than that of the disk's outer parts.

(3) The outside transfer of the angular momentum by viscous forces acting within the disk.

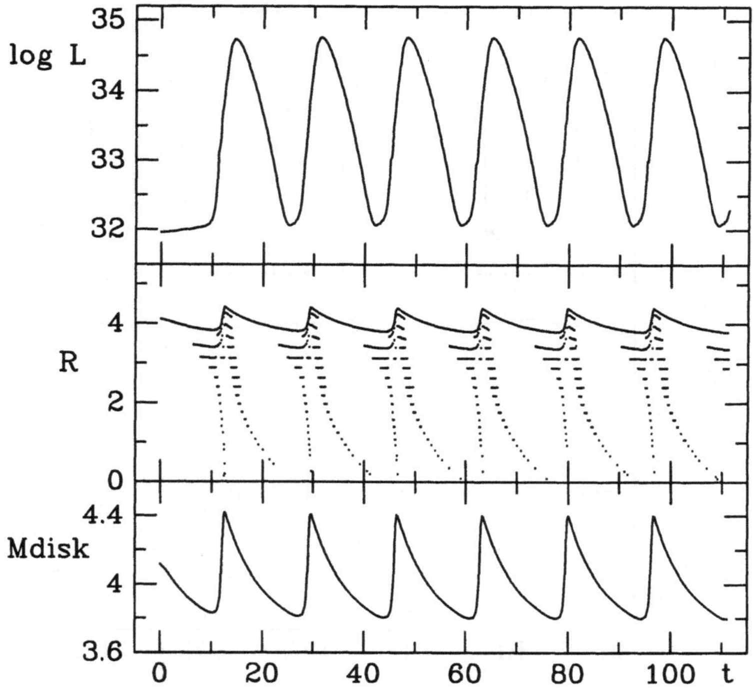

Fig. 3.12. Model outbursts of Type A obtained with model parameters $M_1 = 1.0M_\odot$, $M_2 = 0.4M_\odot$, $\log \dot{M} = 17.1$, and $\alpha = 0.1$. Shown are the bolometric luminosity (in g/s), radius (in 10^{10} cm), and mass of the disk (in 10^{24} g). Time is in 10^5 s. Dots in the middle plot mark regions of the disk undergoing thermal instability: they show how the heating and then cooling fronts propagate within the disk during outburst.

The most important results of such model calculations are (see, for example, Figs. 3.12 and 3.13):

(1) The limit cycle mechanism, resulting from the thermal instability of the disk, is indeed responsible for the dwarf nova behavior.

(2) Two different values of the viscosity parameter α are needed for the two branches of the $\Sigma - T_e$ relation, with $\alpha_{hot}/\alpha_{cool} \approx 4$, for the limit-cycle mechanism to operate coherently on a global scale.

(3) There are two types of outbursts: Type A, or *outside-in*, when the instability begins in the outers parts of the disk and propagates inward, and Type B, or *inside-out*, when it begins in the inner parts of the disk and propagates outward. Those two types differ in their observable characteristics and those differences help to identify them among well observed dwarf novae.

(4) The radius of the disk expands during an outburst and contracts during quiescence. Such variations are indeed observed in the case of few eclipsing dwarf novae, notably in U Gem and IP Peg.

Unlike the observed light curves, the model produced light curves are strictly periodic. Even in the case of Type B outbursts of different amplitudes and du-

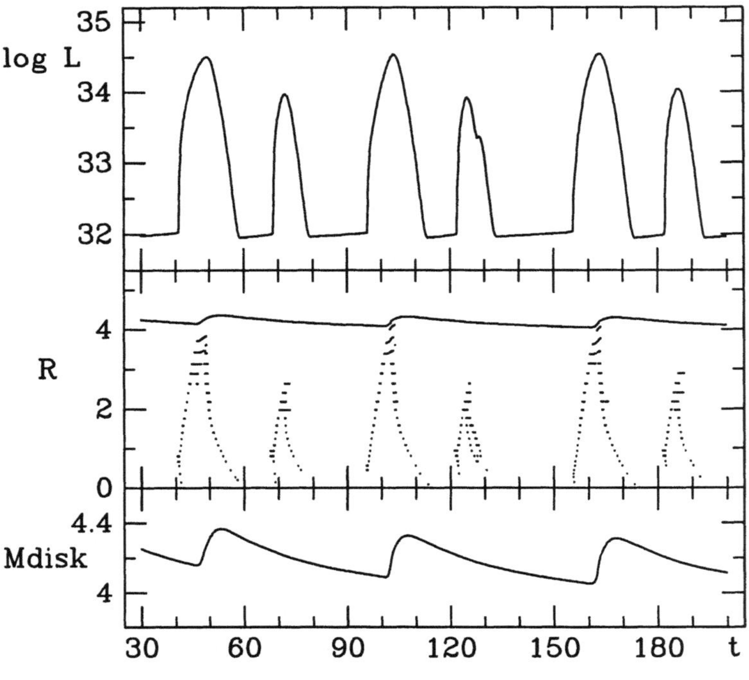

Fig. 3.13. Same as in Fig. 3.12 for a Type B outburst with $\log \dot{M} = 16.5$. Note that some of the *inside-out* instabilities do not propagate all the way to the disk's outer edge, with the resulting outburst amplitude being lower.

rations, the characteristic pattern, consisting of alternating outbursts, repeats periodically (see Fig. 3.13). In order to explain the observed light curves (including those of Z Cam and SU UMa stars) it is necessary to take into account several additional effects and processes, such as:

(1) Heating of the secondary component by the white dwarf and the boundary layer, resulting in modulations of the mass outflow rate.

(2) Heating of the inner parts of disk, affecting their structure and potentially capable of removing the cause of the thermal instability.

(3) Tidal instability of the outer parts of the disk and its effect on global disk behavior.

3.4.3 The Value of α_{hot}

Characteristic time-scales observed during dwarf nova outbursts include the widths (or durations) of outbursts, the rates of decline following outburst maximum, and the UV delays observed during rise, which are correlated with the orbital period (see Fig.. 3.14). Their dependence on the viscous time-scale, defined by α on the hot branch of the $\Sigma - T_e$ relation, provides an important and almost unique opportunity of an empirical determination of α_{hot}.

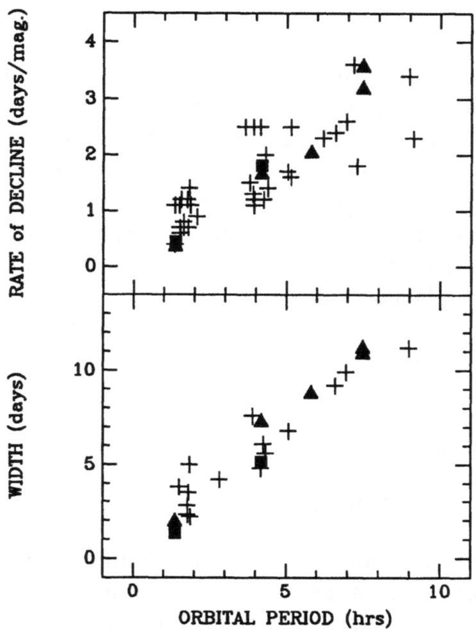

Fig. 3.14. The width of outburst and the rate of decline *versus* the orbital period relations. Crosses are observational data. Filled squares and triangles are model data with $\alpha_{hot} = 0.2$ for, respectively, Type A and Type B outbursts.

It is instructive to begin with the following analytical considerations (for details, simplifying assumptions and approximations – see Smak 1999). The characteristic time scales observed during outbursts can be related to the travel time of the accretion wave. In the α-disk approximation, by integrating v_r from (3.83), we can obtain for the travel time between r_d and the inner radius r_i

$$\Delta t\ (r_d, r_i)\ \sim\ \alpha^{-0.7}\ \dot{M}^{-0.37}\ {M_1}^{-0.46}\ ({r_d}^{1.33} - {r_i}^{1.33}) \qquad (K) \qquad (3.95)$$

Assuming that r_d is the tidal radius $r_{tid} \sim 0.9\ r_{Roche}$, using Kepler Law to replace r_d with P_{orb}, and neglecting r_i, we obtain the following dependence on the orbital period and the viscosity parameter $\alpha = \alpha_{hot}$:

$$\Delta t\ \sim\ \alpha^{-0.7}\ {P_{orb}}^{0.74} \qquad (K) \qquad (3.96)$$

This shows that the observed correlations with the orbital period simply reflect the dependence on the radius of the disk.

By comparing the observed correlations with those obtained from model light curves calculated with different values of α_{hot}, we find $\alpha_{hot} \approx 0.2$. Accordingly, Fig. 3.14 shows model data with $\alpha_{hot} = 0.2$. Similar comparison for UV delays gives $\alpha_{hot} \sim 0.1 - 0.2$.

3.4.4 Problems

4-1. Using (3.94) calculate critical values of the mass transfer rate $\dot{M}_{crit}$ for various combinations of M_1 and r_d. Adopt (a) $\log T_{crit} \approx 3.87$, and then (b) $\log T_{crit} = 3.87 - 0.08 \log(r/10^{10})$, as resulting from data presented in Fig. 3.9.

4-2. Thermal instability can also be avoided in the case of *very low* mass transfer/accretion rates, when T_e *everywhere* in the disk is *lower* than the critical temperature T_e'', defined by point B of the $\Sigma - T_e$ relation. Adopting $\log T_e'' \approx 3.79$ (see Fig. 3.9), calculate $\dot{M}_{crit}''$ corresponding to such a situation. (Note that this will be more complicated than in 4-1).

4-3. Assuming $z_o/r = const.$ integrate (3.83) to obtain a simpler version of (3.95), not containing $\dot{M}$.

3.5 Boundary Layer

3.5.1 Simple Model of the Boundary Layer

The angular momentum h, angular velocity ω, and energy e, of the material at the inner edge of the disk, i.e. at $r = r_i$, are given by the standard, Keplerian formulae. The same parameters describing the equatorial surface layers of the white dwarf, h_1, ω_1, e_1, depend on its rotation. In the extreme case of rotation with the critical, break-up, angular velocity

$$\omega_{crit} = \left(\frac{GM_1}{r_1{}^3}\right)^{1/2} \qquad (K) \tag{3.97}$$

they would also be Keplerian. In the more likely case, however, of a more slowly rotating (in particular: non-rotating) white dwarf, h_1, ω_1, and e_1 are smaller. It is obvious then that between the inner edge of the disk and the surface of the star, the excess mechanical energy of disk's element must be dissipated and its excess angular momentum transferred away, before that element can be accreted onto the stellar surface. This region is called *the boundary layer.*

The importance of the boundary layer can be recognized if we calculate the amount of energy to be dissipated. For simplicity let us assume that $r_i \approx r_1$ (note that this assumption has already been made in Sect.3.3.4; it will be justified later in the present Section). Then for the luminosity of the boundary layer we get

$$L_{BL} = \frac{1}{2}\frac{GM_1}{r_1}\dot{M}\left[1 - \left(\frac{\omega_1}{\omega_{crit}}\right)^2\right] \tag{3.98}$$

By comparing L_{BL} with L_d (3.61), we can conclude that, in the case of a non-rotating white dwarf ($\omega_1 = 0$), the luminosity of the boundary layer is *equal* to that of the disk which extends to infinity; in the case of a finite disk (3.60) it is even *higher.*

Figure 3.15 shows the expected dependence of the angular velocity on r. In the disk ω is Keplerian or nearly Keplerian. In the boundary layer it must

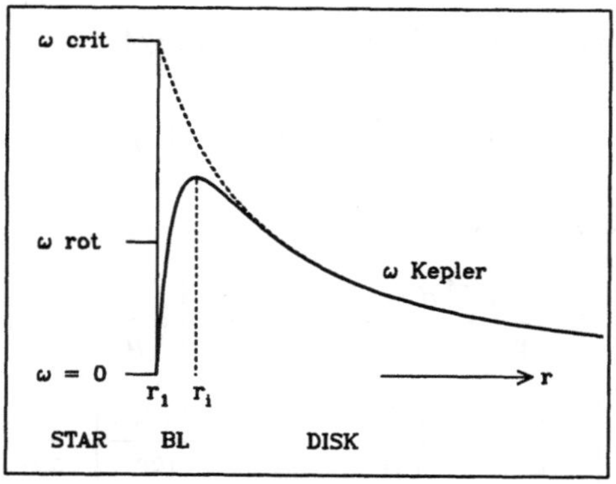

Fig. 3.15. Schematic dependence of the angular velocity on radial distance.

decrease to ω_1. Note that, by definition of r_i, $\partial\omega/\partial r = 0$ and $g_i = 0$ at that point. This implies that the excess angular momentum can be transferred from the boundary layer only to the equatorial surface layers of the white dwarf.

The main properties of the boundary layer can be established by considering the following, very simple model (Lynden-Bell and Pringle 1974). The equations of motion (in the cylindrical system, integrated in z, and with pressure term neglected) are

$$\Sigma \left[\frac{\partial}{\partial r} \left(\frac{{v_r}^2}{2} \right) - \frac{1}{r} \frac{{v_\theta}^2}{2} \right] = 2 \frac{\partial}{\partial r} \left(\mu \frac{\partial v_r}{\partial r} \right) + 2\,\mu \frac{\partial}{\partial r} \left(\frac{v_r}{r} \right)$$

$$+ \frac{\partial}{\partial r} \left[\left(\mu - \frac{2}{3}\lambda \right) \frac{1}{r} \frac{\partial}{\partial r} (r v_r) \right] - \Sigma \frac{GM_1}{r^2} \tag{3.99}$$

$$\Sigma\, v_r \frac{\partial}{\partial r} (r v_\theta) = \frac{1}{r^2} \frac{\partial}{\partial r} \left(\mu r^3 \frac{\partial \omega}{\partial r} \right) \tag{3.100}$$

Assuming that the flow is stationary (but not Keplerian!), by combining (3.27), (3.34), and (3.38) we get for the radial component of the velocity

$$v_r = \bar{\nu} \frac{d\omega}{dr} \,/\, (\omega - \omega_i) \tag{3.101}$$

Let us define new, dimensionless variables x, v, and Ω, all of order of unity:

$$x = \frac{r - r_1}{\delta} \tag{3.102}$$

where $\delta = r_i - r_1$ is the radial extent, or width of the boundary layer (we reserve the term "thickess" for the z-coordinate),

$$v = \frac{\delta}{\bar{\nu}} v_r \tag{3.103}$$

$$\Omega = \omega/\omega_{crit} \tag{3.104}$$

With these definitions, using v_r from (3.101), we get

$$v = \frac{d\Omega}{dx} \,/\, (\Omega - \Omega_i) \tag{3.105}$$

Inserting this into (3.99) we obtain

$$\frac{d}{dx}\left(\frac{v^2}{2}\right) + 2\frac{d}{dx}\left(\frac{1}{r}\frac{d\,(rv)}{dx}\right) - \frac{d}{dx}\left[\left(\frac{\lambda}{\mu} - \frac{2}{3}\right)\frac{1}{r}\frac{d\,(rv)}{dx}\right]$$
$$+ \left[2\frac{dv}{dx} + \left(\frac{\lambda}{\mu} - \frac{2}{3}\right)\frac{1}{r}\frac{d\,(rv)}{dx}\right]\frac{d}{dx}\,[\,\log\,(rv)\,]$$
$$= - \left\{\frac{\delta^3}{\nu^2}\,r_1\left(\omega_{crit}^2 - {\omega_1}^2\right)\right\}\left(\frac{r_1}{r}\right)^2\left[\frac{1 - \Omega^2\left(\frac{r}{r_1}\right)^3}{1 - \Omega_1^2}\right] \tag{3.106}$$

where $\Omega_1 = \Omega(r = r_1)$ and the expression in last square bracket is a dimensionless measure of the imbalance of centrifugal force and gravity which varies from unity at $r = r_1$ to zero at larger distances, where the motion becomes Keplerian.

We now note that each major bracket in (3.106) contains a dimensionless expression which – with one exception – is of order of 1. The only exception is the curly bracket on the right-hand side and it is obvious that it must also be of order unity. This leads to the following expression for δ

$$\delta \approx \frac{\bar{\nu}^{\,2/3}}{[\,r_1\,(\omega_{crit}^2 - \omega_1^2)\,]^{1/3}} \tag{3.107}$$

In the case of a slowly rotating white dwarf we can neglect ω_1^2 (in comparison with ω_{crit}^2) and adopting again the α-approximation for viscosity, insert $\bar{\nu} \approx \alpha\,\omega\,{z_\circ}^2$, from (3.81), to get

$$\delta \approx \alpha^{2/3}\,z_\circ\left(\frac{z_\circ}{r}\right)^{1/3}\left(\frac{\omega^2\,r}{\omega_{crit}^2\,r_1}\right)^{1/3} \tag{3.108}$$

Since all terms on the right-hand side, except $z_\circ$, are either < 1 or $\ll 1$, we finally get

$$\delta = (\,r_i - r_1\,) \ll z_\circ \tag{3.109}$$

The radial extent of the boundary layer turns out to be *much smaller* than its vertical thickness $z_\circ$. With $z_\circ$ being smaller then r, the final conclusion is

$$r_i \approx r_1 \tag{3.110}$$

which remains true also in the case when $z_\circ$ is comparable to r_1 (see below), since even in that case, on account of other terms in (3.108), δ remains very small. An important consequence of (3.110) is that $\omega_i \approx \omega_{crit}$. It was already used in Sect.3.3.4 to derive (3.39)-(3.42).

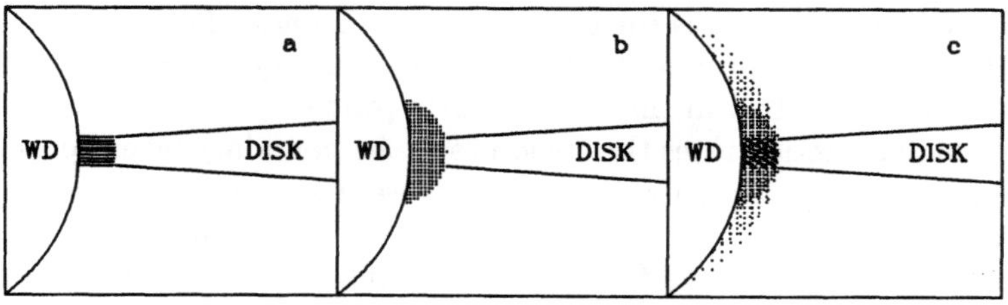

Fig. 3.16. Schematic presentation of models of the boundary layer. (a) Simple model presented in Sect.3.5.1. (b) The hot, optically thin model applicable low $\dot{M}$. (c) The complex model, with density and temperature gradients, applicable to high $\dot{M}$. Dotted regions are hot ($\sim 10^7 - 10^8$K) and *optically thin*. Shaded regions are cool ($\sim 10^5$K) and *optically thick*. Note the large extent of the boundary layer in z over the surface of the white dwarf.

Let us now assume that the boundary layer is geometrically thin ($z_\circ/r_1$ small) and *optically thick* (see Fig. 3.16a). If so, its luminosity can be calculated as

$$L_{BL} = 2 \times 2\pi\, r_1\, \delta\, \sigma {T_e}^4 \tag{3.111}$$

Let us furthermore assume for simplicity that the white dwarf is non-rotating. Taking δ, from (3.108), putting $\omega/\omega_{crit} \sim 1$ (in fact it decreases from ≈ 1 at $r = r_i$ to zero at $r = r_1$), inserting it into (3.111), and comparing it with our earlier expression for L_{BL} (3.98), leads to the following estimate of the effective temperature

$${T_e}^4 \approx \frac{1}{8\pi\sigma}\, \frac{GM_1}{{r_1}^3}\, \dot{M} \left[\alpha \left(\frac{z_\circ}{r_1}\right)^2 \right]^{-2/3} \tag{3.112}$$

With $\alpha \sim 0.1 - 0.2$ and $z_\circ/r \sim 0.01 - 0.2$ this gives

$$T_e \approx (3 - 11) \times 10^5 \left(\frac{M_1}{M_\odot}\right)^{1/4} \left(\frac{r_1}{5.4 \times 10^8}\right)^{-3/4} \left(\frac{\dot{M}}{1 \times 10^{18}}\right)^{1/4} \tag{3.113}$$

The simple model presented above was sufficient to obtain an estimate of the radial extent of the boundary layer, and to show that $r_i \approx r_1$. It is too crude, however, to provide an adequate description of the structure of the boundary layer, or to reliably predict its observable properties. Note, for example, that if we assume that the boundary layer is *optically thin* (and, since it is so hot, it is likely to be optically thin), then its temperature would be much higher.

Due to the time limitations it is impossible to continue now with a detailed description of more realistic models for the boundary layer (e.g. Drew and Kley 1993, Narayan and Popham 1994, Popham 1997 and references therein). We conclude only with a qualitative picture, resulting from such models and from their comparison with observations, particularly with the observed X-ray flux (see, e.g., Patterson and Raymond 1985a):

(1) At low accretion rates the boundary layer is *optically thin*, very hot ($T \sim 10^7 - 10^8$K), nearly uniform, and extending considerably in the vertical direction (see Fig. 3.16b). The flux in this case consists primarily of hard X rays.

(2) At high accretion rates the structure of the boundary layer becomes more complex (Fig. 3.16c). Its central parts are dense, cool, and *optically thick*, with temperature being, roughly, that given by (3.113), so that the flux from that region comes primarily in the form of far UV and soft X-ray radiation. Regions above (and below) those central parts are *optically thin* and are characterized by strong density and temperature gradients; in particualr, the temperature increases with z up to $T \sim 10^7 - 10^8$K. Those regions are then responsible for the hard X-rays.

3.5.2 Heating of the White Dwarf

The boundary layer, being located so close to the white dwarf, is heating its surface layers. Due to geometry (Fig. 3.16) roughly one-half of the flux from the boundary layer must be intercepted by the white dwarf and reprocessed in its outer layers into thermal radiation with $T_e = 10^4 - 10^5$K. But, due also to geometry, this occurs only in an equatorial belt adjacent to the boundary layer (cf. Patterson and Raymond 1985a).

In the case of dwarf novae the accretion through the boundary layer and, in consequence, its luminosity L_{BL}, are strongly modulated during the outburst cycle. As a result the white dwarf is heated during outbursts and then cools down during quiescence. This is well documented with observations in the far UV (in particular with the IUE and HST), their interpretation providing information about the temperature and its distribution on the surface of the white dwarf, and about their variations during the dwarf nova cycle (see Sion 1999ab and references therein).

3.5.3 Origin of Emission Lines

There are no definite models explaining in quantitative terms the origin and, in particular, various regularities and correlations shown by the emission lines observed in the spectra of CV's. There is no doubt, however, that the basic mechanism involves irradiation of the disk by the boundary layer and the white dwarf. This is strongly suggested by the following observational evidence.

(1) The first and most dramatic evidence for irradiation came from observations of emission lines in the spectrum of DQ Her by Chanan et al. (1978). They found that the intensity and profile of the He II λ4686 line vary with the 71-sec period. The phase of these oscillations depends on wavelength in a way which demonstrates that they are produced by irradiation of the disk by the rotating beam from the white dwarf.

(2) Patterson and Raymond (1985ab) found that the emission line intensities are correlated — directly or indirectly — with the X-ray flux. There is a correlation between the X-ray-to-visual flux ratio and the equivalent width of the Hβ line. This correlation could only be a consequence of two other correlations

between F_x/F_v and EW(Hβ) and the accretion rate. But it is more likely, that – to quote Patterson and Raymond – *F_x/F_v is well correlated with EW(Hβ) because the X-rays produce the emission lines.* There is also a correlation between the flux in the He II λ4686 line and the accretion rate. Patterson and Raymond demonstrated that at higher accretion rates the *soft* X-ray flux from the boundary layer is fully sufficient to produce this line *via* photoionization. At lower accretion rates the same role could be played by the *hard* X-ray flux.

(3) In the case of dwarf novae, during early rise to an outburst, the emission lines are replaced with absorption lines which dominate the spectrum at maximum and during early decline reverse again to emission (for references see Robinson et al. 1993). The same phenomenon shows up in the (U-B) – (B-V) color variations. During rise, when the contribution from the disk becomes dominant, and when its Balmer continuum reverses also from emission to absorption, the colors become similar to those of an A-type star and — as the disk becomes hotter — of a B-type star. During the decline, when Balmer continuum in emission reappears again, the colors return to their values at quiescence.

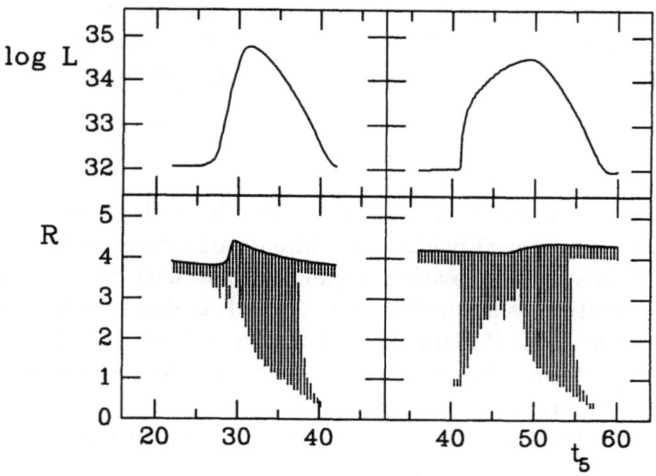

Fig. 3.17. The self-screening of the disk during dwarf nova outbursts of Type A (*left*) and Type B (*right*). Upper panels show the light curves. Lower panels show the disk-radius variations and — as shaded areas — those disk parts which are screened from irradiation. R is in units of 10^{10} cm, t_5 is time in 10^5 s.

The explanation of this behavior is simple and provides a strong argument in favor of irradiation (Smak 1990). Accretion in dwarf novae during their outbursts is non-stationary, with all disk parameters varying with time. The crucial point is that the disk is not always *concave* and, in particular, that its variable shape controls irradiation by the central source. At a given point on the disk surface the condition for irradiation is that the local ratio z/r must be larger than z/r everywhere at smaller distances; otherwise this point is screened from irradiation by the inner parts. Fig. 3.17 shows the shade-patterns obtained from two dwarf nova models for Type A and Type B outbursts. The screening effects appear rather dramatic: large parts of the disk are in shade from early rise to about mid-decline. In the absence of irradiation these parts will produce *absorption* lines.

Irradiation of the disk's atmosphere by the boundary layer and the white dwarf results in

1. photoionization by high energy photons, and
2. heating of the disk atmosphere.

The relative importance of the two effects depends on the spectral energy distribution of the irradiating flux. Major fraction of the flux emitted directly by the boundary layer consists of soft and hard X-rays. They are responsible primarily for the production of higher excitation/ionization lines, for example those of He I and He II. On the other hand the flux reprocessed by the white dwarf comes in the form of thermal radiation with $T_e = 10^4 - 10^5$K. Compared to the local disk's flux, this irradiating flux is usually rather small. Irradiation, however, is very oblique (typically at an angle in the range $80-90°$) and therefore affects effectively the upper layers of the disk atmosphere. This results in a temperature inversion, i.e. in the formation of a chromosphere, where emission lines of lower excitation, in particular – those of hydrogen, must be formed.

3.5.4 Problems

5-1. Assuming that exactly one-half of the flux from the boundary layer is intercepted by the equatorial belt of the white dwarf extending to latitude φ and that this leads to a uniform heating within that belt, derive an expression for the resulting temperature as a function of M_1, r_1, $\dot{M}$, and the intrinsic temperature of the white dwarf $T_{e,\circ}$. Putting $M_1 = 1M_\odot$, $r_1 = 5.4 \times 10^8$, $T_{e,\circ} = 10^4$K, and $\dot{M} = 1 \times 10^{18}$, perform calculations for $\varphi = 30°$ and $90°$ (i.e. for uniform heating of the entire white dwarf.

5-2. Repeat calculations from 5-1 for the main sequence primary in an Algol system, with $M_1 = 5M_\odot$ and $R_1 = 5R_\odot$.

References

1. Cannizzo, J.K. (1993) in: *Accretion Disks in Compact Stellar Systems*, ed. J.C.Wheeler, World Scientific Publ.Co., Singapore, p. 6.
2. Chanan, G.A., Nelson, J.E., Margon, B. (1978) ApJ **226**, 963.
3. Cordova, F.A.-D. (1995) in: *X-ray Binaries*, eds. W.H.G.Lewin, J. van Paradijs, E.P.J. van den Heuvel, Cambridge Univ.Press, p. 331.
4. Drew, J.E., Kley, W. (1993) in: *Accretion Disks in Compact Stellar Systems*, ed. J.C.Wheeler, World Scientific Publ.Co., Singapore, p. 212.
5. Duschl, W.J., Frank, J., Meyer, F., Meyer-Hofmeister, E., Tscharnuter, W.M., eds. (1994) *Theory of Accretion Disks - 2*, Kluwer, Dordrecht.
6. Faulkner, J., Flannery, B.P., Warner, B. (1972) ApJL **175**, L79.

7. Frank, J., King, A.R., Raine, D.J. (1992) *Accretion power in astrophysics*, 2nd ed., Cambridge University Press.
8. Hack, M., la Dous, C. (1993) *Cataclysmic Variables and Related Objects*, NASA SP-507.
9. Hessman, F.V., Robinson, E.L., Nather, R.E., Zhang, E.-H. (1984) ApJ **286**, 747.
10. Honeycutt, R.K., Kaitchuck, R.H., Schlegel, E.M. (1987) ApJS **65**, 451.
11. Horne, K. (1993) in: *Accretion Disks in Compact Stellar Systems*, ed. J.C.Wheeler, World Scientific Publ.Co., Singapore, p. 117.
12. Kato, S., Fukue, J., Mineshige, S. (1998) *Black-Hole Accretion Disks*, Kyoto Univ.Press.
13. King, A. (1995) in: *X-ray Binaries*, eds. W.H.G.Lewin, J. van Paradijs, E.P.J. van den Heuvel, Cambridge Univ.Press, p. 419.
14. Kruszewski, A. (1966) Adv.Astron.Astrophys. **4**, 233.
15. Kruszewski, A. (1967) Acta Astr. **17**, 297.
16. la Dous, C. (1994) Space Sci.Rev. **67**, 1.
17. Lin, D.N.C., Papaloizou, J.C.B. (1996) Ann.Rev.Astron.Astrophys. **34**, 703.
18. Lynden-Bell, D, Pringle, J.E. (1974) MNRAS **168**, 603.
19. Lubow, S.H., Shu, F.H. (1975) ApJ **198**, 383.
20. Marsh, T.R. (1999) This Volume.
21. Meyer, F., Meyer-Hofmeister, E. (1981) A&A **104**, L10.
22. Meyer, F., Meyer-Hofmeister, E. (1983) A&A **121**, 29.
23. Narayan, R., Popham, R. (1994) in: *Theory of Accretion Disks - 2*, eds. Duschl, W.J., Frank, J., Meyer, F., Meyer-Hofmeister, E., Tscharnuter, W.M., Kluwer, Dordrecht, p. 293.
24. Osaki, Y. (1974) PASJ **26**, 429.
25. Osaki, Y. (1985) A&A **144**, 369.
26. Osaki, Y. (1996) PASP **108**, 39.
27. Paczyński, B. (1967) Acta Astr. **17**, 287.
28. Paczyński, B. (1971) Ann.Rev.Astron.Astrophys. **9**, 183.
29. Paczyński, B. (1978) Acta Astr. **28**, 91.
30. Paczyński, B., Sienkiewicz, R. (1983) ApJ **268**, 825.
31. Papaloizou, J.C.B., Lin, D.N.C. (1995) Ann.Rev.Astron.Astrophys. **33**, 505.
32. Papaloizou, J., Pringle, J.E. (1977) MNRAS **181**, 441.
33. Patterson, J., Raymond, J.C. (1985a) ApJ **292**, 535.
34. Patterson, J., Raymond, J.C. (1985b) ApJ **292**, 550.
35. Popham, R. (1977) in: *Accretion Phenomena and Related Objects*, eds. Wickramasinghe, D.T., Bicknell, G.V., Ferrario, L., ASP Conf.Series **121**, 230.
36. Robinson, E.L., Marsh, T.R., Smak, J.I. (1993) in: *Accretion Disks in Compact Stellar Systems*, ed. J.C.Wheeler, World Scientific Publ.Co., Singapore, p. 75.
37. Sellwood, J.A., Goodman, J., eds. (1999) *Astrophysical Disks*, ASP Conf.Series **160**.
38. Shakura, N.I., Sunyaev, R.A. (1973) A&A **24**, 337.
39. Sion, E.M. (1999a) Physics Reports **311**, 353.
40. Sion, E.M. (1999b) PASP **111**, 532.
41. Smak, J. (1981) Acta Astron. **31**, 395.
42. Smak, J. (1989) Acta Astron. **39**, 317.
43. Smak, J. (1990) in: *Structure and Emission Properties of Accretion Disks*, eds. C.Bertout, S.Collin-Souffrin, J.P.Lasota, J.Tranh Thanh Van, Edition Frontières, Paris, p. 247.
44. Smak, J. (1994a) Acta Astron. **44**, 59.

45. Smak, J. (1994b) Acta Astron. **44**, 265.
46. Smak, J. (1999) Acta Astron. **49**, 391.
47. Warner, B. (1995) *Cataclysmic Variable Stars*, Cambridge University Press.
48. Wheeler, J.C., ed. (1993) *Accretion Disks in Compact Stellar Systems*, World Scientific Publ.Co., Singapore.
49. Williams, G. (1983) ApJS **53**, 523.

4 Observations of Cataclysmic Variable and Double Degenerate Stars

Tom R. Marsh

Department of Physics and Astronomy, Southampton University, Highfield, Southampton SO17 1BJ

4.1 Introduction

In this chapter I will discuss cataclysmic variable stars and close pairs of white dwarfs or "double degenerates". These stars are connected through their evolution, but in this case also because they are personal interests of mine. The detailed topics that I discuss are also personal choices and I make no attempt to be comprehensive in my coverage. The reader is referred to Brian Warner's book [104] for such a treatment of cataclysmic variable stars. The theory behind the accretion discs which play such a key rôle in these objects is covered elsewhere in this volume by J.Smak.

Cataclysmic variable stars (CVs) are beautifully set up to allow us to study accretion. The two stellar components of the binary, a white dwarf and a low-mass main-sequence star, are faint, and their semi-detached configuration means that the geometry is entirely specified by the mass ratio and orbital inclination alone. In eclipsing systems we can see the accretion regions occulted by the main-sequence star allowing us to determine their contribution to the light from the system. Much of what we know of CVs has come from such studies. The problem faced in all such investigations is how to deduce the distribution of accretion light from the data. CVs are far too small to be resolved directly – they typically subtend $< 10^{-4}$ seconds of arc – and we learn nothing of their structure from direct imaging except in the case of nova shells. As a result several indirect imaging techniques have been developed, and it is these that will occupy me in this contribution. I begin by looking first at the method of "eclipse mapping", developed by Keith Horne to address the problem of understanding light curves of systems with discs [33]. Next I look at "Doppler tomography", which deals with line emission [50]. The emphasis of these sections is on the disc-accreting systems, and to redress the balance I then consider the magnetic accretors, for which similar techniques have been applied in more recent years. Finally I change tack to another hobby of mine, the evolution of close binaries. I do so by focusing upon near-relatives of CVs in evolutionary terms, close pairs of white dwarfs.

4.2 Eclipse Mapping

To understand eclipse mapping, one must first understand how the light curves of CVs come about, and how they can be computed. I thus begin this section by describing how it is possible to specify the geometry in CVs, and then look in general at how one might invert light curve information; this leads onto the eclipse mapping method. I then describe results, and finish by discussing some significant problems that might affect many such investigations.

4.2.1 Calculating Light-Curves of CVs

The basic geometry of CVs is well established. An example of the sort of data that has inspired this confidence is shown in Fig. 4.1 which shows the light curve of an eclipsing dwarf nova, Z Cha, along with illustrations of the geometry at various phases. This is an example of how a fairly complex set of light sources may appear in a light curve. In this low (quiescent) state of Z Cha the main sources of light are (a) the white dwarf, (b) the bright-spot where the mass transfer stream hits the disc, and (c) the accretion disc itself. The mass donor star is a very cool M dwarf and contributes negligibly at optical wavelengths. The white dwarf and bright-spot are compact sources of similar size which lead to sharp steps in the light curve. The white dwarf is the first of these two to be eclipsed, and the first to come out of eclipse. The steps caused by the white dwarf are of equal size going into eclipse (ingress) and coming out (egress). The bright-spot, which is on the edge of the disc, causes the large "hump" in the light curve, as its projected area reaches a maximum when pointing at us. By

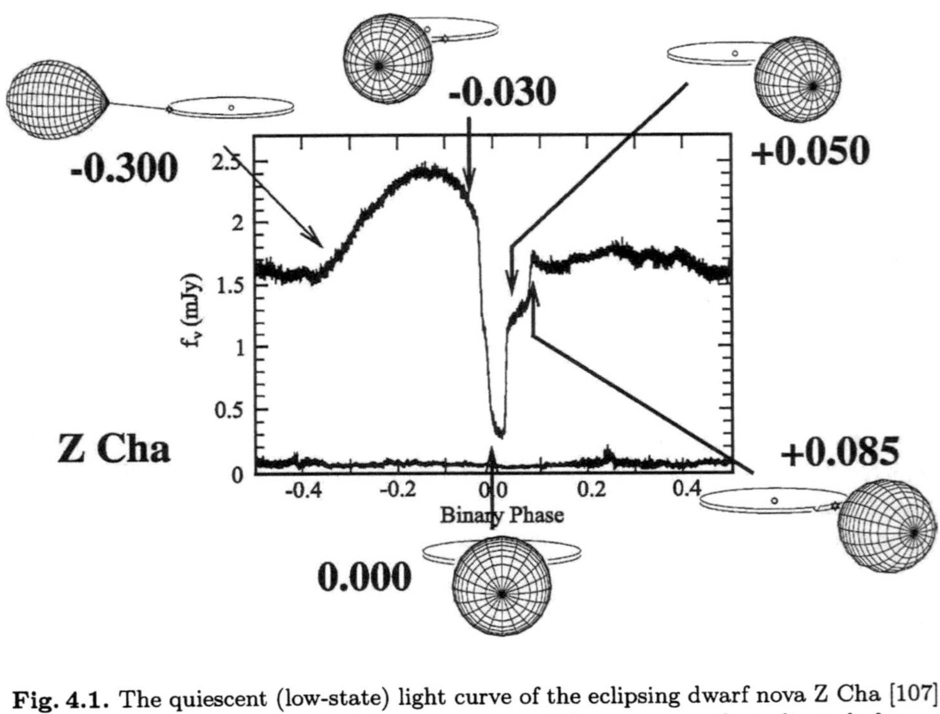

Fig. 4.1. The quiescent (low-state) light curve of the eclipsing dwarf nova Z Cha [107] along with schematic pictures of the appearance of the system at the indicated phases.

the time of eclipse, the bright-spot is fading, and as a result the egress step is smaller than the ingress step. What of the disc? In this system it is faint, but not negligible, because one can see some light at phase 0.0 when both the white dwarf and bright-spot are in eclipse. The disc also reveals itself in the rise in flux in between the egress of the white dwarf and that of the bright-spot.

It is data like those of Fig. 4.1 that have led to the standard model of CVs: a Roche-lobe-filling, near-main-sequence star orbits a white dwarf. Mass transferred from the larger star forms a disc with a bright-spot at its edge, or, if the white dwarf is sufficiently magnetic, it falls in a stream straight onto the white dwarf. Since the secondary star fills its Roche lobe, its surface lies close to the potential surface that touches the inner Lagrangian point. The shape of this surface is entirely specified by the mass ratio $q = M_2/M_1$ where M_1 is the mass of the white dwarf primary star and M_2 is the mass of the main-sequence secondary star. The only other parameter needed to describe the eclipse is the orbital inclination i. If the eclipse of the white dwarf is visible, then its width gives a relation between q and i, and we have only one degree of freedom. In some systems the eclipse of the bright-spot gives another independent q–i constraint, and the geometry is fully specified [107].

Once q and i are specified, the visibility of any point in the system can be determined in the following manner:

(i) First calculate the vector in the rotating frame of the binary which points towards Earth (a function of orbital phase).
(ii) Next, iterate to find the minimum Roche potential along this vector starting from the point of interest. The minimum in this case is restricted to the vicinity of the secondary star.
(iii) If this minimum potential is less than that at the inner Lagrangian point, the point is occulted.

The above procedure can be applied to any model of the accretion brightness, and allows computation of a light-curve for any given surface brightness distribution or "map". The next question is: can we do the reverse? That is, given a light curve, can we deduce the equivalent map?

4.2.2 Light Curve Inversion

The answer to the question posed at the end of the last section is, in general, no. To see why, consider the eclipse of a single point. Its ingress phase picks out a surface of all points with the same ingress phase, corresponding to the edge of the region obscured by the secondary star. The egress phase picks out another such surface. The intersection of these two surfaces is a line. However light is distributed along this line, it will produce an eclipse indistinguishable from that of the original point. Evidently we require more information than just a light curve. In fact we need to make some *assumptions* about the form of the map. This leads to two general approaches, *parameterisation* and *regularisation*.

4.2.3 Parameterisation

In this approach one tries to identify the simplest, perhaps physically motivated, model of the light distribution which can fit the data. This is the basis of the description of Fig. 4.1. Normally such a model comes with several parameters that need adjusting to the particular system in question. Thus one might need to fit the radius and surface brightness of the white dwarf, similar parameters for the bright-spot and also its position, the radius of the disc along with some sort of specification of its intensity distribution. This approach can be successful, and, for example, was used to deduce the extraordinary absorbed spectrum of the white dwarf in the dwarf nova OY Car [37].

Parameterisation has some considerable advantages. It is usually easy to estimate uncertainties upon the fitted parameters. The fitting process is also typically fast. Unfortunately, there are sometimes even more significant drawbacks. If the range of variation of real systems is large, one can reach a state of requiring more and more components, which often become rather *ad hoc*. At the same time, it is easy to build degeneracy into the model, which makes fitting, and the interpretation of any uncertainties, difficult. Finally, it can be hard to know just how to improve a model that is not fitting well.

4.2.4 Regularisation

To reduce the problems of parameterised models, one tries as much as possible to constrain the model, for example by using as few components as possible. In regularisation one does the reverse: the model is designed to be as flexible as possible, within some set of limited, and, it is hoped, broadly applicable conditions. Examples of such a conditions are the common assumptions that the disc is flat and that it lies in the orbital plane.

The flexibility is achieved by dividing up the region to be imaged into very many, M say, separate elements or pixels. The effects of neighbouring pixels are of course very similar and near degenerate, thus in this approach severe degeneracy is built in from the start. The idea at this point is not to find the best fit, but to find a "good enough" fit, usually one of low enough χ^2 to be considered a statistically acceptable description of the data. For example if there are N data points, one may require

$$\chi^2 \leq N \tag{4.1}$$

In general there are infinitely many models that can satisfy such a restriction; they define a volume of the M-dimensional space needed to define the pixels. To select a unique model, one then picks out some other feature felt to be desirable, such as smoothness. This can be done if this property can be expressed as some function of the model values. Let's say this property is encapsulated in a function S, and that we want the model of maximum S, consistent with the restriction above. Then we use Lagrange's method of undetermined multipliers and maximise

$$S + \lambda(\chi^2 - N)$$

with respect to the model and the multiplier λ. This is what is meant by regularisation.

This approach can worry those used to the straightforward χ^2-minimisation approach. After all, does not χ^2 encapsulate all that we know? Well, in fact it does not because we always know, or think we know, something about the likely brightness distribution even before any data is taken. Thus given the strong shear present in accretion discs we do not expect significant structure in the azimuthal direction. Adopting a pure minimum χ^2 approach effectively throws away such knowledge in favour of the data no matter how poor the latter may be. In regularisation the function S encodes our prior knowledge and the resulting images are those that fit the data acceptably, but also match our expectations of smoothness, or whatever, as well as possible.

4.2.5 The Maximum Entropy Method

What do we use for the regularising function S? There are many possibilities. For instance one might try to minimise the difference between each pixel value I_i, and the average of its close neighbours A_i which would give

$$S = -\sum_{i=1}^{M} (I_i - A_i)^2 .$$

Although many such possibilities exist, one function, closely related to the equation for entropy familiar from statistical mechanics, has become popular. This takes the form

$$S = -\sum_{i=1}^{M} p_i \ln p_i \tag{4.2}$$

where

$$p_i = \frac{I_i}{\sum_{j=1}^{M} I_j} . \tag{4.3}$$

Whether this function is the *only* correct choice has been the subject of heated debate. One of the original motivations for this choice was the idea that the image of maximum entropy was one that had the least information, and that it only contained features forced by the data. Since then a more rigorous derivation based upon Bayesian statistics and the imposition of consistency conditions has been developed [20,89]. However some researchers do not attach particular significance to the function $-\sum p \ln p$, regarding it rather as one example amongst a set of possible functions that give the much the same effects, essentially encoding our prior belief that the outcome must be smooth [66,73]. However, whatever the arguments for or against entropy, its use has been facilitated by the wide availability of the powerful MEMSYS code [90] and so I will not consider alternatives any further.

In the absence of data, the entropy (4.2) is maximised if the p_i are all the same. This is easily seen using Lagrangian multipliers by maximising

$$S + \mu \left(\sum_{i=1}^{M} p_i - 1 \right) .$$

The second term imposes the restriction implied by the definition of p_i (Eq 4.3). Maximising with respect to p_i gives

$$-1 - \ln p_i + \mu = 0$$

so the p_i are identical as stated.

Default images. In his original paper [33], Horne used a modified form of the entropy:

$$S = -\sum_{i=1}^{M} p_i \ln \frac{p_i}{q_i} \qquad (4.4)$$

where as before

$$p_i = \frac{I_i}{\sum_{j=1}^{M} I_j}$$

and in addition

$$q_i = \frac{D_i}{\sum_{j=1}^{M} I_j}$$

where the vector $\boldsymbol{D}$ is called the *default image*. To see why it has this name, if we maximise S as before with Lagrangian multipliers we obtain

$$-1 - \ln \frac{p_i}{q_i} + \mu = 0$$

which therefore leads to $p_i = q_i$ and $\boldsymbol{I} = \boldsymbol{D}$. So in the absence of data, the modified form of entropy (4.4) is maximised when the image is the same as the default. This form of entropy therefore measures departures from the default. Viewed in this way the original form (4.2) measures departures from a constant.

The motivation for using default images is that we know that the surface brightness of a disc is likely to be far from constant. Instead it will typically rise steeply towards the centre of the disc. However, we still expect azimuthal symmetry. Thus Horne used default images that were themselves computed from the model images by computing images of perfect azimuthal symmetry with nearly the same radial profile as the model images. Horne showed that this allowed reliable recovery of the radial surface brightness from light-curves, although spot features were azimuthally blurred by the process.

It should be realised that the default image is only a guide for the image. If the data require departure from axi-symmetry, there is nothing to stop the growth of such structure in the final model.

I now move on to the results of applying the maximum entropy method (MEM) to the lightcurves of CVs.

4.2.6 Early Results of Eclipse Mapping

The novalike RW Tri. The first application of eclipse mapping was to the nova-like (steady-state) system RW Tri [35]. The light-curves, MEM fits and radial brightness distributions are plotted in Fig. 4.2. This was an important result as it largely confirmed the theoretical steady-state effective-temperature distribution [see Smak this volume]:

$$T_e^4 = \frac{3GM_W\dot{M}}{8\pi\sigma R^3}\left(1-\sqrt{\frac{R_W}{R}}\right). \tag{4.5}$$

More precisely it confirmed the $T \propto R^{-3/4}$ that holds at large radii. Horne [35] was also able to show from his multi-colour data that the colours of the disc in RW Tri were intermediate between those of black-bodies and stars.

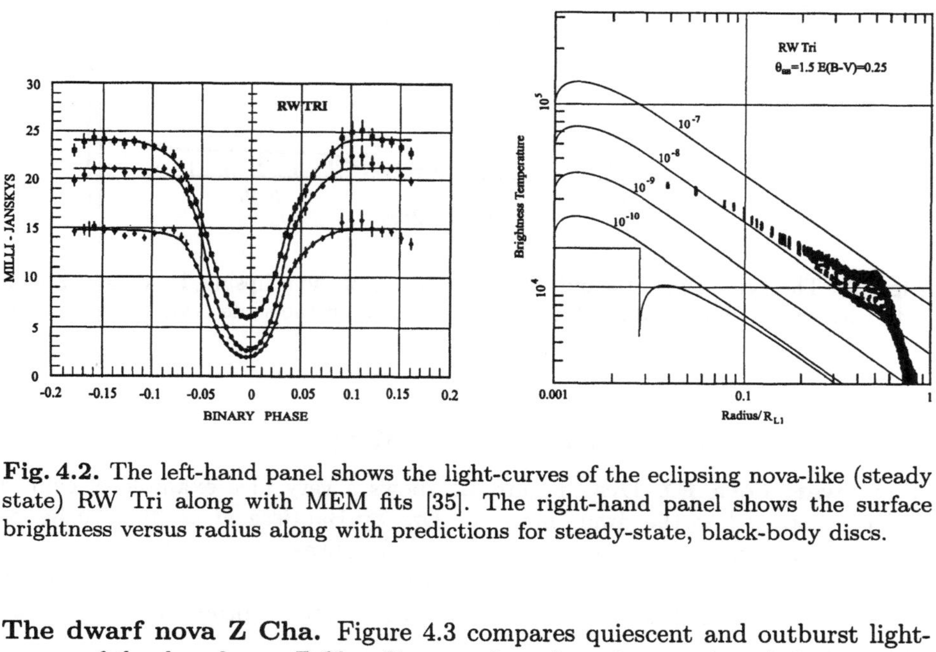

Fig. 4.2. The left-hand panel shows the light-curves of the eclipsing nova-like (steady state) RW Tri along with MEM fits [35]. The right-hand panel shows the surface brightness versus radius along with predictions for steady-state, black-body discs.

The dwarf nova Z Cha. Figure 4.3 compares quiescent and outburst light-curves of the dwarf nova Z Cha. Compared to the quiescent data (which we saw before in Fig. 4.1), the outburst data exhibit a much smoother eclipse, indicative of domination by the disc. Horne & Cook applied the eclipse mapping method to these data [34], obtaining the radial brightness profile shown on the left of Fig. 4.4.

The outburst light-curve corresponds to a radial profile that is close to a steady-state. Wood *et al.* [107] decomposed the quiescent light-curve into white dwarf, bright-spot and disc components. The radial profile corresponding to their disc light-curve is plotted on the right of Fig. 4.4. This reveals a profile dramatically different from steady-state predictions, being much less intense in the centre of the disc. This is much as expected from the disc instability model of dwarf nova outbursts in which mass accumulates at the edge of the disc during quiescence.

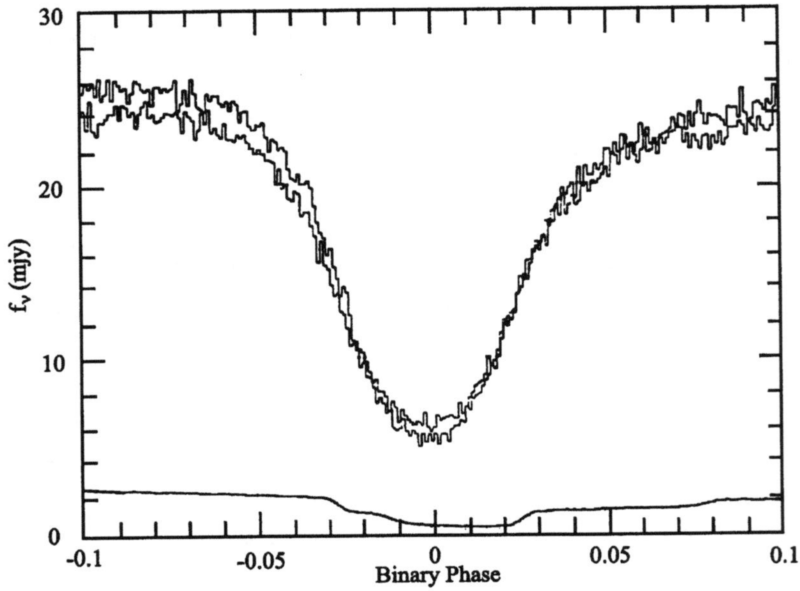

Fig. 4.3. The figure shows the outburst light-curves of the eclipsing dwarf nova Z Cha [34], with the quiescent data of Fig. 4.1 [107] plotted on the same scale (lower curve).

4.2.7 More Recent Results of Eclipse Mapping

The dwarf nova EX Dra. Figure 4.5 shows a very nice example of eclipse mapping applied to the dwarf nova EX Dra by Baptista *et al* [3]. The figure shows the substantial changes in the surface brightness of the disc that occur during the outburst cycles of dwarf novae; EX Dra is a particularly good system for such studies as it outbursts frequently compared to other eclipsing dwarf novae.

Spectrally-resolved eclipse mapping. All the work I have discussed so far has been based upon broad-band photometry. In 1993, Rutten *et al.* [79,81] had the simple but powerful idea of taking high time-resolution spectrophotometry, effectively giving many hundreds of very narrow-band light curves. When each of these is mapped, one can obtain the spectrum over different regions of the disc, exactly the sort of data that is needed for comparison with models of atmospheres of discs [101,102]. This technique is known as "spectrally-resolved eclipse mapping".

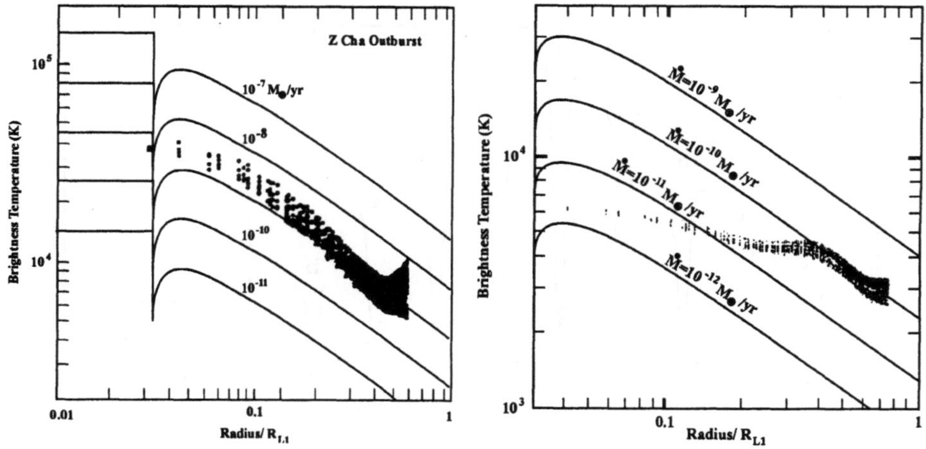

Fig. 4.4. The left-hand panel shows the radial brightness profile of Z Cha in outburst [34], while the right-hand panel shows the profile corresponding to the quiescent disc [107].

Baptista *et al.* [2] applied this method to *HST* data of the eclipsing nova-like variable, UX UMa. Examples of the light curves and the final spectra they derived are plotted in Fig. 4.6.

These spectra clearly show the increase of intensity as one moves from the outer to the inner disc. As one might expect, this increase is most significant at short wavelengths. In the outer disc a strong Balmer jump in absorption is visible.

Note the spectrum at the bottom of Fig. 4.6, marked as "uneclipsed light". This is needed to prevent a common artifact of eclipse mapping in which the code tries to push light to the back of the disc where it does not get eclipsed. This comes about because the eclipses are shallower than expected for an axi-symmetric disc. In some cases this might be a result of significant emission from the mass donor star which I have so far assumed to be negligible. However, this is not the case with UX UMa since the spectrum of the uneclipsed light does not correspond to any normal stellar spectrum.

4.2.8 Flat Temperature Distributions

Horne's first application of eclipse mapping showed that the disc of the novalike variable RW Tri matched steady-state expectations [35]. However, this has by no means always been the case. Rutten *et al.* [78] present the radial profiles of six novalikes, 2 or 3 of which have much flatter profiles than expected for steady-state models. In other words, the inner disc appears fainter than expected relative to the outer disc. In the light curves, this reveals itself in V-shaped rather than U-shaped eclipses.

The nature of the problem is well shown in the eclipsing old nova, V Per [108]. Figure 4.7 shows that steady-state models only match the V-shaped light

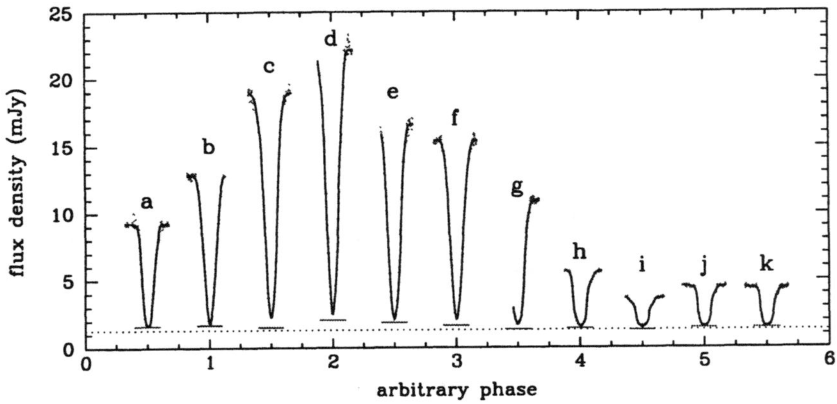

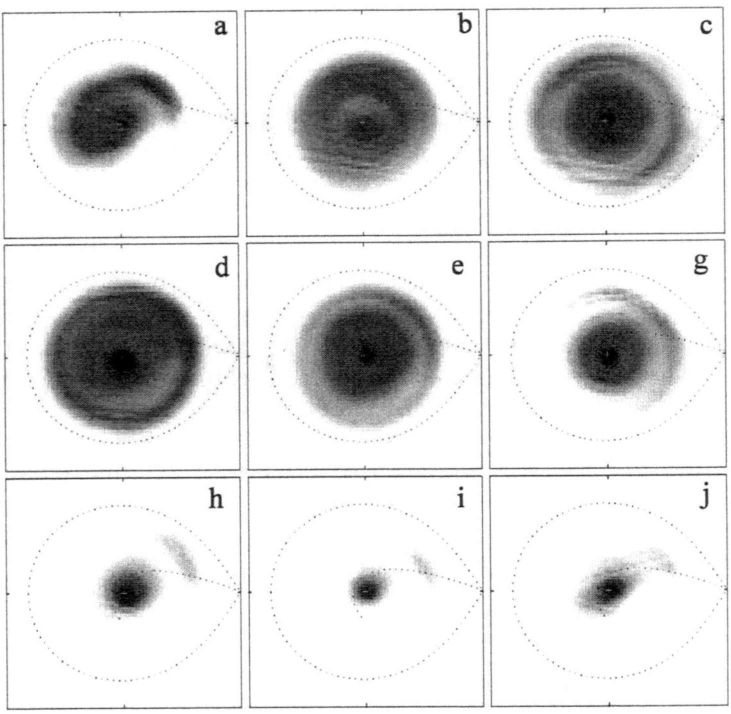

Fig. 4.5. Light curves and MEM images of the eclipsing dwarf nova EX Dra observed at a series of different stages of its outburst cycle [3].

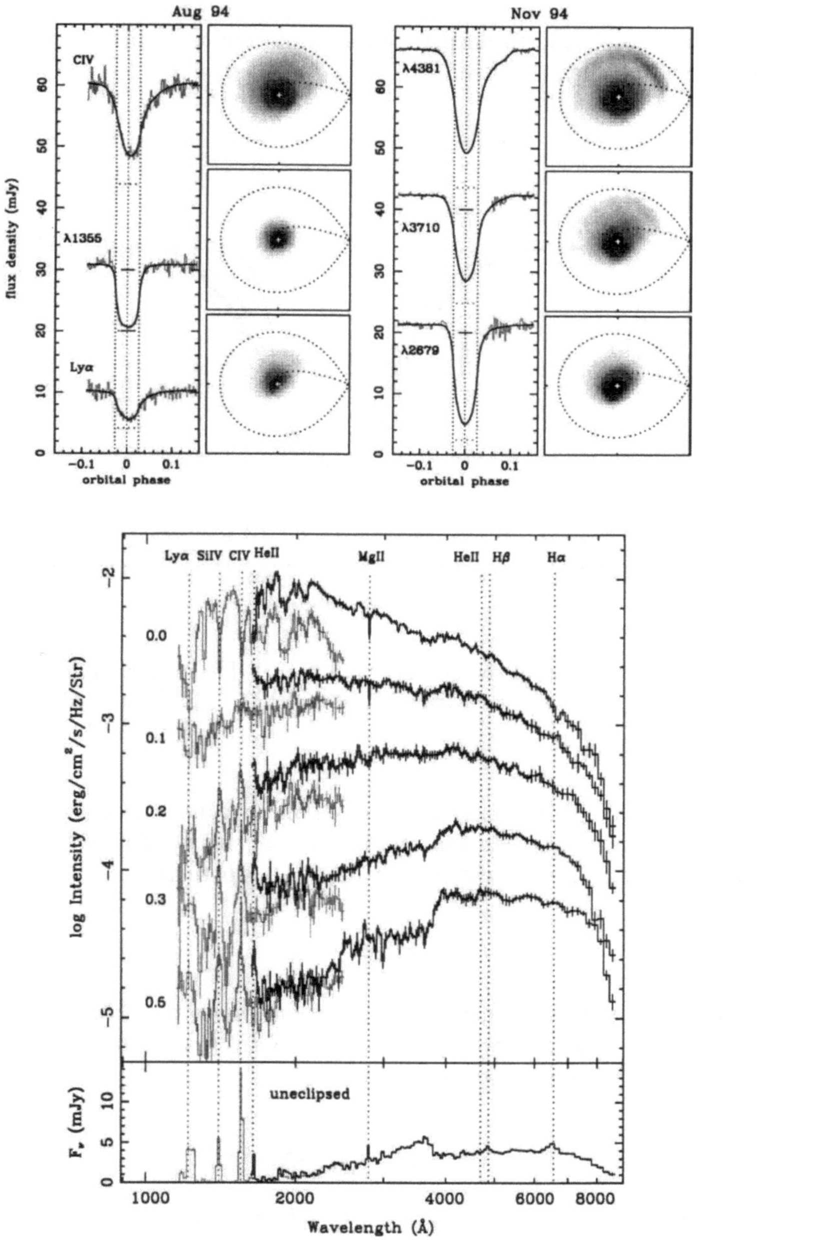

Fig. 4.6. Light-curves of the novalike UX UMa are plotted in the left panel with equivalent MEM images. On the right are plotted the spectra at different radii obtained after all wavelengths have been mapped [2]. Radii are indicated to the left of the spectra and are measured in terms of the distance of the inner Lagrangian point to the white dwarf.

curve at implausibly high mass transfer rates ($> 10^{-6}\,M_\odot\,yr^{-1}$). More realistic mass transfer rates lead to deep U-shaped light curves. This effect translates to the flat radial profiles discussed.

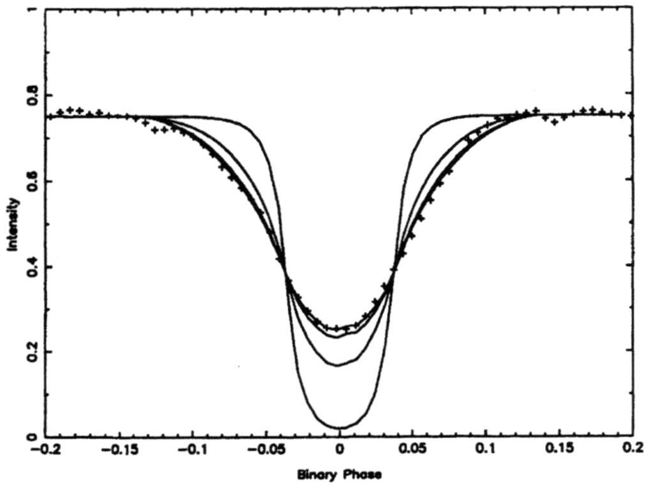

Fig. 4.7. The light-curve of the eclipsing old nova V Per is plotted along with steady-state model light curves for mass transfer rates of 10^{-10}, 10^{-8}, 10^{-6} and 10^{-4} $M_\odot yr^{-1}$ [108].

In my opinion this problem is a serious one because the steady-state prediction is really little more than the conservation of energy. To see this consider the energy liberated as mass flows from radius $R + dR$ to R:

$$\frac{GM_W\dot{M}}{2R^2}dR.$$

(The factor 1/2 appears because half the energy goes into kinetic energy rather than radiation). Spread over an area of $4\pi R\,dR$ (two sides of the disc), this implies an effective temperature T_e given by

$$T_e^4 = \frac{GM_W\dot{M}}{4\pi\sigma R^3}.$$

This is very close to Eq. 4.5, with the only difference a factor of $3(1-\sqrt{R_W/R})/2$ owing to the work done by viscosity which transfers energy liberated close to the white dwarf out to the rest of the disc.

There is no doubt over the conservation of energy, so how can theory and observations be reconciled? There are several possibilities:

- the energy liberated in the inner disc is not radiated but is carried off in some other manner, e.g. a wind.

- the energy liberated in the inner disc is not radiated but is advected onto the white dwarf.
- we don't see the inner disc because the outer disc obscures it.

The second possibility is a straw man that can be knocked down easily: if power was advected onto the white dwarf (such models are popular for black-hole binary stars), it would still have to be liberated at the white dwarf, and we would surely detect it in X-rays. The first possibility is less easily dismissed although it would be remarkable if such a mechanism could remove 90% of the energy as needed to fits observations, and it provides no convincing reason for why some systems match steady-state predictions, and others do not. I believe it is the third, and perhaps most prosaic, explanation that is the correct one.

All the eclipse maps presented so far were calculated assuming that the disc is flat. Although it has been shown that results are little affected if instead the disc has a modest flare [108,82], there is no question that if the outer disc was so thick that the inner disc was invisible, then we could not derive a reliable surface brightness. There is good evidence that exactly this is the case.

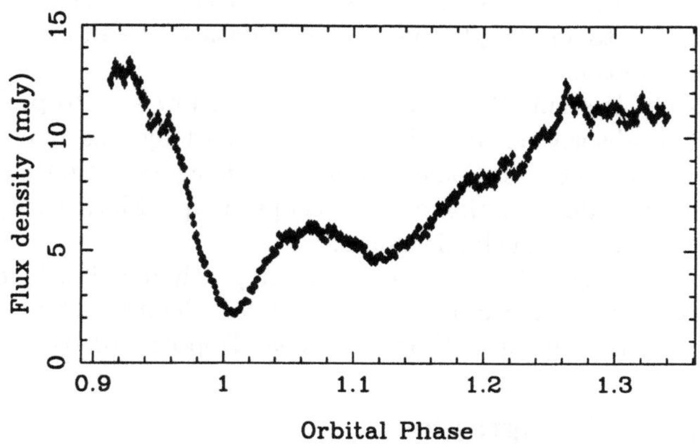

Fig. 4.8. The light-curve of the eclipsing dwarf nova OY Car in super-outburst [8].

Figure 4.8 shows the *HST* light curve of the dwarf nova OY Car in a state of "super-outburst" (unusually long and bright outbursts) [8]. The eclipse starts off normally enough, but during egress there is a substantial drop in flux – a "dip". This is caused by (time-variable) obscuration of the inner disc by the outer disc. In fact it is very likely that there is still obscuration away from the dip phases because there is spectroscopic evidence for absorption at all phases in this data.

Further evidence comes from absorption of the white dwarf's spectrum by cool material in OY Car during quiescence [37]. Finally in UX UMa it is found that the side of the disc nearest to us is very deficient in UV flux compared to the side furthest from us [2]. This is exactly as expected if the outer edge of the disc is absorbing light from the inner disc because this will affect the nearest side of the disc and UV wavelengths most of all.

The apparent ability of discs to be thick is a little surprising. The theoretical disc height to radius ratio H/R is of order the sound speed in the disc C_S divided by the orbital speed V_{orb}:

$$\frac{H}{R} \sim \frac{C_S}{V_{orb}}.$$

For the expected temperatures of the outer discs of CVs, this gives $H/R \sim 1/50$, and so we don't expect the thickness of discs to matter much unless a system is within a few degrees of being edge-on. However, the inclination of UX UMa is only 71°, and yet obscuration seems significant. To date there has been little work on trying to explain this problem.

4.2.9 Eclipse Mapping Summary

Eclipse mapping is a simple and powerful technique that tests our most fundamental predictions for accretion discs. It allows us to obtain images of discs that subtend less than 0.1 milli-arcseconds at Earth. The greatest triumph of eclipse mapping is probably the confirmation of the steady-state temperature distribution, one of the corner-stones of the theory of accretion discs. The extension of eclipse mapping to spectrophotometry gives data that can be tested directly against model atmospheres of discs.

Unfortunately, it seems that real discs are somehow able to puff up at their outer edge. The absorption and obscuration caused by this may prevent our seeing the inner discs at all in some systems, which suggests that caution should be exercised in the interpretation of some eclipse maps. More work is needed to evaluate just how serious such effects are.

With that warning, I leave eclipse mapping, although I will mention very similar techniques when discussing magnetic CVs. Before that however, I turn my attention to the imaging method known as "Doppler tomography".

4.3 Doppler Tomography

Doppler tomography is a technique that allows us to image CVs and related binary stars in the light of their emission lines [50]. Its name comes about because it turns out that the formation of the emission line profiles is closely analogous with the problem of X-ray tomography used in medicine to image internal organs of the human body. As it deals with line emission, Doppler tomography is immediately at a disadvantage compared to eclipse mapping in terms of telling us where most of the energy is emitted in CVs. However, it makes up for this by being a much better conditioned inversion, which means that Doppler tomography can reveal more complex structures than can eclipse mapping. I will discuss several of these. I begin by considering the formation of the line profiles of CVs.

4.3.1 Emission Line Profiles

It has long been known that accretion discs can produce double-peaked line profiles [91,38,92,36]. This can occur if the disc produces line emission from its surface with an intrinsic width much less than the orbital speed. This is only possible if the sound speed C_S is much less than the orbital speed V_{orb}, conditions which we have already seen apply to the discs of CVs.

Figure 4.9 illustrates how the double-peaked profiles come about.

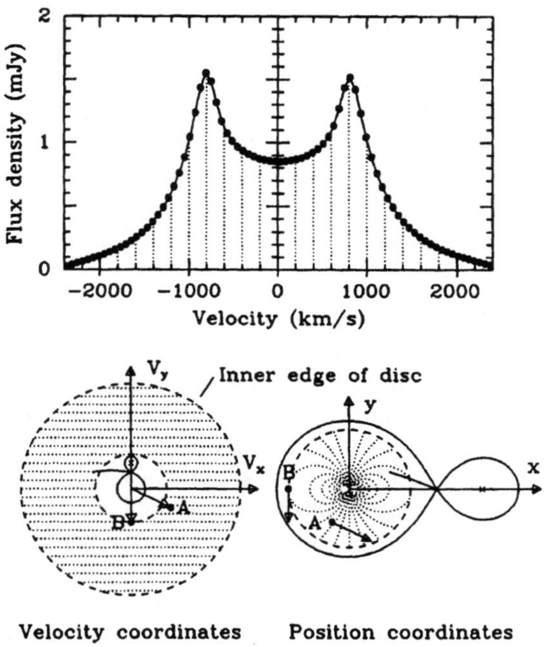

Fig. 4.9. An illustration of how an accretion disc can produce a double-peaked profile. At the lower-right is a schematic illustration of the binary with dashed lines of equal radial velocity (the system is viewed from below, equivalent to orbital phase 0.25). At the lower-left, is the equivalent in velocity coordinates (see text).

When the orbital velocity dominates over the intrinsic line profile, emission from a particular point in the disc will add into the observed line profile at a velocity matching its radial velocity at the time of observation. Thus it is useful to consider lines of equal radial velocity in the system at a particular phase. Fig. 4.9 does this for orbital phase 0.25. A pattern of curved lines can be seen running over the disc. This has a shape identical to a magnetic dipole field pattern, and is a direct result of a so-called "keplerian" velocity field, $V \propto R^{-1/2}$. As we move from high to low velocities, the areas of the crescent-shaped regions of equal radial velocity increase rapidly. This corresponds to the increase in flux moving from the wings of the profile towards zero velocity. However, the increase in area cannot be sustained forever because eventually it is curtailed by the edge of the disc. If we try to go to lower velocity, the area decreases once more, and so two peaks are formed at the projected speed of the outer disc [91].

Throughout most of this section I will be referring to images in "velocity space". This is illustrated in the lower-left of Fig. 4.9. I define coordinate axes with x running from the accretor to the donor, y in the direction of motion of the donor, and z to form a right-handed triad, i.e. $\boldsymbol{z} = \boldsymbol{x} \wedge \boldsymbol{y}$. This definition leads naturally to the definition of V_x and V_y, with one subtlety, which is that these are *projected inertial speeds* rather than rotating-frame speeds ("projected" meaning projection onto our line of sight). Thus the accretor has velocity $(0,-K_1)$, while the donor has velocity $(0,K_2)$, where K_1 and K_2 are the respective projected orbital speeds. Lines of equal radial velocity in velocity space are straight, but their angle depends upon the orbital phase. With the definitions given, the lines of equal radial velocity make angles of $360°\phi$ with respect to the V_y axis at orbital phase ϕ, increasing in the clock-wise direction.

I will come back to discuss velocity space more later, but before proceeding, I first look at the observational evidence in favour of the model.

Evidence for the Doppler-shifted profile model. The most obvious evidence for the Doppler-shifted model is the ubiquity of double-peaked profiles in CVs, particularly in dwarf novae. U Gem [97], Z Cha [48], OY Car [1], IP Peg [49], V2051 Oph [105], and many others show such profiles. However, there are also many systems that do *not* show double-peaked profiles, most notably many eclipsing novalike variables. That theses systems are eclipsing shows that it is not just low inclination that prevents our seeing double-peaks [92]. The absence of double-peaked profiles in many systems is a worry, but luckily there is even better evidence that the model is correct for those systems which do have them. Consider Fig. 4.10 which shows the eclipse of Hβ in the dwarf nova IP Peg [49]. Since the disc rotates in the same direction as the donor, just before mid-eclipse the approaching, blue-shifted side of the disc is occulted. Thus in the middle profile of Fig. 4.10, the blue-shifted side of the profile is absent. The top profile was taken only a few minutes later, by which time the approaching side of the disc has emerged from eclipse while the receding, red-shifted half has been occulted. There can be little doubt of the validity of the Doppler-shifted formation model in this case, and such data can be used to confirm the keplerian velocity field [111,49].

4.3.2 Profile Formation by Projection

When discussing the profile formation model, I introduced the idea of velocity coordinates, in which lines of equal radial velocity were parallel straight lines. This makes the translation from model image to profile particularly straight-forward, as all we need do is *project* the image along the direction appropriate to the orbital phase in question. Projection here means the mathematical transformation of integration along (in this case) lines across a 2D image to produce the 1D profile. The projection of an image at two phases is illustrated in Fig. 4.11.

The artificial image has been created with a spot which can be seen to project into different parts of the profile at different phases. If one saw the two peaks in

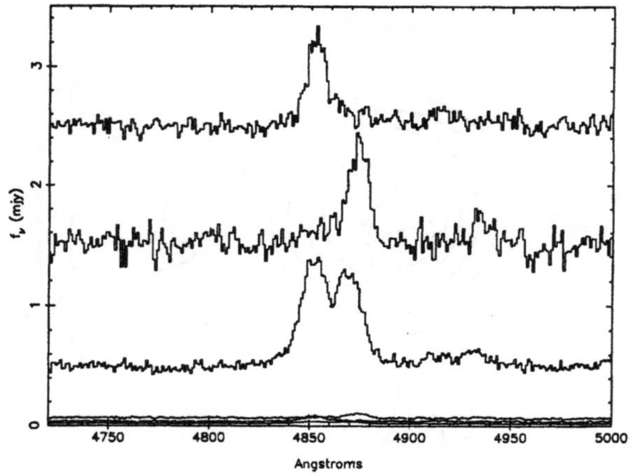

Fig. 4.10. The emission line Hβ in the eclipsing dwarf nova IP Peg as seen at three times. From top to bottom are plotted (i) the profile just after mid-eclipse, (ii) the profile just before mid-eclipse, and (iii) the profile out of eclipse.

the profiles, one could deduce the position of the spot in the image by tracing back along the projection directions. This is in essence how line profile information can be used to reconstruct an image of the system; we will encounter this as *back-projection.*

A series of line profiles at different orbital phases is therefore nothing more than a set of projections of the image at different angles. The inversion of projections to reconstruct the image is known as "tomography", the case of medical X-ray imaging being perhaps the most famous, although it occurs in many other fields too. I now discuss the inversion method known as "filtered back-projection".

4.3.3 Filtered Back-Projection

By the definition of the coordinate axes, the radial velocity V_R of a point at (V_x,V_y) at orbital phase ϕ is given by

$$V_R = \gamma - V_x \cos 2\pi\phi + V_y \sin 2\pi\phi, \tag{4.6}$$

where γ is the mean or systemic velocity of the star. The projection to form the line profile, $f(V,\phi)$, a function of radial velocity V and orbital phase ϕ, from an image in velocity coordinates, $I(V_x,V_y)$, can then be written:

$$f(V,\phi) = \int_{-\infty}^{\infty}\int_{-\infty}^{\infty} I(V_x,V_y)g(V - V_R)\,dV_x\,dV_y, \tag{4.7}$$

where g is a function representing the line profile from any point in the image, including effects such as instrumental blurring. Note that g is assumed to be

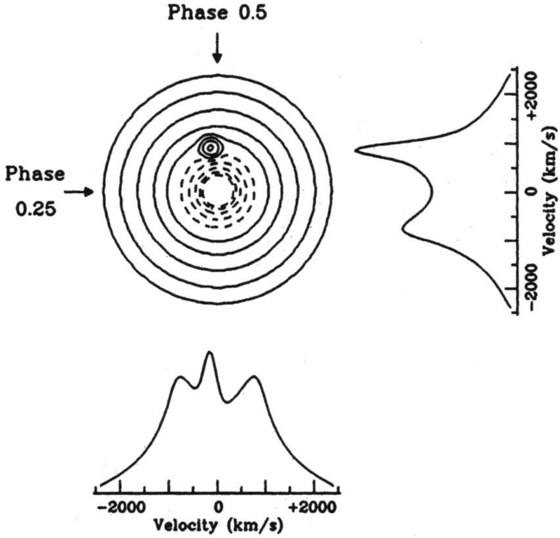

Fig. 4.11. A model image and the equivalent profiles formed by projection at angle appropriate to orbital phases 0.25 (right-most profile) and 0.5 (lower profile).

the same at every point. It is the linear dependence of V_R upon V_x and V_y (4.6) which gives the straight line projection in velocity coordinates.

The inversion of Eq. 4.7 is detailed in the appendix. The process can be summarized in the following two steps. First the line profiles are filtered in velocity to derive modified profiles, $\tilde{f}(V, \phi)$. For mathematically exact inversion, the filter is $|s|/G(s)$, where $G(s)$ is the Fourier transform over V of $g(V)$ and s is the frequency in inverse velocity units. If the local line profile $g(V)$ is gaussian then so too is $G(s)$, dropping to zero at large s. Thus the filter will strongly amplify high frequencies, and the image will likely be corrupted by noise. This may be familiar when it is realised that division by $G(s)$ is just the standard (and noise-sensitive) Fourier deconvolution; the presence of $|s|$ in this case only exacerbates the problem. Thus in practice, the filter is modified to $|s|W(s)$, where $W(s)$ is a (typically gaussian) "window" function to cut off high frequencies and therefore limit the propagation of noise into the final image. The penalty for this is that the final image is a blurred version of the true image.

The second step is that of *back-projection*:

$$I(V_x, V_y) = \int_0^{0.5} \tilde{f}(\gamma - V_x \cos 2\pi\phi + V_y \sin 2\pi\phi, \phi)\, d\phi. \qquad (4.8)$$

An intuitive understanding of back-projection is very useful when trying to make sense of Doppler maps. There are two ways of imagining the process. The first one is perhaps the most obvious from Eq. 4.8 which implies that each point in the image can be built by integrating along a sinusoidal path through "trailed" spectra (spectra viewed in 2D form with axes of velocity and phase). The particular sinusoid is exactly that which a spot at the particular place in the image would produce in the trailed spectra. This view is illustrated in Fig. 4.12.

Back-projection is named for another, perhaps more useful, way of regarding this operation. In effect, Eq. 4.8 means that the image is built up by smearing each filtered profile along the same direction as the original projection which formed it (see Fig. 4.13).

This way of looking at back-projection shows very clearly why small numbers of spectra cause linear artifacts in Doppler maps, and one should always be wary of such features. Similarly, any anomalies, such as unmasked cosmic rays or unmasked eclipses are liable to cause streaks across Doppler maps.

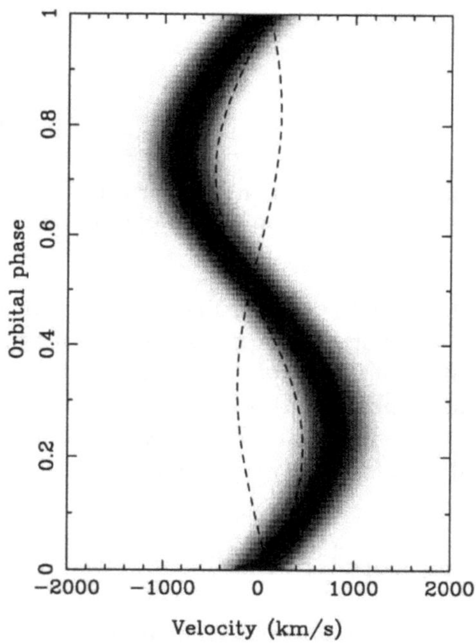

Fig. 4.12. An illustration of back-projection as integration through trailed spectra. If a track coincides with a sinusoidal component, there is a large contribution to the integral and a spot will appear in the final image.

4.3.4 MEM Inversion

The inversion from line profiles to image can be regarded in the same sense as the eclipse mapping problem, and MEM can be applied in a very similar way [50]. In practice, filtered back-projection often does a perfectly adequate job, and is comparatively easy to code and fast to apply. MEM seems to be better at suppressing the "spokes" characteristic of limited numbers of spectra, and can

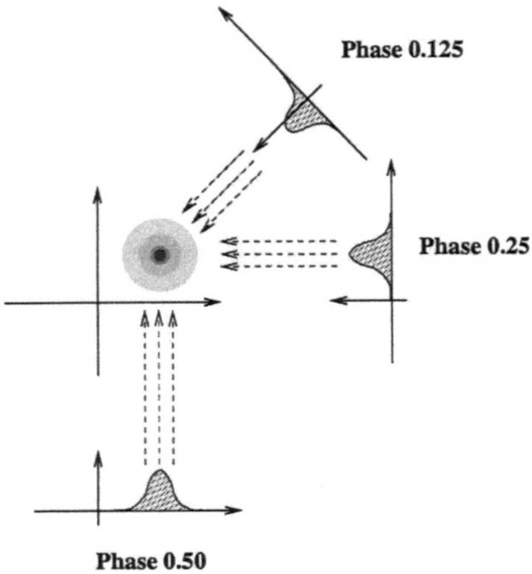

Fig. 4.13. An illustration of back-projection. Each profile is smeared across the image at an angle matching the original direction of its projection.

be easily extended to cope with blended lines (see [57] for example), variable local line profiles, and many other deviations from the simple model discussed. Moreover, just as MEM is routinely applied to sharpen blurred images (without allowing excessive noise amplification), so it can go further towards the effective implementation of the exact filter $|s|G^{-1}(s)$ and thus produce somewhat sharper features than the windowed filter $|s|W(s)$ allows.

4.3.5 Understanding Doppler Maps

Figure 4.14 shows some of the key components of a CV represented in velocity space. The donor star is assumed to co-rotate with the binary, which means that it appears with the same shape in velocity as in position coordinates, although rotated by 90° owing to the relation $\boldsymbol{v} = \boldsymbol{\Omega} \wedge \boldsymbol{r}$ between velocity and position for "solid-body" rotation. This reassuring property is somewhat misleading, since the disc, which is very definitely not co-rotating with the binary, ends up being turned inside out so that the inner disc is at large velocities while the outer disc appears as a ring at low velocity. The gas stream is plotted twice: once with its true velocity and once with the velocity of the disc along its path. The positions of all these components is fully specified if the projected orbital velocities of the two stars, K_1 and K_2, and the orbital phase are known. The overall scale is set by $K_1 + K_2$; their ratio $q = K_1/K_2 = M_2/M_1$ defines the detailed shape of the stream and Roche lobe. The orbital phase sets the overall orientation of the image, and if it is not known the image will be rotated by an unknown amount relative to the "standard" orientation of Fig. 4.14.

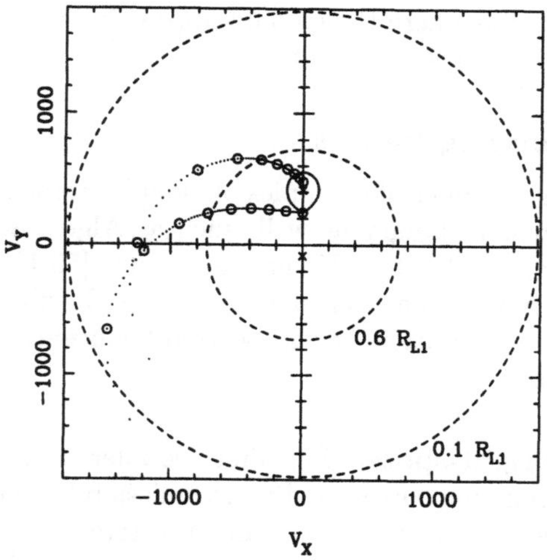

Fig. 4.14. A schematic of some key components in velocity coordinates.

Although velocity coordinates simplify the picture of line profile formation, it is simple enough to invert into position coordinates – indeed this is how I originally computed Doppler images [47]. However, I abandoned this approach for two reasons. First, the translation between velocity and position is often not known. In fact, perhaps it is never known, given that it is likely that deviations from keplerian flow occur. This means that position maps would require recomputation each time system parameters were updated. Second, the same place in the system can produce emission at more than one velocity. This is not an abstract possibility, but happens in almost every system that has been imaged. There are many examples of bright-spot emission from the gas stream while the disc at the same location produces emission at a completely different velocity. If such data is imaged into position coordinates on the basis of keplerian rotation, a spot of emission would be produced at a completely spurious location in the disc, but it is hard to see what else could be done. Sticking to velocity coordinates is a reminder of these potential difficulties of interpretation. Only in eclipsing systems is there potential for disentangling such effects.

While we cannot translate the data into position space, there is no difficulty in translating any theoretical model into velocity coordinates. Indeed, ideally, the theory–data comparison should be made by predicting trailed spectra, doing away with the need for Doppler maps altogether. However, Doppler maps still have a rôle in that theoretical models are not good enough to predict all the peculiarities of real systems, and comparison is easier in the half-way house of velocity space.

The idea of translating to velocity space also applies to how one should think about Doppler maps. Rather than trying to translate features of maps mentally from velocity to position coordinates, one should try to think of various components and imagine where they would appear in velocity space. With that said, I now turn to look at some results.

4.3.6 Doppler Imaging Results

There are now a large number of examples of Doppler tomography, covering CVs along with other types of binary as well, such as Algols and X-ray transients, the latter being very similar to CVs in many ways [53,13]. Some of the most spectacular images have come from magnetic CVs; I will look at these in the third section of this contribution. In this section I concentrate upon results from disc accreting systems.

Disc rings and bright-spots. The first Doppler maps published revealed broadened rings of emission corresponding to emission from discs and also bright spots of emission caused by the gas stream/disc impact [51,31,52]. It has since become clear that usually the spot emission comes from the gas stream rather than the disc. An exception to this rule is seen in the system U Gem where the bright-spot appears half-way between the predicted stream and disc positions, which may indicate that we are seeing emission from downstream of the impact [51].

An example of a Doppler map of this variety is shown in Fig. 4.15 which shows images of several lines of the 46-minute period, helium accretor GP Com. This system does not have a well-determined orbital phase, so this is a case where there is rotational uncertainty in the images, albeit the same amount for each image. GP Com shows the broadened disc rings, and an obvious bright-spot at the edge of the disc (and therefore near the peak of the disc ring in velocity coordinates). In addition it displays emission at zero velocity. This is a "forbidden region" for most CVs – there is no single part of the system that is stationary relative to the centre of mass. GP Com has an unusually extreme mass ratio – 50:1 – which means that the accretor is almost stationary and may be the origin of the low velocity spot.

The images of GP Com are very similar in nature to those of other short period/extreme mass ratio systems such as WZ Sge [94], GD 552 [31], LY Hya [96] and X-ray novae [53,12]. In more equal mass ratio systems the disc component is displaced noticeably downwards, as expected because the radial velocity amplitude of the accretor is larger, but again the images have the same overall structure.

In systems with well-determined parameters the position of the bright-spot in Doppler images can be used to measure its distance from the white dwarf, which can serve as an estimate of the outer disc radius. So too can the disc ring itself. This is superior to simply measuring double-peak separations from mean spectra since the latter may include unrecognised contributions from the bright-spot and/or red star.

The existence of bright-spots of emission was well-known before the advent of Doppler tomography, as in many systems the 'S'-wave is obvious, and measurable, see for example Stover et al.'s analysis of U Gem [97]. It is fair to ask

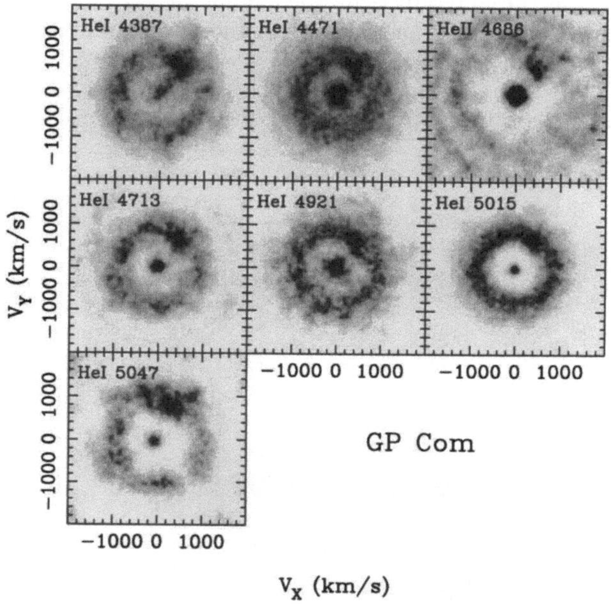

Fig. 4.15. Doppler images of the helium accretor, GP Com showing a broad ring of emission from the disc and two spots of emission, the outermost of which is from the gas stream/disc impact region.

then, what extra does Doppler tomography bring to this topic? Chiefly, Doppler tomography gives an objective description of the emission without requiring intuitive development of models. If the emission can be described by a disc and single spot, this is not much of an advantage, but it becomes vital in more complex cases, such as is shown in Marsh et al.'s study of U Gem [51] and, as we will see below, in the discovery of spiral shocks in IP Peg. Doppler tomography can make an important contribution even when the emission distribution is simple. A good example of this is Johnston et al.'s analysis of the X-ray nova V616 Mon [41], where they modelled the emission pattern in terms of a simple parametric model of the disc plus "spot". They described the spot with a very large-scale asymmetry, and their best-fit parameters for this model led them to deduce an extreme mass ratio for the system. Later Doppler tomography revealed that the spot was in fact very localised, and, combined with study of the secondary star, was consistent with a very different mass ratio.

Emission from the secondary star. It is common to see emission from the secondary star [52,23,24] in Doppler images. The exact location of this is informative as to whether the emission is intrinsic to the secondary stars or generated by irradiation. Irradiation, for example, favours emission on the side of the secondary closest to the white dwarf, and this can be seen in Doppler images. A nice example of this was presented by Schwope [88] for the star QQ Vul where one can see the effect of irradiation in producing emission and suppressing photospheric absorption at the same tie (Fig. 4.16).

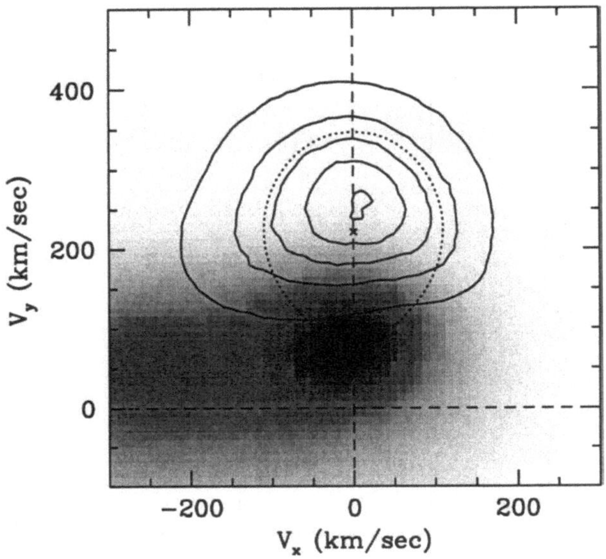

Fig. 4.16. Doppler images of the secondary star in QQ Vul shows emission concentrated on the side facing the white dwarf (grey-scale) whereas photospheric absorption (contours) is mostly on the opposite face [88].

Standard Doppler tomography assumes that all elements are seen equally at all phases, which is clearly not the case for emission from the secondary star. Rutten and Dhillon [80] developed code which can cope with this and therefore can image the secondary star correctly. The potential of this method, at least when applied to Balmer emission, may be limited by intrinsic broadening other than Doppler — there is good evidence for this in detached white dwarf/M-dwarf binaries [58]. CaII lines are likely to prove a better bet in such work.

The mystery of missing discs. While dwarf novae often show double-peaked lines, it is well known that many nova-like variables fail to do so. Indeed, there is a group of such objects called the "SW Sex" stars, with the presence of single-peaked lines as one of its defining criteria, in spite of most of these stars being of sufficiently high inclination to eclipse. It is no surprise then that these lead to very different Doppler images from those discussed so far, as can be seen in the Doppler map atlas published by Kaitchuck et al. [43]. Novalikes often show a single blob of emission to the lower-left of the centre of mass, and usually too low to be consistent with the gas stream. There is evidence for the gas stream over-flowing the disc which can cause emission in this region of Doppler maps [27], but there are still features for which a convincing explanation is lacking in my opinion, such as the relative weakness of eclipses in the lines. The reader is referred to [28,17] for further discussions of these systems.

Spiral shocks. As I mentioned above, Doppler tomography is seen at its best on complex emission patterns, which produce spectra which defy simple intuitive inversion. In the 1980s there were several theoretical papers describing "spiral shocks" in discs driven by the tidal field of the secondary star [85,93]. It was noted by Robinson et al. [77] that these should be visible in Doppler images, but that, so far, there was no sign of them. It was a great surprise, perhaps even a "shock", then when Steeghs et al. [95] (Fig. 4.17) found evidence for very strong spiral shocks in the dwarf nova IP Peg during outburst. Spiral shocks have since been confirmed in several other outbursts of IP Peg, and, with hindsight, were probably present in earlier outburst spectra of the system [52].

The spiral shocks of IP Peg are persistent: Figure 4.18 shows them still going strong six days into an outburst.

They may even last into quiescence, although they are expected to be more tightly wound and difficult to detect [95]. Still, this is something that should be searched for.

More difficult to understand is why these shocks are so strong in IP Peg, but relatively vestigial in other systems for which suitable data exists, although it has to be remarked that there is very little outburst spectroscopy overall, and IP Peg is unusual in having particularly strong line emission during outburst.

There is much to be done in this area. Above all perhaps, we need to cover a full outburst of IP Peg rather than comparing different stages of different outbursts as we have been forced to do so far. The way in the spiral shocks evolve may have much to tell us about the evolution of the disc during outburst. In addition, these shocks need to be searched for in other systems to see if we can understand when and why they appear.

4.3.7 Extensions to Doppler Tomography

There are a number of possible extensions to Doppler tomography, although they all have additional possible pit-falls too. I have already touched on Rutten & Dhillon's Roche tomography [80] which describes the secondary star accurately, although in practice intrinsic broadening may limit its usefulness. Bobinger et al. [9] describe a method for simultaneous Doppler and eclipse mapping of emission lines. In this a single image is computed to fit both spectra and light curves of the lines, with a keplerian velocity field used to translate between position and velocity space. It is difficult to evaluate how much spectra or line flux dominate the final maps, but it is clear that spectral information does alleviate the degenerate nature of eclipse mapping. Of course, the need to assume a particular velocity field is a dis-advantage. An attempt has been made to avoid this

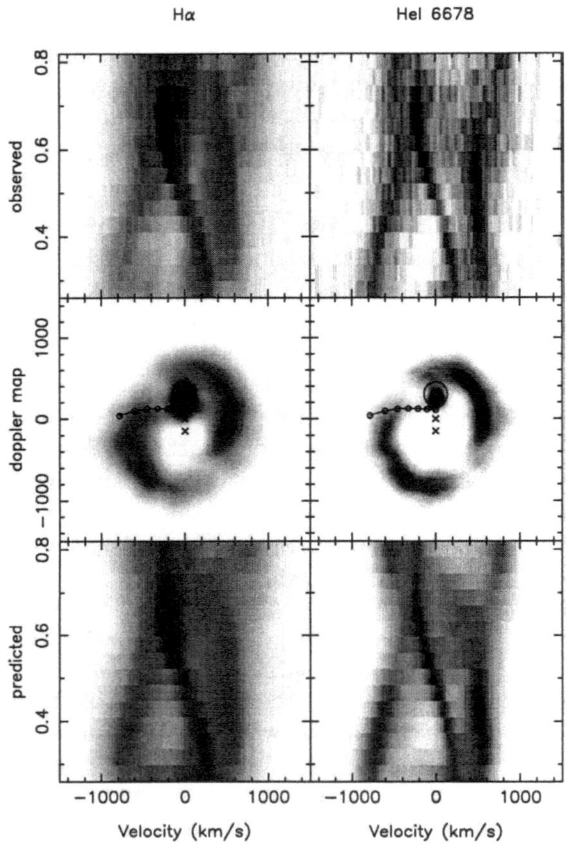

Fig. 4.17. Trailed spectra (data at the top, fits at the bottom) and Doppler maps of IP Peg during outburst show a two-armed spiral pattern [95].

by simultaneously adjusting a spatial image and a position–velocity map to fit spectra of eclipsing systems [7]. In this method, spectra out of eclipse serve to fix the velocity space image as usual, which is then translated to position space through the eclipse information. The technique was able to recover a $V \propto R^{-1/2}$ relation from spectra of V2051 Oph, but as developed it could not handle the difficult case of the same place producing emission at more than one velocity.

4.3.8 Doppler Mapping Summary

Doppler mapping provides us with a fairly objective interpretation of emission lines in cataclysmic variables. It is seen at its best when unravelling the spiral shocks seen in IP Peg where the large-scale of the features makes the spectra hard to understand from intuition alone. The significant caveat is that Doppler tomography is based upon a few simple assumptions which are certainly vio-

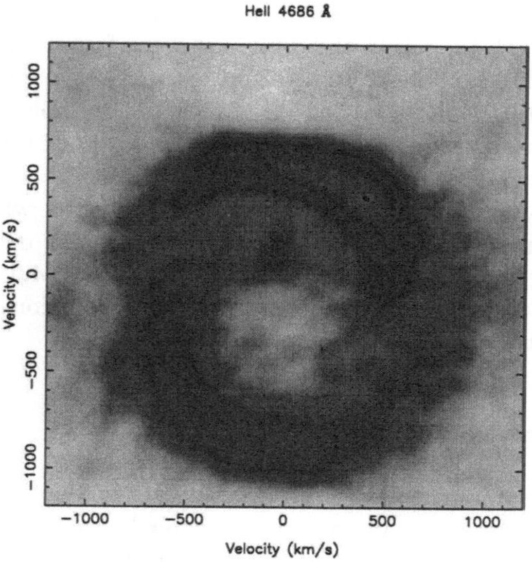

Fig. 4.18. Spiral shocks are still strong six days after the start of an outburst [63].

lated in some, and probably most, systems. It is up to individual researchers to evaluate the significance of departures from the ideal case.

A brief literature search reveals published Doppler maps for some 11 dwarf novae, 17 old novae and nova-likes, 9 intermediate polars and 12 polars, but compared to over 300 systems listed in Ritter & Kolb [75], this is is not so very many. I don't feel that we have reached the stage when we can say what an average Doppler map of a dwarf nova is like. Still less have we explored in a consistent way the evolution of Doppler maps with time or in the light of different emission lines. The commonly-observed Balmer lines may not be so favourable because of intrinsic broadening which may limit the resolution attainable to something much worse than is possible from purely instrumental limitations. Many well-known systems – RX And, AH Her, RU Peg to name a few – have no published Doppler maps, and outburst images are limited to IP Peg, SS Cyg and EX Dra. There is thus plenty of work to do in this area.

4.4 Magnetic Cataclysmic Variables

I have so far concentrated upon systems which accrete via discs around the compact objects. There is however a very different class of systems in which the disc is either truncated with a large hole at its centre or entirely absent, owing to the strong magnetic field of the accretor. These systems are divided into two types, the AM Her stars or "polars" and DQ Her stars or "intermediate polars". The AM Her stars are the most magnetic. They are so magnetic that they have no disc and the white dwarf is locked to the orbit of the binary star. Thus all variations occur on the orbital period. In the DQ Her stars, the white dwarf is not magnetic enough to lock to the orbit and spins faster than the binary. The magnetic field still disrupts the disc, but at least in some systems, not entirely, and so one pictures them as having discs with central holes. The white dwarf spin is seen in photometry of DQ Her stars along with combinations of the spin and orbital frequencies.

I will start by concentrating upon the AM Her stars which have received intensive study in recent years as a result of many new discoveries from the ROSAT X-ray satellite. For reviews of these systems see [104,6,15]. In this section I will concentrate on aspects connected with the earlier sections, namely attempts to map the structures in these systems.

4.4.1 The Accretion Column

Figure 4.19 shows the nature of accretion in an AM Her star. Mass overflows from the mass donor as usual through the inner Lagrangian point, but rather than orbiting the white dwarf, to result in formation of a disc, the material locks onto the white dwarf's field and flows down onto one or both magnetic poles.

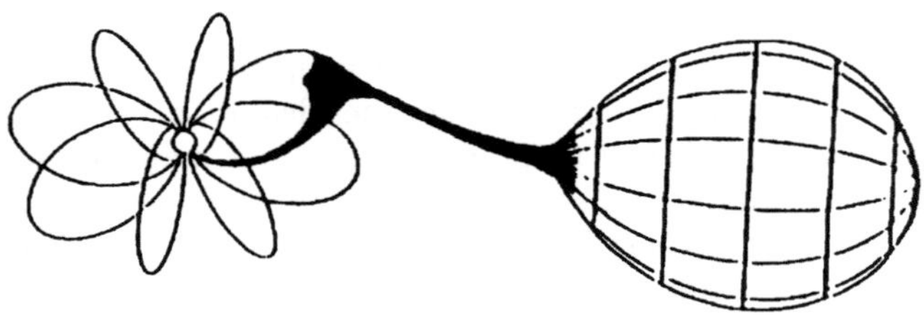

Fig. 4.19. A schematic picture of an AM Her star [15].

Accretion in AM Her stars is very different from disc dominated systems. Rather than spending many orbits slowly spiralling down onto the white dwarf, the gas flows more-or-less in free-fall onto the white dwarf. In doing so it accelerates to close to the escape velocity of the white dwarf (of the order of 5,000 Km/s) but must at some point come to a halt in order to accrete. It does so in a strong shock a little way above the white dwarf's surface, converting most of the kinetic into thermal energy. Protons with velocities of 5,000 Km/s have energies of 130 keV. This energy is transferred to the electrons and radiated as bremsstrahlung at X-ray wavelengths (Fig. 4.20).

At the same time, the electrons are in a very strong magnetic field and radiate cyclotron radiation as well. Most of this comes out at optical and infrared wavelengths. Cyclotron radiation is particularly effective immediately after the shock; bremsstrahlung is increasingly important as the gas cools following the shock. The cyclotron radiation is strongly polarised. When seen face on, circular

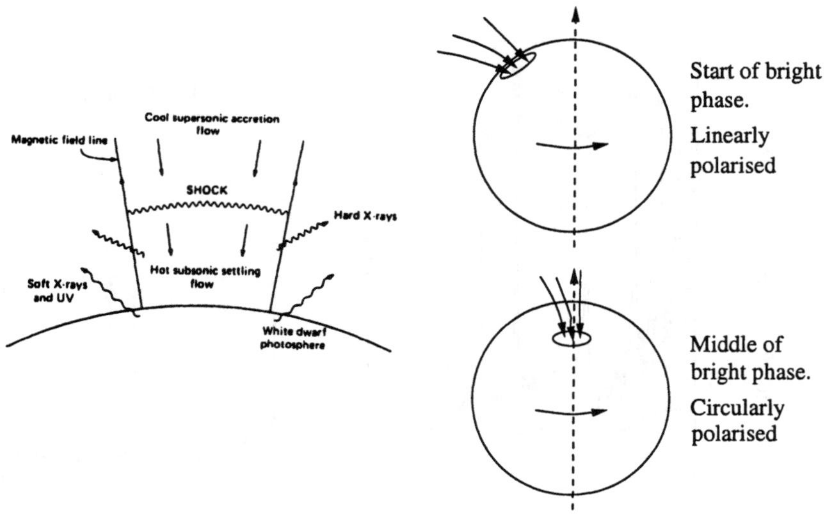

Fig. 4.20. On the left is a schematic picture of the base of the accretion column in a magnetic cataclysmic variable star [15]. On the right are two views of the accretion spot.

polarisation dominates, whereas when seen on the limb of the white dwarf, a pulse of linearly polarised radiation is seen (Fig. 4.20).

The above discussion only scratches the surface of the radiation sources in AM Her stars. There is evidence to suggest that the accretion flow is blobby in nature and that the blobs can carry energy below the photosphere of the white dwarf; this eventually escapes as soft X-ray emission. Adding to this, much of the radiation after the accretion shock will be intercepted by the white dwarf and then re-radiated. Radiation from the spot can also be expected to impinge on other structures in the system such as the accretion stream prior to the shock and the mass donor star.

4.4.2 Magnetic Fields

There are two main indicators of the strength of the magnetic fields in AM Her stars. In their high states, cyclotron emission is significant at optical wavelengths, and it is possible to detect broad spectral humps from cyclotron harmonics (Fig. 4.21). With a broad spectral range it is possible to identify which harmonics the humps correspond to. The magnetic field then follows more-or-less from the standard relation

$$\omega_C = \frac{eB}{\gamma m_e}$$

where ω_C is the fundamental cyclotron angular frequency, e is the charge on the electron, m_e its mass, γ the Lorentz factor and B the magnetic field. For $\gamma \approx 1$

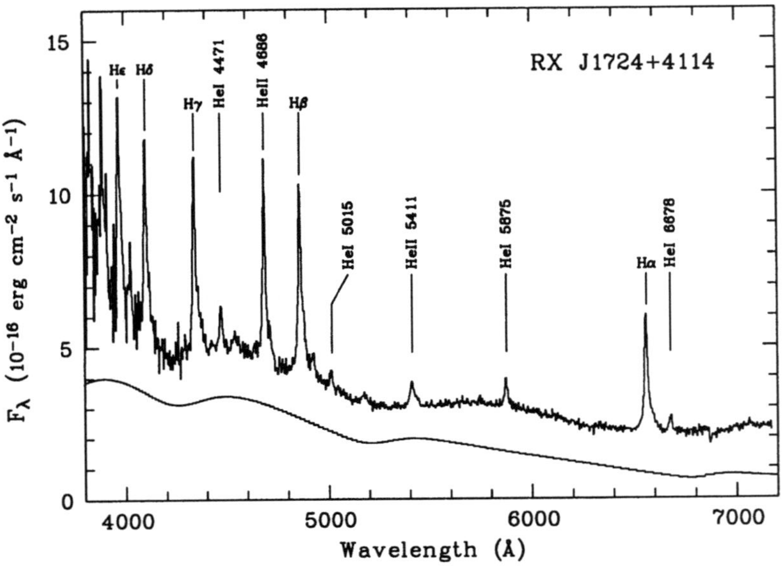

Fig. 4.21. The optical spectrum of the AM Her, RXJ1724.0+4114 showing cyclotron humps.

this leads to

$$\lambda_C = 10.7 \left(\frac{B}{10\,\mathrm{MG}} \right)^{-1} \quad \mu\mathrm{m}$$

where λ_C is the wavelength equivalent to ω_C.

AM Her stars occasionally go into extended low states, during which the photospheres of the white dwarfs can dominate their spectra. Then Zeeman-split absorption lines can be seen, allowing another method of measuring the field. The two methods are not exactly the same since the cyclotron radiation comes almost exclusively from the magnetic poles whereas the Zeeman lines can come from the whole photosphere and some assumed field pattern is needed to deduce the polar field strength. The accretion column is the site of optical line emission, which raises the potential for measurement of magnetic fields from Zeeman splitting and associated circular polarisation in the lines. Only recently has this been seen in the high field system AN UMa [86].

Fields in AM Her stars range mostly from 10 to 100 MG, leading to fundamental cyclotron humps at wavelengths from 1 to 10 μm; the exception is AN UMa which has a field of 230 MG, the current record [86]. The cyclotron emission weakens at high harmonics, part of the reason, presumably, why it is not often seen in DQ Her stars. The field distribution in AM Her stars is very different from that of isolated magnetic white dwarfs where roughly equal numbers of systems are seen in each decade of field strength from 1 to 1,000 MG. It may be that very magnetic white dwarfs are only rarely produced during the evolution of CVs, or that very magnetic white dwarfs speed the evolution of CVs, reducing their apparent numbers.

4.4.3 Light Curves

In their high states, the light curves of AM Her stars are dominated by their accreting pole or poles, with significant contributions from the rest of the white dwarf's photosphere and the gas stream. The energy is radiated close enough to the white dwarf that it can appear and disappear from our view as the white dwarf rotates. The light curves depend, amongst other things, upon the inclination of the system and the latitude and longitude of the accretion spot, and therefore one can hope to measure these parameters from light curves.

Even more can be learned from polarimetry which shows other distinct characteristics as the accretion spot changes orientation. For instance when the accreting pole is at the limb of the white dwarf, linear polarisation is seen with an electric vector parallel to the limb. Consider then the case of an edge-on system with a spot on the equator compared with the same system with a spot at latitude 45°. In both cases, the spot will be visible for half a spin (and therefore orbital) period and therefore not obviously distinguishable from light curves alone. However, the first case will show little change in polarisation angle, while the latter will switch by 90°.

Figure 4.22 shows polarimetry of the star ST LMi. In this case the accreting pole is on the hemisphere furthest from us leading to a short period of visibility; the variation of polarisation can be understood approximately as outlined earlier. An important early result of such analyses was the discovery by Cropper [14] of a clustering in the longitudes of the accreting poles which implied *exact* synchronism between the white dwarf and the binary orbit. This is in spite of a tendency for the white dwarf to spin-up owing to continued accretion of matter. The only plausible explanation for this is dipole/dipole interaction between the white dwarf and mass donor star. On the other hand, to reach synchronism requires a dissipative process such as the induction of currents in the secondary star as the white dwarf's magnetic field sweeps by it. An interesting recent application of these ideas has been made by [61] who suggest that the loss of the secondary star's magnetic field, which may occur at very short orbital periods when it becomes a brown dwarf, could cause loss of synchronism, followed by a short spin-up phase during which the star becomes first a DQ Her star and then a "propeller" – a system that does not accrete but expels matter. This could lead to the observed lack of AM Her stars at the shortest orbital periods.

Absorption by the accretion stream can have a very important effect upon the light curves of AM Her stars. This is especially obvious at wavelengths of high opacity such as the EUV.

Figure 4.23 shows a good example in the eclipsing system HU Aqr. The eclipse appears to vary in width according to the wavelength, and is particularly broad at EUVE and ROSAT wavelengths. The HST UV data is distinctly different but there is a definite dip just before eclipse. The interpretation of this data is that we are seeing the material in the stagnation region where the stream threads

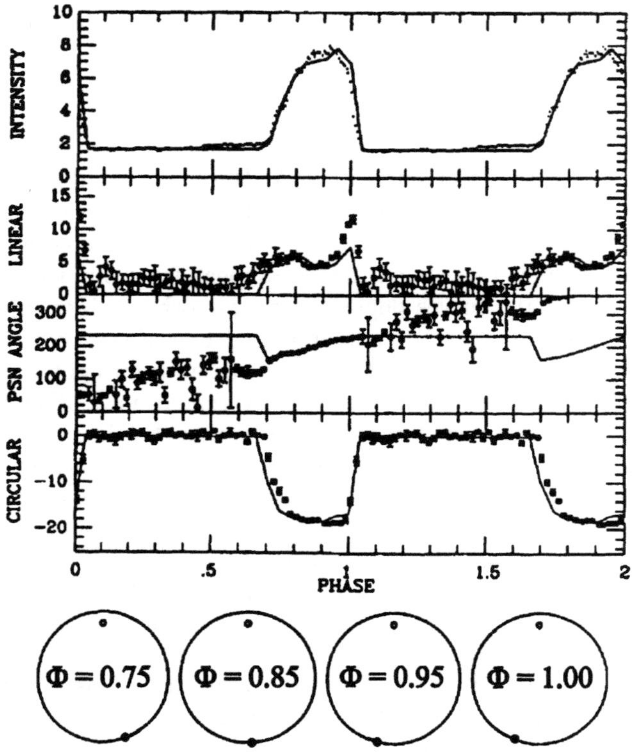

Fig. 4.22. Polarimetry of the AM Her system ST LMi shows a distinct bright phase as the accreting pole (which is placed in the lower hemisphere) comes into view [16].

onto the magnetic field pass in front of the accreting spot just prior to the true eclipse by the companion star. At EUVE wavelengths the stream obscuration is total, whereas it only appears as a dip in the UV. The position of the dip is observed to vary with time and this can be used to measure the change in location of the stagnation spot, which is expected when the mass transfer rate changes [87].

Modelling of the light curves of AM Hers has become sophisticated with essentially similar methods to that of eclipse mapping applied to a situation of very different geometry. Thus assuming a particular flow geometry, as can be fixed by assuming a dipole field geometry and knowing the locations of the accreting spots, one can adjust the brightness along the stream, both ballistic and magnetically-dominated, to match observations [25,26].

This has been taken still further in order to fit polarimetric data [72]. In this case a model of cyclotron emission is required rather than just the usual simple sum over intensity.

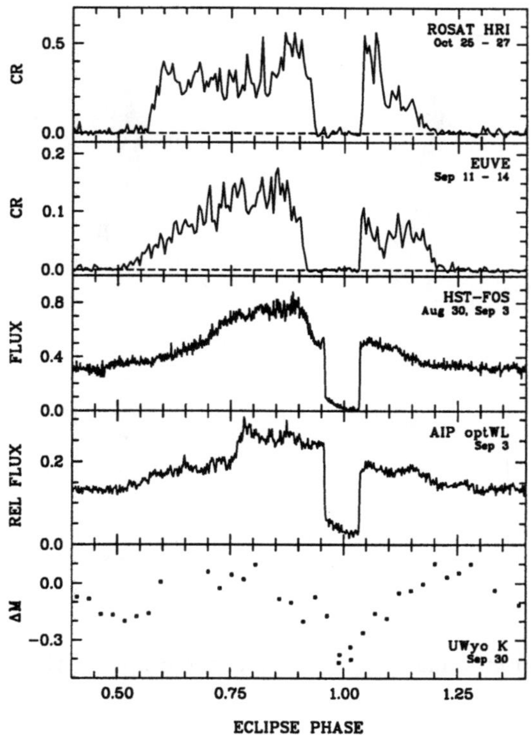

Fig. 4.23. Light curves of the eclipsing AM Her star HU Aqr at different wavelengths [87].

4.4.4 Optical Spectra

In recent years the application of tomography to AM Her stars by Schwope and collaborators has produced some remarkable results. There are obvious risks in naively applying tomography to such systems – we know that there are flows out of the orbital plane, violating one of the core assumptions built into standard tomography. However, in practice such regions do not seem to contribute strongly enough to the line emission to make much difference. In any case the main effect of such out-of-plane motion is a blurring of the reconstruction rather than anything more serious [50].

The system HU Aqr provides a good example of what tomography can tell us – Fig. 4.24.

The contrast with the disc accreting systems is dramatic. The main structures visible are the heated face of the secondary star – there being no disc to shield it – and the initial part of the gas stream. There is also a distinct change between low and high state data. As expected, it seems that the stagnation point moves outwards in the low state to appear as a bright region on the stream. In addition there is also some evidence for emission from the magnetically-dominated part of

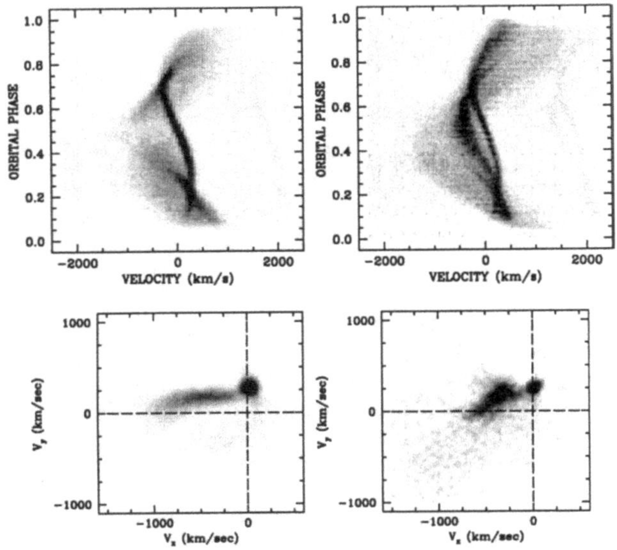

Fig. 4.24. Trailed spectra and Doppler maps of the eclipsing AM Her star HU Aqr [87].

the flow, and relatively simple models of the field threading process can explain the location of this emission in velocity space.

4.4.5 DQ Her Stars

Some researchers differentiate between DQ Her stars and "intermediate polars" on the basis of spin period, with the term "DQ Her star" being reserved for the short spin period systems such as GK Per ($P_{\rm spin} \approx 6\,$mins) and DQ Her itself ($P_{\rm spin} \approx 140\,$secs), while the intermediate polars are longer period systems with, typically, $P_{\rm spin} \sim 0.1 P_{\rm orb}$. Although they could well have different magnetic field strengths, it is not at all clear that there is anything that fundamentally distinguishes the two, and so I will use the term DQ Her to refer to all such systems, distinguished by their coherent pulsations on periods other than the orbital period.

DQ Her stars are not as well understood as their more magnetic counterparts. This is probably mostly because they are more complex. DQ Her stars show pulsations so they must have an inner magnetically dominated region. On the other hand, some of them show disc-like features such as double-peaked lines (most obviously DQ Her itself), and so they may have disc-like structures. On the other hand, X-ray data of DQ Her stars often show a signal at the "beat" frequency $\omega - \Omega$ between the spin and orbital frequencies ω and Ω, which indicates that the accretion spot "knows" where the secondary star is [103,67,4].

This suggests that there is stream overflow – matter streaming over the disc and hitting the magnetosphere directly. Such components are variable, so it looks as though sometimes there is stream overflow and sometimes there is not.

To add to the complication, despite stream overflow, much of the accretion may occur as material from the inner edge of the disc connects to the field. This is not as clean a process as occurs in the AM Her stars and will lead to an accretion "curtain" rather than a spot. This accretion curtain will obscure the accretion areas on the white dwarf for a wide range of inclinations. The obscuration is most marked at X-ray wavelengths and means that the X-rays peak when the accreting regions are pointing *away* from us. The situation described is illustrated in Fig. 4.25.

The precise behaviour is likely to vary markedly with details of geometry. Add to this the difficulty in many DQ Her stars of detecting the mass donor star, and thus the orbital phase, and a rather confusing picture emerges.

The spin periods of DQ Her stars. Any discussion of DQ Her stars would be incomplete without presenting a plot of their spin versus orbital periods, shown in Fig. 4.26.

There has been a great deal of work trying to explain this diagram, which was partly responsible for the intermediate polar/DQ Her division discussed earlier because when there were fewer systems known, it appeared that the systems that cluster below the $P_{\rm spin} = 0.1 P_{\rm orb}$ line were distinctly different from the shorter spin period systems such as DQ Her itself. What needs explaining is how the white dwarfs manage to attain an equilibrium spin period. On the one hand accretion of angular momentum acts to spin them up, while on the other the magnetic field drags on the inflowing material, which, at least when it is far out, will tend to slow the spin down.

Material orbiting the white dwarf only slows the latter down if it is outside the co-rotation radius where the white dwarf's spin period matches the orbital period at that radius. A key factor is the relative size of this radius compared to the magnetospheric radius at which the magnetic and ram pressures balance. The latter radius is dependent upon the strength of the magnetic field but also upon the mass transfer rate, which may vary. This can lead to some interesting consequences. For instance, suppose that a white dwarf is spun up under high mass accretion which subsequently decreases. It may happen that the magnetospheric radius then becomes larger than the co-rotation radius. A possible consequence is that most material is flung away from the white dwarf rather than accreted by it. The white dwarf loses energy in the process. Exactly this may be occurring in the peculiar system AE Aqr in which the observed rotational power loss implied by the spin-down of the white dwarf greatly exceeds the total electromagnetic energy that we observe from the system [110]. Such systems are the "propellers" I referred to earlier when discussing the evolution of AM Her stars. AE Aqr is therefore *not* near a state of equilibrium at present.

Wynn & King [109] have modelled the process of accretion in magnetic systems by introducing a drag term for material crossing field lines. This was used

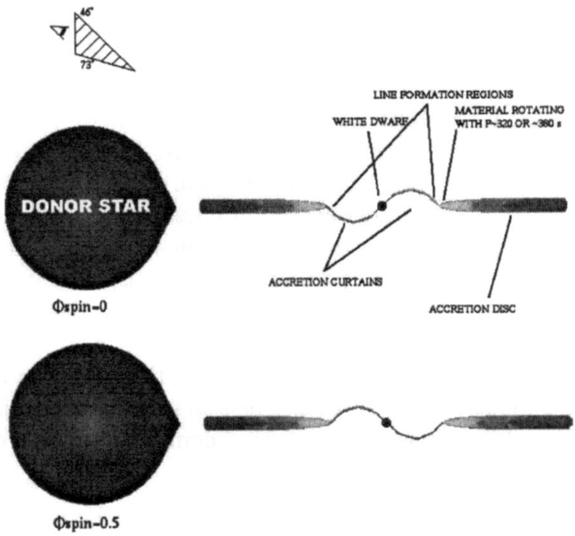

Fig. 4.25. Schematic picture of a DQ Her star (from [62]).

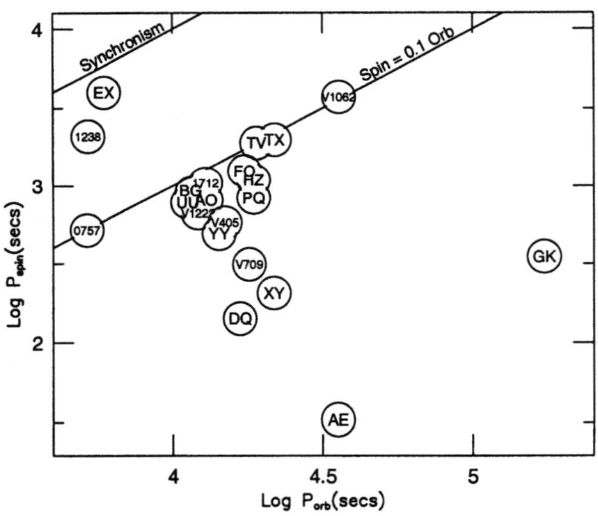

Fig. 4.26. Spin versus orbital periods of DQ Her stars (from [29]).

to interpret AE Aqr, and more recently in understanding EX Hya, which has a spin period 2/3 as long as the orbital period, and which is explained by having a magnetospheric radius comparable to the distance to the inner Lagrangian point [44]. These models are fascinating for the variety of new phenomena they predict, but I think it is fair to say that they are only weakly backed up by observations. Part of the problem here is that the models predict dynamics but not the brightnesses of the various components and it may be that components specific to the DQ Her nature of the stars are not especially prominent. The reader is referred to King & Wynn [44] for more details.

Tomography of DQ Her stars. Tomography of DQ Her stars is in its infancy, and presents new problems. If we are looking for features specific to DQ Her stars, as we surely should be, we must tackle the issue of multiple periods. Normal tomography will give us an average view of the system. A better approach is only to image the system with spectra taken when the relative orientations of the white dwarf and binary are the same, or approximately the same [56]. This can be achieved by selecting all spectra of similar "beat" phase (i.e. phased on the $\omega - \Omega$ frequency). Since this involves selecting only a fraction of the spectra, it is clear that acquiring the data for tomography of DQ Her stars is significantly more challenging than for either AM Her stars or non-magnetic CVs. The situation is even more difficult than this suggests however. Much of the variation in DQ Her spectra may be caused by optical depth effects in the accretion curtain near the white dwarf. This occurs on the white dwarf's spin period – the orbital phase is relatively unimportant – and suggests that tomography is best carried out by selecting spectra of constant spin phase. Such data show evidence for emission from curtains on one or both poles [30]. We have two interpretations here for the same data. In one case maps are calculated by selection of the same beat phase, and in the other on the same spin phase. It is not clear which is to be preferred, and probably both contain an element of truth but not all of it.

4.4.6 Magnetic CV Summary

In the past few years many new AM Her stars have been discovered from ROSAT data. The new discoveries have led to impressive advances in the field. Optical spectra and light curves have probed the ballistic stream and threading region of the flow far from the white dwarf; X-ray light curves, optical photometry, spectroscopy and polarimetry have revealed the accretion shock and magnetic white dwarf. There is much scope for more work of this nature. For example, there are a small number of AM Her stars which are not quite synchronised and can be expected to change structure as the white dwarf slowly rotates relative to the binary; these stars could reveal fascinating aspects of how gas threads onto the magnetic field. Many AM Her stars are optically faint and will require the use of the new generation of large aperture telescopes. Advances in X-ray satellites are finally making X-ray spectroscopy of reasonable resolution a reality which will have a substantial impact upon our understanding of accretion columns in these stars.

Progress on the related DQ Her stars has been slower. These are intrinsically complex stars, and lack most of the diagnostics available in the AM Her case. In many cases for instance, we do not even know the orbital phase for these stars for sure. It is also likely that DQ Her stars are not homogeneous in the sense that some have discs whereas others might not. With the diamagnetic blob models of Wynn & King [109], I have the impression that theory has crept rather ahead of observations. Nevertheless, the variety of possible phenomena, and the likely evolutionary connections between DQ Her and AM Her stars make it imperative that we continue to study these enigmatic objects.

4.5 Double-Degenerate Binaries

In the last section I change subject to look at a set of non-interacting binary stars consisting of close pairs of white dwarfs. These are also known as "double degenerates". My interest in these stars is for what they can tell us about the evolution of binary stars. These systems were predicted to exist in substantial numbers some time ago [39,106], moreover they were suggested as potential progenitors of Type Ia supernovae. By the early 1990s, after several searches had turned up just one system [76,18,10] in addition to another found by chance [83], this possibility was looking less likely. However, since that time several more systems have been found and, while other objections can be raised to double-degenerates as the progenitors of Type Ia supernovae [112], the intrinsic numbers are probably not a problem. Indeed, these systems are probably at least a factor of 10 more common than CVs.

4.5.1 Progenitors of Type Ia Supernovae

Type Ia supernovae (SNe) are very bright explosions of stars that have attracted much attention in recent years as a result of their use in cosmology. Observations of Type Ia SNe at redshifts $z \sim 1$ appear to show that the expansion rate of the universe may be increasing with time, indicative of a significant cosmological constant [70,74]. This has led to increased interest in the explosions themselves, if only to check that the cosmological result is not an artifact of the evolution of Type Ias with redshift.

Type Ia SNe are characterised by an absence of hydrogen and helium in their spectra at any time. They also occur in all types of galaxy, in contrast to the better-known Type II SNe which are not seen in elliptical galaxies. These two features have led to general agreement that Type Ia SNe are caused by explosions of the degenerate material of white dwarfs [11]. Since white dwarfs do not explode on their own, these white dwarfs must be accreting within binary systems. Explosions may occur when a white dwarf exceeds its Chandrasekhar limit, with carbon igniting at the centre. Indeed at one time the perceived similarity of Type Ia SNe was used as an argument in favour of Chandrasekhar-limit explosions. This has since become less compelling as it is now known that there are significant variations amongst Type Ia SNe. Other possibilities exist. Hydrogen dumped onto a white dwarf can explode once 10^{-5} – $10^{-4} M_\odot$ of material has accumulated; we know such explosions as classical novae. If the same process takes place with helium, much more material is needed ($\sim 10^{-1} M_\odot$) before an explosion occurs. As a result it is much bigger, and could even trigger the rest of the underlying white dwarf to explode as well. These are known as Edge-Lit

Detonations (ELDs), but whether they occur or not depends upon the accretion rate, with higher rates reducing the chance of explosive ignition.

Modelling of such explosions is difficult, and it is not yet clear which initial circumstances will lead to something approximating the explosions that we see. This makes it hard to decide what sort of binary system produces Type Ia SNe. Some, however, can probably be ruled out. For instance, CVs suffer from the violent thermonuclear runaways of hydrogen referred to above. These are seen as classical novae which if anything erode the mass of the white dwarf and prevent it ever exploding as a Type Ia SN. The unstable burning that leads to classical novae occurs at accretion rates below about $10^{-7} M_{\odot}\ yr^{-1}$. Much above this rate the white dwarf cannot cope and develops an extended envelope, becoming essentially a red giant once more. Even if such an object did explode, it would not be seen as a Type Ia SN. We are left with a small range around from 1 to $4 \times 10^{-7}\,\mathrm{M_{\odot}\,yr^{-1}}$ in which steady accretion and hydrogen burning can take place with little expansion of the white dwarf. In the early 1990s a new class of luminous (10^{29} – 10^{31} W) X-ray source – the Super-Soft Sources (SSSs) – were recognised [99,19] – see [42] for a review. Apart from their high luminosities which are close to the Eddington limit for a $1\,\mathrm{M_{\odot}}$ object, these sources are characterised by very soft X-ray spectra which peak at around 40 eV. They are now widely believed to be white dwarfs undergoing steady accretion and nuclear burning. It is the relatively large area of a white dwarf compared to a neutron star that explains their soft spectra compared to the better-known low mass X-ray binaries. Since the potential energy/unit mass at the surface of white dwarfs is relatively small, SSSs are dominated by fusion power rather than accretion power. This means that their high luminosities require lower accretion rates than if only accretion was taking place, and this explains why we can see the X-rays at all [100]. SSSs are now one of the leading candidate progenitors of Type Ia SNe [112]. The high accretion rates of these objects are thought to stem from thermal-timescale mass transfer from stars somewhat more massive than the white dwarfs, e.g. 1.3 – $2.5\,\mathrm{M_{\odot}}$ [42]. They are in fact "failed" CVs in the sense that their mass ratios do not allow the low mass transfer rates of CVs since mass transfer from one star to a less massive one shrinks the binary orbit. The SSSs are an attractive class of progenitor, except that the small range of stable burning accretion rates may make it difficult for enough Type Ia SNe to be produced by this route (but see [21]), and they may not last long enough to produce SNe in elliptical galaxies. Thus we should keep in mind alternatives, with double-degenerates probably the leading contenders.

All binary stars are predicted to emitt gravitational radiation, losing orbital energy and angular momentum in the process. This process makes the binary star's orbit tighter and will ultimately lead to merging. Two stars of masses M_1 and M_2 in a circular orbit of period P hours will merge on a timescale of

$$t_m = 1.0 \times 10^7 M_G^{-5/3} P^{8/3}\,\mathrm{yr} \tag{4.9}$$

where M_G is known as the "chirp mass" (in solar units), and is given by

$$M_G = M \left(\frac{\mu}{M} \right)^{3/5} ,$$

where $M = M_1 + M_2$ is the total mass and μ is the reduced mass given by $1/\mu = 1/M_1 + 1/M_2$. This loss is insignificant for most binary stars, but becomes important for periods of a few hours or less. If the two stars are white dwarfs, once the orbital period is of the order of 100 s, the lighter white dwarf will undergo rapid Roche lobe overflow onto its companion, and we have the potential for explosions to occur. Two carbon-oxygen (CO) white dwarfs may lead to a combined object which exceeds the Chandrasekhar mass, while CO+He double degenerates (DDs) may lead to ELDs. The merger is a violent process which will heat the material. While this is needed to ignite the material it can reduce the degeneracy and make the subsequent explosion less violent. In fact it is not clear that an explosion will occur at all, with a slower transition to a neutron star, or perhaps relatively sedate burning as an R CrB star amongst alternative outcomes. Nevertheless, since the details of the explosions are not well understood, DDs must remain as potential progenitors.

Until the 1990s, DDs suffered from another problem – they were not common enough. What should they look like? As pairs of detached white dwarfs, DDs are to be found amongst the white dwarfs. Systems capable of leading to Type Ia SNe need to have short enough periods to merge within the age of the Galaxy. This generally implies an orbital period $P < 10\,\mathrm{h}$, and therefore substantial radial velocities for the component stars. This has been the principle behind all searches for DDs. However, as mentioned at the start of this section, these searches [76,18,10], which covered 44, 25 and 54 white dwarfs respectively, only uncovered one DD (WD 0957-666) in addition to another found by chance (L870-2, [83]), and neither of these had a period short enough (at least when first measured – see [64]) to be representative of a Type Ia progenitor.

The failure to find any possible progenitors was taken to be a serious blow against DDs as Type Ia progenitors [76] on the basis of an estimated local Type Ia rate of $6.8 \times 10^{-14}\,\mathrm{pc}^{-3}\,\mathrm{yr}$ [98]. However, there were problems with the analysis: first, as Robinson & Shafter recognised [76] and was later confirmed theoretically [40], a significant fraction of DDs may merge so quickly that they would be too hot to show the absorption lines needed for detection. More simply, the Type Ia rate they used was probably too high, since multiplying by an effective volume for the Galaxy of $5 \times 10^{11}\,\mathrm{pc}^3$, gives a total rate in the Galaxy of one every 30 years, whereas it is more commonly thought to be about 10 times lower than this [112]. Instead binary "population synthesis" studies, in which primordial binaries are evolved to estimate the probabilities of various outcomes, suggest that only 1 in ~ 400 white dwarfs need be Type Ia progenitors to match the observed SN rate [40]. Coincidentally, these models predict a total merger rate of order 1 every 30 yr [22], but most of these presumably do not lead to Type Ia supernovae. Judged against this fraction, the failure of the early searches which covered ~ 100 systems is hardly surprising. The question remains however, how many DDs are there?

4.5.2 Finding Double-Degenerates

The fraction of 1 in 400 mentioned above refers to systems with short enough periods to merge within a Hubble time and which have enough mass to exceed the Chandrasekhar limit [40]. However, there are *many* more of longer period and lower mass. Indeed the fraction of white dwarfs that are DD without qualification is predicted by the same models to be more like 1 in 10. Finding them is a little harder than finding *bona fide* Type Ia progenitors since longer period and lower mass systems have smaller radial velocities, but it is still not difficult. We now know some 15 of these systems.

The breakthrough started when Bergeron, Saffer & Liebert [5] published a spectroscopic study of white dwarfs. They fitted blue spectra to determine gravities and effective temperatures of their targets, and then translated these to masses using cooling models (Fig. 4.27). The possibility of deducing masses from spectra alone is one of the great advantages of white dwarfs.

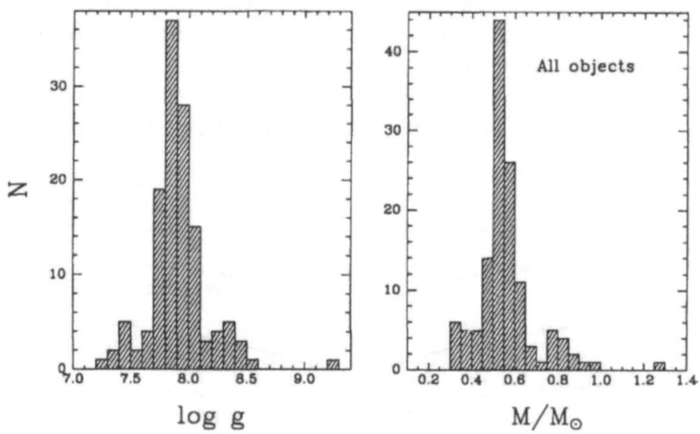

Fig. 4.27. The gravity and mass distributions of the white dwarfs studied by Bergeron, Saffer & Liebert [5].

Of 129 targets, 14 had masses below $0.45\,M_\odot$, well below the typical mass of white dwarfs of $\approx 0.55\,M_\odot$. There has not been enough time for stars below $0.8\,M_\odot$ to become white dwarfs and it is thought that the minimum mass of white dwarfs from stars of higher mass is similar to the observed peak at $\sim 0.55\,M_\odot$, although this is dependent upon uncertain details of mass loss on the red giant branch (RGB) and asymptotic giant branch (AGB). There is however a simple way of producing white dwarfs of lower mass – by mass loss within a binary star. If a star within a binary comes into contact with its Roche lobe before reaching the tip of the RGB it may end as a white dwarf of low mass and mainly helium composition.

In the case of low mass white dwarfs, we know that any companion is either an M dwarf or brown dwarf or another compact object, since any other star would be brighter than the white dwarf that we see. Stars of these types cannot accrete at high rates and therefore it is likely that they were overwhelmed in the most recent mass transfer phase that led to the emergence of the low mass white dwarf. If so then the most recent phase would have been a *common envelope phase*. During the common envelope phase both stars orbit within a single envelope. They lose energy and angular momentum to the envelope until the latter is dissipated and the binary emerges considerably tighter than before mass transfer started. This leads to the following conclusion: if low mass white dwarfs are indeed the products of binary evolution, and assuming that the two stars did not merge, they will now be *close* binary stars and probably straightforwardly detectable as such.

With this motivation Marsh, Dhillon & Duck [54] studied 7 of the low mass white dwarfs found by Bergeron, Saffer & Liebert [5], and found that 5 of them were binary stars, a startling success rate compared with the earlier searches for Type Ia SN progenitors [76,18,10]. There were several reasons for this. First, the selection of white dwarfs of low mass paid off as they are indeed predominantly made within binary stars. Second, the first observing run was spaced in time to raise the sensitivity to long period systems. Third, and probably most importantly, we observed Hα at moderately high resolution $R \sim 8,000$ which allowed us to use the sharp non-LTE core shown by Hα but not the other Balmer lines in white dwarfs (Fig. 4.28).

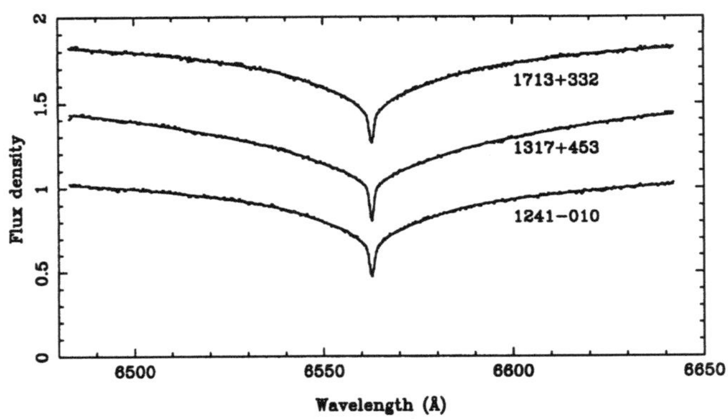

Fig. 4.28. Mean spectra of three double degenerates corrected for orbital motion along with the multi-gaussian fits used to measure their velocities [54].

Subsequent work [55,64,84] has raised the total number of DDs of known orbital period to the 15 quoted above. Along with these several sub-dwarf B (sdB)/white dwarf binaries have been discovered [84,65,68] – these systems are probably immediate precursors to DDs since it is thought that the sdB star will evolve into a white dwarf. Amongst the systems found are several with periods short enough to merge well within a Hubble time. The current record-holder is WD 0957-666, which with $P = 88\,\text{min}$ will merge 200 million years from now

[64]. WD 0957-666 consists of two white dwarfs with masses of $\approx 0.33\,M_\odot$ each and so is unlikely to become a supernova — what it will become is an interesting question. One system, WD 1704+481, has an orbital period of $1/7^{th}$ of a day and appears to consist of a CO/He white dwarf pair. If Type Ia SNe are ELDs then WD 1704+481 is a possible progenitor [60].

Whatever these systems do become on merging, it looks as though their numbers are not a problem, as I now look at in a little more detail.

4.5.3 The Space Density of DDs

While an improvement upon the situation a decade ago, the total number of DDs known pales beside the large number of CVs (> 300) with measured orbital periods and yet intrinsically DDs are more numerous by far. The space density of CVs is of order $3 \times 10^{-5}\,\mathrm{pc}^{-3}$, although most of them are probably very faint and may not actually appear very much like the brighter CVs that we know.

The space density of DDs is probably best obtained by considering the fraction of white dwarfs that are DD. I have already quoted a number of 1 in 10 obtained from theory, and several observations support this. First, there are the 14 low mass white dwarfs out of 129 from Bergeron, Saffer & Liebert [5], most of which are DD. Second, Saffer et al. [84] carried out a radial velocity survey of 107 white dwarfs, finding evidence for variability in 13 of them. Some of their detections are probably false alarms [59], but at least 7 of them are confirmed. Maxted & Marsh [59] carried out a smaller but more sensitive survey, and managed to constrain the DD fraction to the range 2 to 19%. This needs improving, but so far all lines of evidence are consistent with a 10% fraction.

The space density of white dwarfs has been measured from proper motion surveys [46,45] and from associations with nearby stars [69]. The total of these is near $10^{-2}\,\mathrm{pc}^{-3}$, making the DD space density $10^{-3}\,\mathrm{pc}^{-3}$, of order 30 times that of CVs. Of course most of them are several billion years old and very faint, but radial velocity studies are possible at least to an age of a billion years, about 10% of the age of the Galactic disk, and therefore there should be of order 400 DDs within 100 pc that can be detected as such. At this distance a white dwarf of this age will have $V \sim 16$, well within reach of relatively modest telescopes.

4.5.4 Gravitational Radiation from DDs

No matter how old they are, DDs will continue to emitt gravitational waves, and their high space density supports the suggestion that gravitational radiation from DDs may dominate the low frequency gravitational wave spectrum ($\nu < 10^{-2}$ Hz) that is potentially observable with space-based interferometers such as LISA [32,71]. Given the space density of DDs, there are of order 500 million in the Galaxy and since interferometers like LISA have little directionality, they could be a major source of noise for observations of other sources. The majority will be too closely spaced in frequency to be resolvable, but at high frequencies it should be possible to see individual sources.

As described earlier, optical searches for Type Ia candidates are insensitive to systems formed at very short periods. Gravitational wave observations avoid this problem entirely. To see this more quantitatively, assume for simplicity that all DDs have identical masses, so that the present day merger rate is given by

$$m = \int_0^\infty b(P, t_m)\, dP$$

where $b(P,t)\, dP\, dt$ is the birth rate of systems with periods from P to $P + dP$ between times t to $t + dt$ ago. Here $t_m(P)$ is the time for a DD of period P to merge, as given in Eq 4.9. Of course at long P, t_m exceeds the age of the Galaxy and the birth rate drops to zero, and so DDs of long period are not directly relevant to the merger rate.

We cannot observe $b(P,t)$ directly but instead can constrain the distribution

$$f(P,t)\, dP\, dt,$$

representing the number of existing DDs with periods between P to $P+dP$ and ages between t to $t + dt$. The two are related by

$$f(P,t) = \left(\frac{P}{P'}\right)^{5/3} b(P',t),$$

where P' is the orbital period at birth, a time t ago, and is given by

$$P'^{8/3} = P^{8/3} + \alpha t,$$

where α is a constant depending upon the stellar masses. Optical searches can constrain $f(P,t)$ over the entire range of periods of interest, but only over restricted ranges of t since very young and very old white dwarfs have weak absorption lines making radial velocity measurements difficult. A lower limit t_l for $f(P,t)$ translates to a lower limit on the value of P for $b(P,t)$. We therefore cannot constrain $b(P,t)$ observationally at short periods because these systems do not last long enough to get into the radial velocity surveys. On top of this to measure $b(P,t_m)$, we must actually measure $f(P,t_m)$ and $P \to 0$, but given the $(P/P')^{5/3}$ factor and that P' tends to constant as $P \to 0$, such short period systems are extremely rare. These uncertainties can only be avoided by relying upon the results of theoretical models of $b(P,t)$, effectively extrapolating from the long periods which we can measure. This is something that we would prefer to avoid given the problems with theory discussed in the next section.

Potentially gravitational wave observations can avoid this problem entirely since they can detect the short period systems which will merge within the next few thousand years, irrespective of their age. Thus in one stroke they can be used to determine the current merger rate. In order to determine the merger times we will need to know the values of M_G; this can be done if it is possible to measure both periods and period derivatives of the shortest period systems. If space-based interferometers can be realised, there is considerable potential for learning about DDs.

4.5.5 Double Degenerates as a Test Population for Binary Star Evolution

The difficulty of finding DDs is down to their similarity to single white dwarfs. Luckily, there are enough of them that simply searching known white dwarfs is feasible. Most importantly, although it is hard work finding DDs, one can quantify the selection effects that go into their discovery in a way impossible with the CVs except at the most rudimentary level.

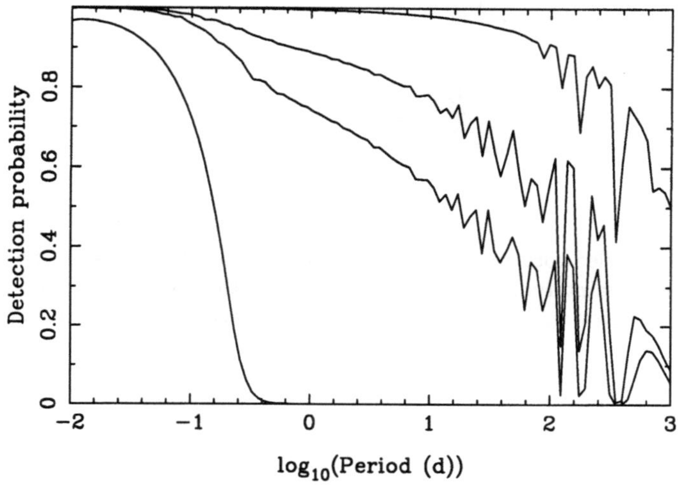

Fig. 4.29. Detection probabilities (99.99% confidence level) versus orbital period are plotted for four sampling strategies. The lowest curve shows the case of a single pair of spectra, each of precision 5 Km/s separated by 10 minutes. The next curves show the effects of adding identical pairs of spectra every year for periods of one year, two years and finally ten years. The calculations are based upon stellar masses of 0.5 $M_{\odot}$.

Figure 4.29 for instance shows examples of the detection probability versus orbital period for various hypothetical radial velocity observations. What is significant about this figure is first that it can be calculated at all, and second that good detection probability can be maintained up to periods > 100 d, covering the entire range of theoretical interest. Add to this that it is often possible to see both components of DDs and that both masses and ages (as white dwarfs) can be estimated from spectra alone (avoiding the need to know the orbital inclination), and it can be seen that DDs furnish us with a rich a set of constraints for testing theory.

The advantages of DDs have already made themselves felt. The fainter component has been detected in six systems so far, making measurement of their mass ratios possible. Fig. 4.30 shows an example of such a system.

The surprise of the observed mass ratios is that most of them are close to, or greater than, unity. The observed mass ratio is defined as the mass of the brighter

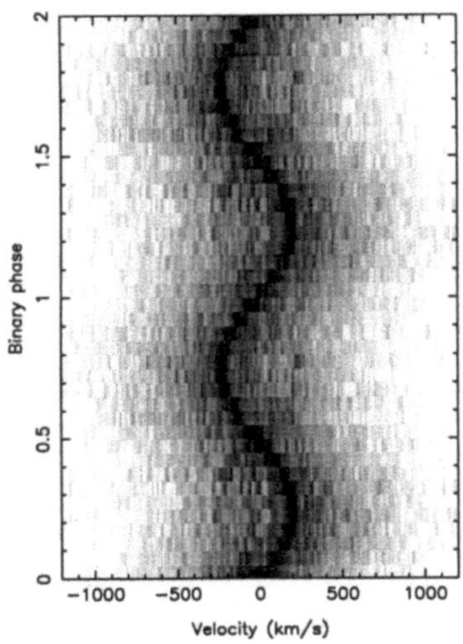

Fig. 4.30. The trailed spectra of the DD WD 0957-666 in the left panel shows two components, with one fainter and in anti-phase with the other. The right-hand panel shows the result of subtracting a fit to the brighter component to emphasize its companion [64].

star divided by the mass of its companion. To first order the brighter white dwarf should be the younger. In population synthesis models the binary often goes through two common envelope phases. The first of these shrinks the binary and the the younger white dwarf therefore forms in a tighter binary than did its companion. As a result the younger white dwarf should be the least massive and the mass ratios should be less than 1. For this reason theoretical distributions peak around $q = 0.5$ to 0.6 [40,22]. The observed mass ratios therefore suggest that, on the contrary, the first mass transfer event does not usually shrink the binary, and may even expand it since two systems have mass ratios significantly in excess of unity. Algol-type mass transfer can do this and therefore may be more significant compared to common envelope evolution than theory originally suggested. This is a significant discrepancy: Han [22] was only partially successful in trying to modify theory specifically to match the observed mass ratios – see Fig. 4.31.

Another comparison of data and theory is presented in Fig. 4.32 which shows orbital periods and primary masses (from model atmosphere fits).

Theory does not convincingly match the observations. In particular too many white dwarfs of very low mass are predicted. There is no obvious way in which such white dwarfs could be hidden, and yet it is also difficult to see why they should not be produced.

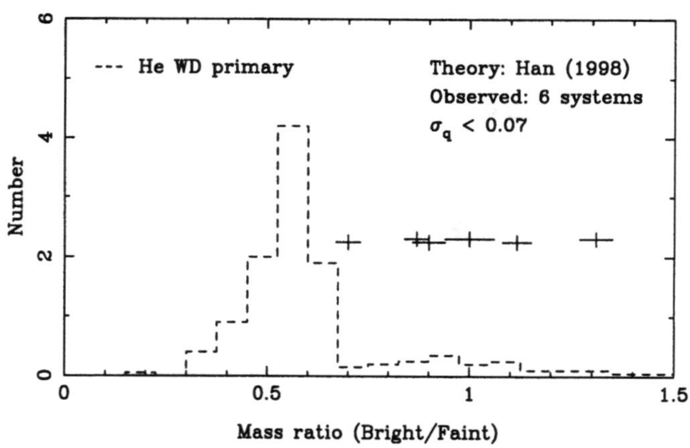

Fig. 4.31. The observed mass ratios are compared with the distribution for helium white dwarf primary stars from Han's models [22]

4.5.6 Double Degenerates Summary

Double degenerates are simple systems which, although hard to detect, are intrinsically common. The lack of strong selection effects and, in many instances, the accuracy of system parameters, makes them a good class of binary for testing binary star evolution. Indeed, this has already proved effective with the high fraction of near equal mass-ratio systems and the absence of very low mass white dwarfs particular problems for theory at present. Whether DDs are Type Ia progenitors remains unknown, but it can no longer be convincingly argued that there are not enough of them. Even the lack of any systems with total masses above the Chandrasekhar mass is probably not a problem as the systems known at present are dominated by the searches of low mass white dwarfs.

4.6 Conclusions

In these lectures I have endevoured to cover recent work on cataclysmic variables and double degenerate binary stars. Indirect imaging techniques have been applied widely in the field of CVs and have sharpened the comparison between theory and observation of these systems. It would be nice to replace these with physically accurate models involving only a few parameters, but it does not look as though this will be possible in the near future at least. Several discoveries have only been possible as a result of indirect imaging: in particular spiral shocks in outbursting dwarf novae and the line emission from stream-threading in AM Her systems.

There is much work to be done in continued application of these techniques to a larger range of systems and more intensively to individual systems in a range of states. While some cataclysmic variables are very similar to each other, their differences can often be more surprising. What makes one star have especially

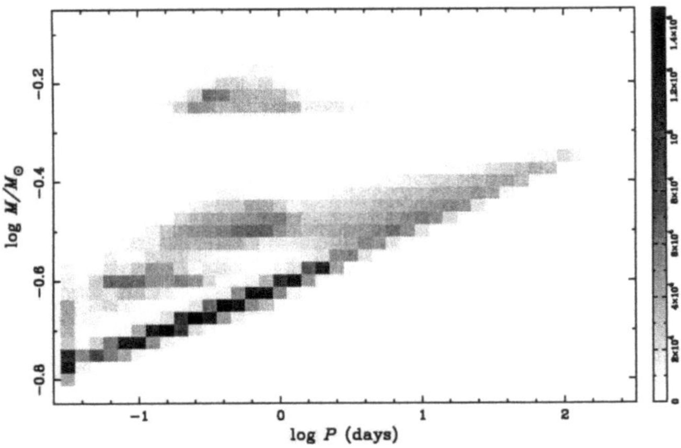

Fig. 4.32. The observed periods and primary masses are compared to theory [84]; stars are sdB/white dwarf binaries mostly plotted with a nominal mass of 0.5 $M_\odot$.

strong line emission and spiral shocks in outburst? Why do some show much stronger continuum and/or line-emission from the bright-spot than others? Why is the ballistic stream obvious in some AM Her stars but not others? Some of these differences may be trivial, but others may be significant, and only continued observation can help.

A significant worry in interpreting observations is the thickness of the discs in non-magnetic CVs. Just how they can be as thick as they sometimes appears is not clear. Many peculiarities of behaviour may be a result of this phenomenon, about which, so far, theory has had little to say. Although it may well be a detail in terms of the overall structure of CVs, its effect upon eclipsing systems may be disproportionately significant since so much of our knowledge of CVs has come from these systems.

The evolution of CVs is an area that may well benefit from large-scale optical surveys over the next few years. I hope that it can be connected to the evolution of other large populations of binary stars as only then can theories be truly tested. I have described one such population – the double-degenerates – and have given examples of how it challenges theory. Larger telescopes will bring many more of these systems within our reach and should tech us much about how they came to be.

Appendix

In this appendix I show that, as stated in section 4.3.3, Eq. 4.7 can be inverted by application of the filter $|s|/G(s)$ followed by the back-projection of Eq. 4.8. I define the Fourier transform $F(s)$ of a function $f(x)$, and its inverse by

$$F(s) = \int_{-\infty}^{\infty} f(x) e^{-i2\pi s x}\, dx.$$

$$f(x) = \int_{-\infty}^{\infty} F(s) e^{i2\pi sx}\, ds.$$

The frequency s here is measured in cycles per unit x. Now take the Fourier transform over V of the line profile equation, 4.7:

$$F(s,\phi) = \int_{-\infty}^{\infty} f(V,\phi) e^{-i2\pi sV}\, dV \tag{4.10}$$

$$= \int_{-\infty}^{\infty}\int_{-\infty}^{\infty} I(V_x,V_y) \int_{-\infty}^{\infty} g(V-V_R) e^{-i2\pi sV}\, dV\, dV_x\, dV_y \tag{4.11}$$

$$= G(s) \int_{-\infty}^{\infty}\int_{-\infty}^{\infty} I(V_x,V_y) e^{-i2\pi sV_R}\, dV_x\, dV_y. \tag{4.12}$$

Dividing through by $G(s)$, multiplying by $|s|$ and taking the inverse Fourier transform gives the filtered line profiles

$$\tilde{f}(V,\phi) = \int_{-\infty}^{\infty} \frac{|s|F(s,\phi)}{G(s)} e^{i2\pi sV}\, ds$$

$$= \int_{-\infty}^{\infty}\int_{-\infty}^{\infty} I(V_x,V_y) \int_{-\infty}^{\infty} |s| e^{-i2\pi s(V-V_R)}\, ds\, dV_x\, dV_y. \tag{4.13}$$

Finally, back-project these filtered profiles according to Eq. 4.8, that is compute the integral

$$\int_0^{0.5} \tilde{f}(V_R,\phi)\, d\phi,$$

where

$$V_R = \gamma - V_x \cos 2\pi\phi + V_y \sin 2\pi\phi. \tag{4.14}$$

Putting dashes on various symbols to avoid confusion later, then the back-projection integral becomes

$$\int_0^{0.5} \tilde{f}(V_R,\phi)\, d\phi = \int_{-\infty}^{\infty}\int_{-\infty}^{\infty} I(V_x',V_y') \int_0^{0.5}\int_{-\infty}^{\infty} |s| e^{-i2\pi s(V_R-V_R')}\, ds\, d\phi\, dV_x'\, dV_y'$$

$$= \int_{-\infty}^{\infty}\int_{-\infty}^{\infty} I(V_x',V_y') \int_0^{1}\int_0^{\infty} s e^{-i2\pi s(V_R-V_R')}\, ds\, d\phi\, dV_x'\, dV_y'$$

$$= \int_{-\infty}^{\infty}\int_{-\infty}^{\infty} I(V_x',V_y') \delta(V_x'-V_x)\delta(V_y'-V_y)\, dV_x'\, dV_y'$$

$$= I(V_x,V_y). \tag{4.15}$$

The third line above follows from the second after transforming from polar coordinates s and ϕ to cartesian $s_x = s\cos 2\pi\phi$ and
$s_y = s\sin 2\pi\phi$, and using Eqs. 4.14 so that

$$\int_0^{1}\int_0^{\infty} s e^{-i2\pi s(V_R-V_R')}\, ds\, d\phi = \int_{-\infty}^{\infty}\int_{-\infty}^{\infty} e^{-i2\pi[-(V_x-V_x')s_x+(V_y-V_y')s_y]}\, ds_x\, ds_y \tag{4.16}$$

and then the integrals over s_x and s_y separate to give the two Dirac δ-functions of the penultimate line of Eq. 6.1 since

$$\delta(x) = \int_{-\infty}^{\infty} e^{\pm i2\pi sx}\, ds.$$

This justifies the assertions of Sect. 4.3.3.

References

1. Bailey, J., Ward, M., 1981, MNRAS, 194, 17P.
2. Baptista, R. et al., 1998, MNRAS, 298, 1079–1091.
3. Baptista, R., Catalan, M.S., Costa, L., 2000, MNRAS, in press.
4. Beardmore, A. P. et al., 1998, MNRAS, 297, 337–347.
5. Bergeron, P., Saffer, R.A., Liebert, J., 1992, ApJ, 394, 228–247.
6. Beuermann, K., Schwope, A.D., 1994, ASP Conf. Ser. 56: Interacting Binary Stars, 119.
7. Billington, I., 1995, PhD Thesis, Oxford University.
8. Billington, I. et al., 1996, MNRAS, 279, 1274–1288.
9. Bobinger, A. et al., 1999, A&A, 348, 145–153.
10. Bragaglia, A., Greggio, L., Renzini, A., D'Odorico, S., 1990, ApJL, 365, L13–L17.
11. Branch, D. et al., 1995, PASP, 107, 1019.
12. Casares, J., Charles, P.A., Marsh, T.R., 1995, MNRAS, 277, L45–L50.
13. Casares, J. et al., 1997, New Astronomy, 1, 299–310.
14. Cropper, M., 1988, MNRAS, 231, 597–608.
15. Cropper, M., 1990, Space Science Reviews, 54, 195–295.
16. Cropper, M., Horne, K., 1994, MNRAS, 267, 481.
17. Dhillon, V.S., Marsh, T.R., Jones, D.H.P., 1997, MNRAS, 291, 694.
18. Foss, D., Wade, R.A., Green, R.F., 1991, ApJ, 374, 281–287.
19. Greiner, J., Hasinger, G., Kahabka, P., 1991, A&A, 246, L17–L20.
20. Gull, S.F., Skilling, J., 1984, IEEE Proc., 131(F), 646–659.
21. Hachisu, I., Kato, M., Nomoto, K.'i., 1999, ApJ, 522, 487–503.
22. Han, Z., 1998, MNRAS, 296, 1019–1040.
23. Harlaftis, E.T., Marsh, T.R., 1996, A&A, 308, 97–106.
24. Harlaftis, E., 1999, A&A, 346, L73–L75.
25. Harrop-Allin, M.K., Hakala, P.J., Cropper, M., 1999, MNRAS, 302, 362–372.
26. Harrop-Allin, M. K. et al., 1999, MNRAS, 308, 807–817.
27. Hellier, C., Robinson, E.L., 1994, ApJL, 431, L107–L110.
28. Hellier, C., 1996, ApJ, 471, 949.
29. Hellier, C., 1999, ASP Conf. Ser. 157: Annapolis Workshop on Magnetic Cataclysmic Variables, 157, .
30. Hellier, C., 1999, ApJ, 519, 324–331.
31. Hessman, F.V., Hopp, U., 1990, A&A, 228, 387–398.
32. Hils, D., Bender, P.L., Webbink, R.F., 1990, ApJ, 360, 75–94.
33. Horne, K., 1985, MNRAS, 213, 129–141.
34. Horne, K., Cook, M.C., 1985, MNRAS, 214, 307–317.
35. Horne, K., Stiening, R.F., 1985, MNRAS, 216, 933–948.
36. Horne, K., Marsh, T.R., 1986, MNRAS, 218, 761–773.
37. Horne, K. et al., 1994, ApJ, 426, 294–307.

38. Huang, S., 1972, ApJ, 171, 549.
39. Iben, I., Jr., Tutukov, A.V., 1984, ApJS, 54, 335–372.
40. Iben, I., Jr., Tutukov, A.V., Yungelson, L.R., 1997, ApJ, 475, 291.
41. Johnston, H.M., Kulkarni, S.R., Oke, J.B., 1989, ApJ, 345, 492–497.
42. Kahabka, P., van den Heuvel, E.P.J., 1997, ARAA, 35, 69–100.
43. Kaitchuck, R. H. et al., 1994, ApJS, 93, 519–530.
44. King, A.R., Wynn, G.A., 1999, MNRAS, 310, 203–209.
45. Knox, R.A., Hawkins, M.R.S., Hambly, N.C., 1999, MNRAS, 306, 736–752.
46. Liebert, J., Dahn, C.C., Monet, D.G., 1988, ApJ, 332, 891–909.
47. Marsh, T.R., 1985, PhD Thesis, Cambridge University.
48. Marsh, T.R., Horne, K., Shipman, H.L., 1987, MNRAS, 225, 551–580.
49. Marsh, T.R., 1988, MNRAS, 231, 1117–1138.
50. Marsh, T.R., Horne, K., 1988, MNRAS, 235, 269–286.
51. Marsh, T. R. et al., 1990, ApJ, 364, 637–646.
52. Marsh, T.R., Horne, K., 1990, ApJ, 349, 593–607.
53. Marsh, T.R., Robinson, E.L., Wood, J.H., 1994, MNRAS, 266, 137.
54. Marsh, T.R., Dhillon, V.S., Duck, S.R., 1995, MNRAS, 275, 828.
55. Marsh, T.R., 1995, MNRAS, 275, L1–L5.
56. Marsh, T.R., Duck, S.R., 1996, New Astronomy, 1, 97–119.
57. Marsh, T.R., 1999, MNRAS, 304, 443–450.
58. Maxted, P. F. L. et al., 1998, MNRAS, 300, 1225–1232.
59. Maxted, P.F.L., Marsh, T.R., 1999, MNRAS, 307, 122–132.
60. Maxted, P., Marsh, T.R., Moran, C.K.J., Han, Z., 2000, MNRAS, in press.
61. Meyer, F., Meyer-Hofmeister, E., 1999, A&A, 346, L13–L16.
62. Morales-Rueda, L., Still, M.D., Roche, P., 1999, MNRAS, 306, 753–765.
63. Morales-Rueda, L., Marsh, T.R., Billington, I., 2000, MNRAS, 313, 454–460.
64. Moran, C., Marsh, T.R., Bragaglia, A., 1997, MNRAS, 288, 538–544.
65. Moran, C. et al., 1999, MNRAS, 304, 535–539.
66. Narayan, R., Nityananda, R., 1986, ARAA, 24, 127–170.
67. Norton, A. J. et al., 1992, MNRAS, 254, 705–718.
68. Orosz, J.A., Wade, R.A., 1999, MNRAS, 310, 773–783.
69. Oswalt, T.D., Smith, J.A., Wood, M.A., Hintzen, P., 1996, Nature, 382, 692–694.
70. Perlmutter, S. et al., 1999, ApJ, 517, 565–586.
71. Postnov, K.A., Prokhorov, M.E., 1998, ApJ, 494, 674.
72. Potter, S.B., Hakala, P.J., Cropper, M., 1998, MNRAS, 297, 1261–1268.
73. Press, W.H., Teukolsky, S.A., Vetterling, W.T., Flannery, B.P., 1992, .
74. Riess, A. G. et al., 1998, AJ, 116, 1009–1038.
75. Ritter, H., Kolb, U., 1998, A&AS, 129, 83–85.
76. Robinson, E.L., Shafter, A.W., 1987, ApJ, 322, 296.
77. Robinson, P.F.L., Marsh, T.R., Smak, J., 1999, Accretion Disks in Compact Stellar Systems, Edited by J. Craig Wheeler. World Scientific, ISBN 981-02-1273-9 (1993)., 75–116.
78. Rutten, R.G.M., van Paradijs, J., Tinbergen, J., 1992, A&A, 260, 213–226.
79. Rutten, R. G. M. et al., 1993, Nature, 362, 518–520.
80. Rutten, R.G.M., Dhillon, V.S., 1994, A&A, 288, 773–781.
81. Rutten, R.G.M., Dhillon, V.S., Horne, K., Kuulkers, E., 1994, A&A, 283, 441–454.
82. Rutten, R.G.M., 1998, A&AS, 127, 581–586.
83. Saffer, R.A., Liebert, J., Olszewski, E.W., 1988, ApJ, 334, 947–957.
84. Saffer, R.A., Livio, M., Yungelson, L.R., 1998, ApJ, 502, 394.
85. Sawada, K., Matsuda, T., Hachisu, I., 1986, MNRAS, 219, 75–88.

86. Schmidt, G. D. et al., 1999, ApJ, 525, 407–419.
87. Schwope, A. D. et al., 1998, Wild stars in the old west: Proceedings of the 13th North American Workshop on Cataclysmic Variables and Related Objects, eds. Howell, S. Kuulkers, E., Woodward, C., astro-ph/9708228, 44–59.
88. Schwope, A. D. et al., 2000, MNRAS, 313, 533–546.
89. Shore, J.E., Johnson, R.W., 1980, IEEE Tran. Info. Theory, IT-26(1), 26–37.
90. Skilling, J., Bryan, R.K., 1984, MNRAS, 211, 111.
91. Smak, J., 1969, ActaAstron, 19, 155.
92. Smak, J., 1981, Acta Astronomica, 31, 395.
93. Spruit, H.C., 1987, A&A, 184, 173–184.
94. Spruit, H.C., Rutten, R.G.M., 1998, MNRAS, 299, 768–776.
95. Steeghs, D., Harlaftis, E.T., Horne, K., 1997, MNRAS, 290, L28–L32.
96. Still, M.D., Marsh, T.R., Dhillon, V.S., Horne, K., 1994, MNRAS, 267, 957.
97. Stover, R.J., 1981, ApJ, 248, 684.
98. Tammann, G.A., 1982, Supernovae: A Survey of Current Research, 371–403.
99. Truemper, J. et al., 1991, Nature, 349, 579–583.
100. van den Heuvel, E.P.J., Bhattacharya, D., Nomoto, K., Rappaport, S.A., 1992, A&A, 262, 97–105.
101. Wade, R.A., 1984, MNRAS, 208, 381–398.
102. Wade, R.A., Hubeny, I., 1998, ApJ, 509, 350–361.
103. Warner, B., 1986, MNRAS, 219, 347–356.
104. Warner, B., 1995, .
105. Watts, D. J. et al., 1986, A&A, 154, 197–213.
106. Webbink, R.F., 1984, ApJ, 277, 355–360.
107. Wood, J. et al., 1986, MNRAS, 219, 629–655.
108. Wood, J.H., Abbott, T.M.C., Shafter, A.W., 1992, ApJ, 393, 729–741.
109. Wynn, G.A., King, A.R., 1995, MNRAS, 275, 9–21.
110. Wynn, G.A., King, A.R., Horne, K., 1997, MNRAS, 286, 436–446.
111. Young, P., Schneider, D.P., 1980, ApJ, 238, 955–963.
112. Yungelson, L., Livio, M., 1998, ApJ, 497, 168.

5 Evolution of Close Binaries

Bert C. De Loore

Astrophysical Institute, Vrije Universiteit Brussel, Belgium

Abstract. The evolution of a star depends primarily on its mass, chemical composition, so it is important to know the distribution of stellar masses when the stars are born. Stars can be born as singles, in pairs, or in multiple systems. A large fraction of the stars belongs to binary systems. Hence it is essential to know how stellar masses are distributed at birth, how many stars are formed, not only as singles, but also what fraction originates as binary systems, and what is then the mass ratio and orbital period distribution of these binaries. Since massive stars are intrinsically the brightest ones, it is interesting to try to discover their properties in distant stellar groups. Until now observations of massive stars are reasonably complete for a restricted portion of our own Galaxy, within a sphere with a radius of about 3 kpc from the Sun. Conclusions from these observations are then extended to the whole Galaxy. Before starting the study of evolution of close binary systems, we examine first the evolution of single stars. So we begin to deal with the stellar structure processes, stellar evolution in general, we investigate how evolution occurs differently for stars of low, intermediate and large mass. Comparison of evolutionary computations with observations allows restricting a number of uncertainties of the value of parameters in the equations of stellar structure and evolution. The evolution of close binaries differs from the evolution of single stars, since the presence of a companion sets limits on the stellar expansion during its evolution. We examine the processes that have to be added to the stellar structure equations to compute the structure of the components of binary systems, and to calculate their evolution. Here also we study in detail the uncertainties and their effects on the results of the computations. An overview is presented of the existing evolutionary computations for various masses, with various assumptions on the processes, stellar wind mass losses for massive components, convection criteria, the behaviour of semiconvection, the accretion process, standard accretion models, full mixing models, the formation of common envelope systems, the formation of discs. Comparison of computations with observations of binaries that are representative for a given class allows to draw general conclusions and to discard a number of uncertainties. The combination of the results of massive single stars and massive binaries allows predictions concerning the evolution of certain stellar populations (Wolf-Rayet stars, luminous blue variables, X-ray binaries, ...), i.e. predictions on their progenitors, and their evolutionary history. We investigate uncertainties in binary statistics (frequency, mass ratio and period distribution) and try to determine how these uncertainties affect the results. Observations of massive single stars and of massive close binaries, allow to present a consistent evolutionary model for both, that is able to explain the stellar content and the overall stellar properties for the solar neighbourhood.

5.1 Stellar Evolution During the Successive Nuclear Burning Phases

Before we discuss the evolution of close binary stars we review very briefly the evolution of single stars. They may be divided into low mass stars, intermediate mass stars and massive stars. Each group has a different evolutionary history. Low mass stars develop after core hydrogen burning a helium core where the electrons are degenerate. This degeneracy is lifted by the helium flash. Intermediate mass stars ignite helium in non-degenerate conditions, but develop a highly electron-degenerate carbon-oxygen core after central helium exhaustion. The massive stars evolve through all phases of nuclear burning without encountering degeneracy. The temperature can increase beyond 6 10^8 K igniting central carbon burning in a quiescent way, and producing Ne and Mg. The mass limits can vary according to assumptions adopted in the construction of stellar models, e.g. mass-loss, overshooting. The treatment of the convection in the interior stellar layers and the ensuing mixing is very important in this respect. Indeed, the extent of convection in the stellar interior determines the mass of the different burning cores and the occurrence of electron degeneracy. In the conventional way the boundary of the central convective regions is determined by the Schwarzschild criterion. When overshooting of convective cores is included the separation masses between the groups of stars are larger. In the case of the Schwarzschild criterion we consider as low mass stars, stars with masses below 2.3 $M_\odot$, intermediate stars have masses between 2.3 and 9 $M_\odot$, and massive stars are larger than 9$M_\odot$; in the case of large overshooting (Roxburgh criterion) these vales are respectively 1.6 and 6$M_\odot$.

5.1.1 The Zero-Age Main Sequence

The term main-sequence means the phase of central hydrogen burning. For lower mass stars (M $< 1.3 M_\odot$) hydrogen burning in the core occurs via the proton-proton cycle, for larger masses via the CNO-tri-cycle. The mass fraction of hydrogen is large (X = 0.7), so the MS phase represents an important fraction of the lifetime of a star (about 90 percent). As hydrogen in the stellar central parts is converted into helium, the star moves to the right in the HRD. The ZAMS is the curve given by (L, T_{eff}) - values that correspond to homogeneous stars that just started hydrogen burning. When the pp-chain is the primary hydrogen burning mode, i.e. for masses below 1.3 $M_\odot$, the central temperatures T_c are lower than $2.10^7 K$. The nuclear energy-generation rate is moderately temperature-dependent; in the temperature regime 6.60 <log T_c < 7.38, $\epsilon_{pp} \approx T^4 to T^6$. In stars with larger central temperatures (7.08 < log T_c < 7.7) the CN and CNO bi-cycles are dominant, the nuclear energy-rate, $\epsilon_{CNO} \approx T_n$, with n between 13 and 20. In the case of the CNO- cycles the energy-generation rates are so temperature-dependant that the energy is produced in the immediate vicinity of the centre. This leads to extremely huge temperature gradients and the core becomes convective. Summarized: in stars on the upper part of the ZAMS the nuclear energy is produced according to the CN and CNO bi-cycles ; they have convective cores. In stars on the lower part of the ZAMS the energy is produced via the pp cycle; these stars have radiative cores. The discriminating mass is $\approx 1.3 M_{\odot}$, the discriminating central temperature is $\approx 210^7$ K. Also in the outer layers there are differences: in the lower ZAMS stars, with cool atmospheres, the adiabatic gradient is reduced by ionization zones, and convective zones are present. In ZAMS stars with $T_{eff} > 8300$ K the ionization zones are very thin, there is no convective energy transport ; the outer layers are in radiative equilibrium. Figure 5.1 shows the effect of the metallicity on stellar models with masses between 0.2 and 100 $M_{\odot}$ in the HRD. The Z-values range between 0.00001 and 0.03. Figure 5.2 shows how the central temperature and the density vary with the stellar mass.

5.1.2 General Characteristics of ZAMS Models

- If M > 1.3 - 1.4 $M_{\odot}$ the cores are convective, the outer layers are in radiative equilibrium. If M< 1.3 - 1.4 $M_{\odot}$ the cores are not convective, the envelopes are convective.

- If M> 1.7 $M_{\odot}$ the envelopes are radiative (envelope convection is ineffective for hot stars).

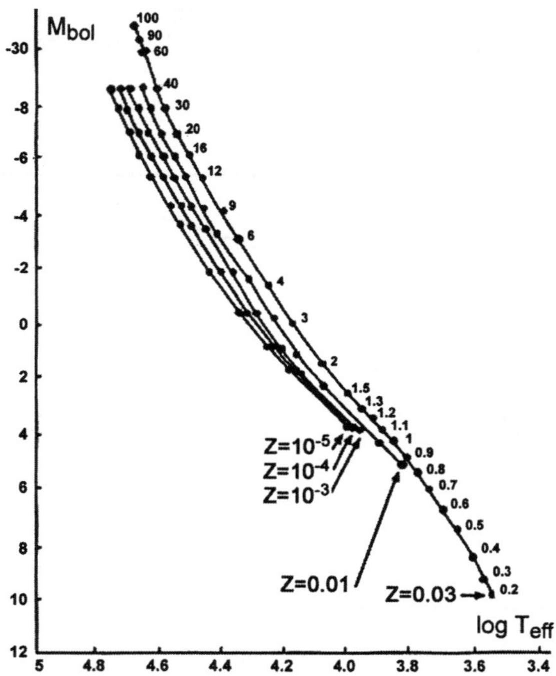

Fig. 5.1. The ZAMS position for stars with masses between 0.2 and 100 $M_{\odot}$ for metallicities between Z = 0.00001 and 0.03. The masses are given in $M_{\odot}$. (After Vanderlinden, 1982)

- The large convective core mass fraction q_{core} for larger mass stars is the consequence of radiation pressure. Larger masses imply larger core masses.

- Stars with masses 0.4 $M_\odot < M < 1.3 M_\odot$ have convective envelopes and radiative cores. This is a consequence of the proton-proton cycle for the energy generation which is less temperature dependent than the CNO tri-cycle.

- In the mass range 1.3 - 1.7 $M_\odot$ small convective cores and shallow or moderately well-developed convective envelopes may occur.

- Stars with $M < 0.4\ M_\odot$ are completely convective (Limber, 1958).

- The ZAMS is terminated at $M \approx 0.04\ M_\odot$

- The central density ρ_c increases with decreasing mass. It attains a maximum at $\approx 1.3\ M_\odot$; then it decreases with decreasing mass. This behaviour is determined by the change-over from the pp- cycle to the CNO tri-cycle at ≈ 1.3 $M_\odot$

- The central temperature increases with increasing mass.

- Radii increases with increasing mass.

- For $M < 0.08\ M_\odot$ the evolution is so slow that these stars are still at the ZAMS. During the lifetime of the Galaxy they barely have evolved.

- ZAMS population II stars are believed to be formed at the early beginnings of the Galaxy; the more massive stars among them have already evolved from the ZAMS ; therefore the computations for population II stars are restricted to low-mass stars.

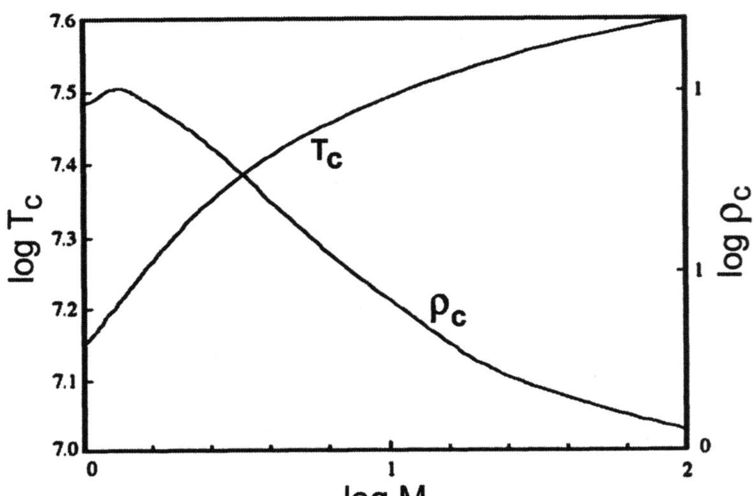

Fig. 5.2. T_c and ρ_c for ZAMS models for various masses. The masses are given in $M_\odot$

5.1.3 Shell Sources

At the end of a burning phase, the central parts of the star are formed by the "ashes"of the nuclear burning. The nuclear reactions stop since the nuclear fuel is exhausted. But just outside this inert core the temperature may have become sufficiently high to start nuclear burning in a shell around this core, using the remains of the nuclear fuel in a thin shell. This process is called nuclear shell burning. It is evident that nuclear shell burning generally occurs at composition discontinuities. So, during core hydrogen burning ^{4}He nuclei are formed in the central regions of stars at temperatures of a few tens of million degrees. These helium layers are surrounded by regions still containing a considerable amount of hydrogen. In thin shells of matter thermonuclear fusion processes may be active, producing energy; these shells are called active shell sources. If the temperature drops below the threshold for nuclear reactions (e.g. if expansion occurs) the nuclear energy production in the shell comes to an end and the shell becomes inactive. Inactive shell sources may be reactivated by rising temperatures .

5.1.4 Hydrogen Burning

The computations of stellar evolution start from a chemically homogeneous model, i.e. at each point in the star the X- Z-and Y values are the same. During a time interval Δt, the changes

ΔX = dX/dt.Δt, and ΔY = dY/dt.Δt are determined at each point of the star and so a stellar model is computed. This new model serves then as input for the next static model and so on. Also mass losses may be considered, e.g. mass transfer and mass loss for the evolution of close binaries, or mass losses by stellar winds for massive stars. Low-mass stars (M< 1.6 - 2.3 $M_\odot$) where H- burning occurs mainly via the pp-chain develop radiative cores surrounded by convective envelopes, extending to the base of the photosphere. Lower-mass stars have smaller cores. At the start the core is chemically homogeneous. The stellar energy losses are supplied by nuclear reactions. These occur at a nuclear time-scale which is much larger than the dynamical time scale. This allows the star to adopt its structure quasistatically to the decreasing hydrogen fuel supply. Matter is not mixed, thus the change of the hydrogen content by weight, X_H, is proportional to the energy generation ϵ_H. During a time step Δt, the change of the hydrogen content ΔX_H is given by $\Delta X_H \approx \epsilon_H \Delta t$. A graph of the hydrogen abundance versus the mass M_r gives the so called hydrogen profile.

The chemical evolution, (the change of X_H in each mass shell can be followed for a sequence of evolutionary models. Figure 5.3 depicts the change of the hydrogen profile during core hydrogen burning for a 1 $M_\odot$ star. When the end of the main sequence is attained X_H= 0 in the stellar centre. Larger masses (M >1.6-2.3$M_\odot$) evolve in a different way. If the central temperature is larger than 2.10^7, H-burning occurs via the CNO tri-cycle in a convective core. Inside this core matter is mixed so fast, compared to the local nuclear fusion processes, that the composition throughout the core is homogeneous, even though the nuclear burning, (proportional to the temperature), is much more efficient in the central

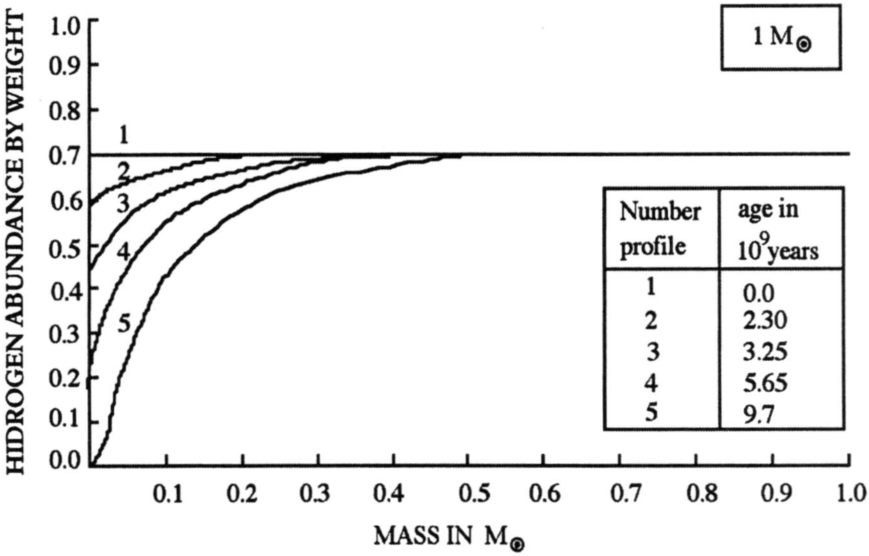

Fig. 5.3. The H-profile of a low mass star of 1 $M_\odot$ versus the mass Mr during the MS phase. Initial chemical composition: X = 0.7, Z = 0.02. The table shows the number of the successive profiles, and the corresponding evolutionary times

regions than in the surrounding layers. The energy generation ϵ is now given by $\epsilon = \rho X Z_{CNO} T^n$ with $n \approx 18$, and Z_{CNO} the mass fractions of C, O and N. The energy production is now linear in X, instead of quadratic. The change of the hydrogen abundance is proportional to the energy generation, just as for lower mass stars, but now we have to calculate the average of the energy production $< \epsilon_H >$ over the whole core, since matter is mixed. Hence $\Delta X_H \approx < \epsilon_H > \Delta t$.

During core hydrogen burning the convective core decreases. In Fig. 5.4 the evolution of the H- profile during the MS stage for a 20 $M_\odot$ star is shown. At the end of H-burning the convective core has completely vanished ; a small helium core is present, of about 5 percent of the total mass. During helium burning the convective core increases again. In the envelope the composition has not changed, since these layers are too cool for nuclear reactions. For all massive stars the behaviour is the same; the more massive the star, the more the H profiles are shifted to the right, i.e. to larger M/M_{tot}-values.

5.1.5 Evolution in the HRD

1. Low mass stars: M< $1.3M_\odot$.
   When H- burning starts the evolutionary models in the HRD are situated at the ZAMS. The energy generation rate $\epsilon \approx \rho X^2 T^n$, hence the energy

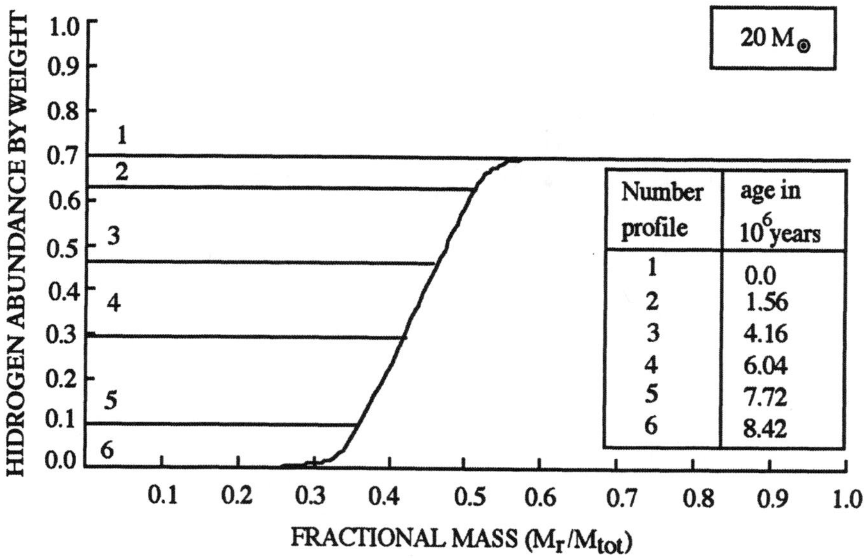

Fig. 5.4. The hydrogen profile of a 20 $M_\odot$ star versus the fractional mass Mr/Mtot during the main sequence phase. Initial chemical composition:X = 0.7, Z = 0.02. The table in the figure shows the number of the successive profiles, and the corresponding evolutionary times

generation decreases with X, unless ρ and/or T increase. Lower energy generation means lower pressure, and the star would contract, leading to larger X. According to the virial theorem a part of the released potential energy is used to increase the temperature; the remaining part goes into radiation. So ρ and T increase; this leads to a larger energy output, hence a larger L. The net effect is a higherr luminosity, and a small increase in the radius of the core and the envelope. The star moves in the HRD upward and slightly to the left (lower curves of Fig. 5.5). Stars on the main sequence convert hydrogen into helium in the core; the core may become helium-dominated. Hydrogen burning occurs mainly in a relatively thick shell around a small, increasing helium core, which is almost isothermal. When the end of the main sequence stage a chemically inhomogeneous structure develops. The star is now centrally condensed. The radius increases, T_{eff} decreases. The right upward motion continues. The evolution accelerates.

2. More massive stars.

In stars with central temperatures larger than 2 10^7K, the energy is produced by the CN- and the CNO bicycle. The masses are larger than $1.3M_\odot$. These stars have convective cores, matter is mixed. The chemical composi-

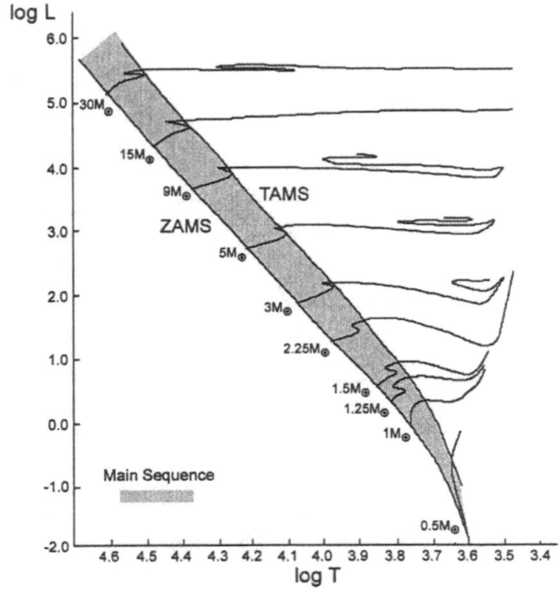

Fig. 5.5. The zero age main sequence, ZAMS and the end points of hydrogen burning, the termination of the main sequence, the TAMS. The hatched area is the main sequence band. L in $L_\odot$

tion in these cores is uniform. Intermediate mass are dominated by Kramers' opacity. For larger masses, electron scattering is dominant. The energy generation rate $\epsilon \approx \rho X Z_{CNO} T^n$; Z_{CNO} is the mass fraction of C,N and O required for the nuclear reactions. N has a typical value of ≈ 18. Since in this case $\epsilon \approx X$,the effect of a decrease of X on the energy production rate will be less important than for smaller mass stars; the contraction of the core is also less pronounced. The luminosity increases and the star expands. The expansion rate is larger than the luminosity increase, hence the star moves slightly upward to the right. When the initial hydrogen abundance is reduced to about 0.05, the energy release rate can no longer support the core; the core contracts. The luminosity does not vary very much, hence the effective temperature increases: the motion of the star in the HRD is upwards to the left (see Fig. 5.5). This is the end of the MS stage. During the main sequence stage helium nuclei are produced at temperatures below the temperature required for helium ignition, a hydrogen-rich envelope surrounds this helium core.

5.1.6 Red Giants

When a shell source is developed, the core becomes inert, it may however contribute to the luminosity by gravitational contraction. Shell sources have the tendency to remain at the same place, so the envelope expands as the core contracts The shell luminosity L_s remains nearly constant, and consequently the surface temperature decreases. The star moves in the HR diagram nearly horizontally to the right. The hydrogen shell is shallow, not far from the stellar centre. The motion is upward towards the red giant region. The envelope may become convectively unstable, hence the chemical profile of the envelope is uniform. Stars of different masses have different evolution schemes: In low-mass stars degeneracy starts before core helium ignition begins and this degeneracy is only lifted by the helium flash (a sudden ignition of helium); this leads to expansion of the core. Finally the core readjusts its structure and helium burning starts at normal, non-degenerate conditions. In higher-mass stars helium burning starts before degeneracy becomes dominant, since a temperature gradient is developed by the increasing importance of gravitation. After core contraction the hydrogen burning shell produces most of the energy. The produced helium layers are added to the core. This core exceeds the Chandrasekhar-Schönberg limit. This is the maximum mass fraction an isothermal, non-degenerate core can have in a star in equilibrium. The energy is produced in a nuclear shell surrounding the core. The core contracts rapidly, the central temperature rises, the shell burns faster and the envelope expands. In the HRD the star moves to the right and slightly downwards. Finally these stars contract and become white dwarfs. More massive stars undergo at the end of their evolution explosions phases, and finally become neutron stars. The most massive collapse and become black holes.

5.1.7 Zero Age Main Sequences

We define as zero-age main sequence any sequence of homogeneous stellar equilibrium models of various masses M. The zero age main sequence (ZAMS) is a curve. We use the term "zero age", to avoid confusion with the main sequence (MS) itself, which is a domain. We will consider H-, He- and C-zero age main sequences, chemically homogeneous stars consisting of a certain element (H, He, C, ...) burning H, He or C in their central parts. H-, He- and C-zero age main sequences are depicted in Fig. 5.6.

1. Hydrogen Zero Age Main Sequence

   The computation of chemically homogeneous hydrogen burning equilibrium models is rather easy, and H-ZAMS models are available for all kinds of chemical compositions. From the computational results simple, but important relations may be derived, MR and ML-relations. L increases strongly, and R increases slowly with increasing M.

   Mass luminosity relation : $L \approx M^b$; $b \approx 10/3$ in the mass range 1 to 40 $M_\odot$

   Mass radius relation : $R \approx M^a$; $a \approx 0.6$ for lower masses, and $a \approx 0.8$ for larger masses

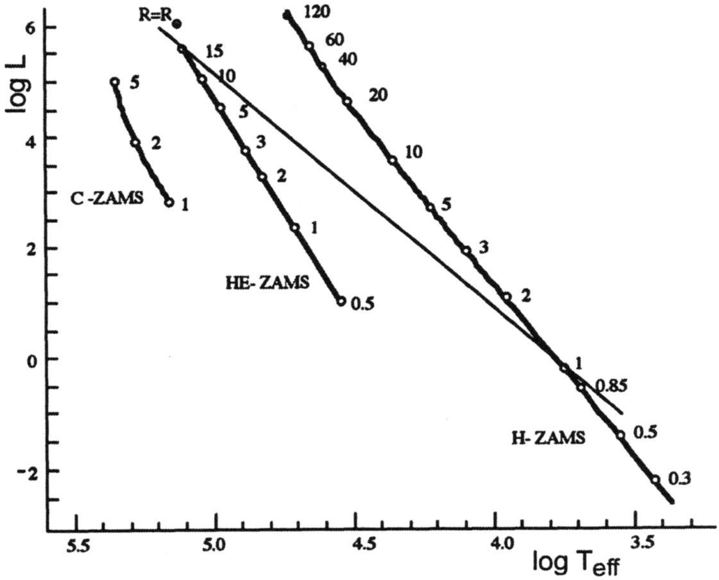

Fig. 5.6. The H-ZAMS, the He-ZAMS and the C-ZAMS (thick lines). The models are labeled with their mass (in $M_\odot$). The curve in fine line shows the locus of constant radius for $R_\odot$

2. The Helium Main Sequence (He MS)
   The He-MS is the set of chemically homogeneous equilibrium models consisting almost completely of helium, at the stage of core He burning. They can represent remnants of initially more massive stars which have lost their H- rich envelope, and developed a central He- core. In the HRD the He-MS is positioned far left of the H-MS. When we compare stars of the same mass at the H-MS and the He-MS it turns out that the helium stars have larger luminosities and smaller radii. The helium burning reactions are very temperature dependent; consequently helium burning occurs in a small central sphere. Owing to the large energy flux a convective core is produced.

   Mass luminosity relation : $\log L = 3.15 \log M + 2.03$

   Mass radius relation : $\log R = 1.13 + 2.26 \log M + 0.78 (\log M)^2$

3. The Carbon Main Sequence (C-MS)

   The C-ZAMS is the set of homogeneous central carbon burning models. The stars contain pure 12C or a mixture of 12C and 16O, the end products of helium burning, and a small quantity of heavy elements. The C-ZAMS in the HRD is positioned at $logT_{eff} > 5$. C-ZAMS models have smaller radii and larger luminosity than stars of the same mass at the H-ZAMS, or the He-ZAMS. C-ZAMS models have large convective cores.

Mass luminosity relation: $\log L = 2.8 + 4.8 \log M - 2.56 (\log M)^2$

Mass radius relation: $\log R = -1.366 + 1.58 \log M - 0.9 (\log M)^2$

Mass luminosity relations for models of the H-, He- and C- ZAMS are shown in Fig. 5.7.

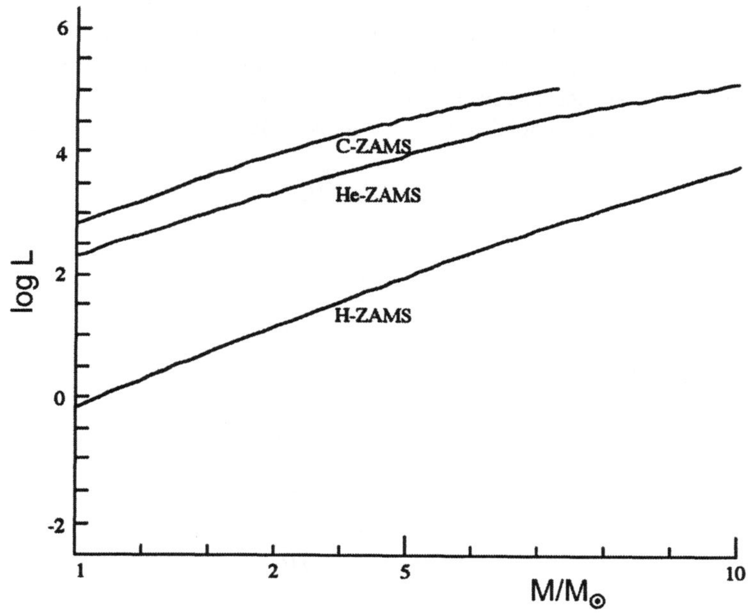

Fig. 5.7. Mass- luminosity relations for models of the H-ZAMS, the He-ZAMS and the C-ZAMS

5.1.8 The Equations of Stellar Structure

The physical state of the stellar interior depends on position and on time. A star has a well determined spatial structure at any time but also changes with time. For simplicity we consider the star as spherically symmetric, not deformed by rotation, and not influenced by a magnetic field. In this case the structure can be described by four physical principles: the equation of motion of the mass layers, conservation of mass and energy and the energy transport mechanism. These principles yield five differential equations describing the stellar structure.

We first list the parameters used to describe the physical conditions inside stars:
r : we consider spherical models, hence all parameters only depend on the distance r from the star's center
ρ : the density of the stellar matter in g/cm^3
M_r : the mass, inside a sphere with radius r. The total stellar mass, M_R, is denoted by M
T : the temperature of the stellar material
P : the total pressure in the stellar interior.
The pressure, exerted by the particles and the photons of the stellar plasma, is related to the density of the material, its temperature and its chemical composition. This relation is the equation of state.
L_r: this parameter denotes the amount of energy flowing across a sphere with radius r. The energy at the stellar surface L_R, is denoted by L. The total luminosity of the sun, $L_\odot$, is used as basic unit. First, we give the equations as functions of the distance r. Afterwards, we give the transformed equations as a function of thhe mass coordinate, M_r, more suitable for the computation of stellar models.

$$\frac{dP}{dr} = -G\frac{M_r\rho}{r^2} \tag{5.1}$$

$$\frac{dM_r}{dr} = 4\pi r^2\rho \tag{5.2}$$

$$\frac{dL_r}{dr} = 4\pi r^2\rho[\epsilon - T\frac{dS}{dt}] \tag{5.3}$$

$$\frac{L_r}{4\pi r^2} = F_{rad} + F_{conv} \tag{5.4}$$

$$\frac{dT}{dr} = \frac{-3}{4ac}\frac{\kappa\rho}{T^3}\frac{L_r}{4\pi r^2} \tag{5.5}$$

$$\frac{dT}{dr} = \frac{\gamma - 1}{\gamma}\frac{T}{P}\frac{dP}{dr} \tag{5.6}$$

$$\dot{M} = \dot{M}(t) \tag{5.7}$$

ϵ is the energy production per unit of mass

κ is the opacity

$T\frac{dS}{dt}$ in the equation for the luminosity is the entropy-term.

All quantities(T, L, P, M) are expressed as a function of r. Since for stars the mass is a more important factor as the radius, we use M_r as independent variable.

We have then as set of equations for the stellar structure:

$$\frac{dr}{dM_r} = \frac{1}{4\pi r^2\rho} \tag{5.8}$$

$$\frac{dP}{dM_r} = -G\frac{M_r}{4\pi r^4} \tag{5.9}$$

$$\frac{dL_r}{dM_r} = \epsilon - \frac{TdS}{dt} \tag{5.10}$$

$$\frac{dT}{dM_r} = \frac{-3}{4ac}\frac{\kappa\rho}{T^3}\frac{L_r}{16\pi^2 r^4} \tag{5.11}$$

$$\frac{dT}{dM_r} = \frac{\gamma - 1}{\gamma}\frac{T}{P}\frac{dP}{dM_r} \tag{5.12}$$

together with the equation of state, expressions or tables for the opacity κ and for the nuclear energy ϵ.

5.2 General Considerations on the Evolution of Close Binary Systems

5.2.1 Characteristics

Binaries are characterized by the masses M_1 and M_2 of their components, the orbital period (P) the eccentricity e and the spins of the components. The semi-major axis A is given by Kepler's law

$$(\frac{2\pi}{P})^2 A^3 = G(M_1 + M_2) \tag{5.13}$$

and the total orbital angular momentum is determined by

$$J = \frac{M_1 M_2}{(M_1 + M_2)}\frac{2\pi A^2\sqrt{(1-e^2)}}{P} \tag{5.14}$$

We define the primary as the component which was originally the most massive component; the originally less massive star is denoted as the secondary. This evolutionary definition may differ from the observational viewpoint where primary means the visually most luminous component. The mass ratio q is the ratio of the mass of the secondary to the mass of the primary.

5.2.2 The Roche Model

The evolution of a star which is a component of a binary is different from the evolution of a single star with the same mass and the same chemical composition, because of the existence of a point in between the two components where the effective gravity vanishes and where matter from one star may flow freely towards its companion.

5.2.3 Circularization - Synchronization

Eccentricities of shorter period binaries (of the order of days), with intermediate masses with components of sizes comparable to the orbital separation are small but the eccentricities of binaries with larger separation are larger. The spins of the components seem to be synchronized with respect to their orbital revolution (Plavec, 1970; Levato, 1976).

About the observed spin periods one is less certain. The projected rotational velocity of the optical star of the HMXB Vela X-1 (HD 77581) is about 125 km/s and its most probable radius $\approx 28 - 35R_{\odot}$. This means that the spin period is $\approx 11 - 14$ days. Comparing this with the orbital period P = 8.9 days, we may conclude that the binary is not far from synchronization. Penny (1996) examined rotational velocities of a number of OB+OB binaries. She concluded that the binary components of HD 215835 (O6n + O6n, P = 2.1 days) seem to rotate at 75 percent of the synchronous rotation; the components of HD 165052 (O6.5V + O, P = 6.14 days) and of HD 149404 (O8.5I + O7III, P = 9.8 days) rotate at 2 to 3 times the synchronous rate. The binaries HD 93403 (O5f + O7.5, P = 15.1 days) and HD 152248 (O7f + O, P = 5.97 days) have almost reached synchronism. So, many possibilities remain open. That circularization and synchronization in shorter period binaries prevail follows from the fact that for given masses and fixed angular momentum, the mechanical energy of a binary attains its minimum when both stars have circular orbits and synchronized spins (Lynden-Bell and Pringle, 1974). Since all physical systems tend to such a stable equilibrium binaries will tend to achieve circularization and synchronization. Zahn (1977) gives typical time scales for the case of small eccentricities:

$$T_{sync} = \frac{1}{6q^2k_2}\frac{1}{MR^2}(\frac{a}{R})^6 t_F \tag{5.15}$$

$$T_{circ} = \frac{4}{63q(1+q)k_2}(\frac{a}{R})^8 t_F \tag{5.16}$$

k_2 is the apsidal motion constant and I the spin moment of inertia of the star with mass M. For information on the apsidal motion constants, see e.g. Schwarzschild, 1958, p 154. A very uncertain parameter is the time scale for frictional dissipation t_F. The time scales depend strongly on the binary period (smaller periods, smaller time scales) and on the stellar radius (larger radii, smaller time scales). In massive stars with radiative envelopes and short orbital periods, these time scales are smaller than the nuclear lifetime of the primary (order of a few days). In binaries with larger periods synchronous rotation and circularization may never be achieved. Some binaries have difficulties to reach their equilibrium configuration; their orbit remains eccentric. On the other hand, asynchronous spins may cause turbulence in stellar envelopes, enhancing considerably the viscosity; hence the tendency to reach synchronization increases (Tassoul, 1987, 1988, 1990; Tassoul and Tassoul, 1992). The general conclusion is that binaries with initial periods of the order of days and with a small initial eccentricity will reach their equilibrium configuration (synchronized and circularized) within a time scale shorter than their evolutionary time scale. For binaries with larger initial periods, and/or large initial eccentricities it is an open question if they will become circularized or not.

5.2.4 Roche Equipotential Surfaces

We consider a binary with a circular orbit, where spin and orbit are synchronized. Already at the beginning of core hydrogen burning, the two components are centrally condensed and their gravitational field may be represented by that of a point source. In a Cartesian co-ordinate system, - the origin at the centre of mass, - z-axis along the spin axis and x-axis along the line joining the stellar centres, the equipotential surfaces are given by:

$$\Phi = -\frac{GM_1}{(\sqrt{(x-\mu a)^2 + y^2 + z^2})} - \frac{GM_2}{(\sqrt{(x-(1-\mu)a)^2) + y^2 + z^2})} - \\ -\frac{1}{2}{\Omega_B}^2[x^2 + y^2] = ct \tag{5.17}$$

$$\mu = \frac{M_2}{M_1 + M_2} \tag{5.18}$$

$$\Omega_B = \frac{2\pi}{P} \tag{5.19}$$

Figure 5.8 shows the intersection of the equipotential surfaces with the orbital plane (z = 0). The function Φ has five saddle points, the Lagrangian points L. In these points the centrifugal and gravitational forces (modified by the Coriolis acceleration) are in balance. The second and third Lagrangian points L_2 and L_3 are potential minima. They are situated on the x-axis at the right and the left of the binary centres. Very important is the first Lagrangian pointL_1, the point of gravitational balance on the x-axis between the two components. The equipotential surface across L_1 is the critical or Roche surface (Roche lobe). The Roche radius R_c is defined as the radius of a sphere with the same volume as the Roche lobe. The Roche radius as a function of the mass ratio q can be approximated by an interpolation formula given by Eggleton (1983):

$$\frac{R_c}{a} = \frac{0.49}{0.6 + q^{2/3}ln(1 + q^{-1/3})} \tag{5.20}$$

5.2.5 Roche Lobe Overflow

If the envelope of the primary rotates synchronously with the binary orbital motion, hydrostatic equilibrium applies in the co-rotating frame and the pressure equation is given by

$$\nabla P = -\rho \nabla \Phi \tag{5.21}$$

Surfaces of equal pressure and density coincide with equipotential surfaces. The shape of a binary component may be represented by an equipotential. Owing

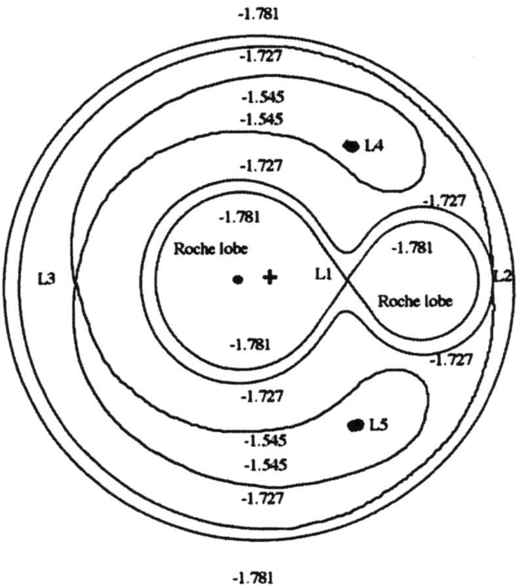

Fig. 5.8. The equipotential surfaces and the 5 points of Lagrange for a binary. The equipotential surfaces are labeled with their potential energy. The critical equipotential surface, the critical Roche lobe in 8-form, is indicated. The mass centre is +

to stellar expansion, the outer layers may reach the critical Roche lobe, and attain L_1. InL_1 hydrostatic equilibrium is no longer valid since $\nabla\Phi = 0$. Considering for simplicity, an isothermal gas near L_1, the one-dimensional equation of motion may be written as:

$$(v^2 - c_s^2)\frac{dv}{dx} = \nabla\Phi \tag{5.22}$$

v is the velocity of the gas, c_s the local isothermal sound speed. In order to retain a monotonic increasing velocity field near and across L_1 (where $\nabla\Phi$=0), $v \approx c_s$, hence matter travels across L_1 and flows at hypersonical speed towards the companion. The process is called the Roche lobe overflow (RLOF). Matter can flow from one component to the other. The component losing matter due to RLOF is called the mass loser or mass donor, and similarly, the component that accepts (accretes) this matter is designated as the mass gainer or the mass accretor. Near the stellar centres the equipotential surfaces are almost spherical; with increasing distance from these centres the deviation from the spherical shape becomes stronger. For a well-determined value of the potential the equipotential surfaces have a common point between the stellar centres. Although Roche lobes are not strictly spherical they do not differ by much from a sphere. The radius of a sphere of the same volume as the volume of the Roche lobe is called the Roche radius R_R.

5.2.6 Discussion of the Roche Model

The Roche model is a good approximation. Condition is that circularization and synchronization are maintained, also during the process of RLOF. Tidal interaction will stabilize the orbit. Even in the case where the envelope of the primary rotates asynchronously with the orbit, a saddle point (zero effective gravity point) exists between the two components (Kruszewski, 1963). Also here it is possible to define a critical surface from where funnelled mass loss takes place. For spin periods not too different from the orbital period, the critical radius differs not very much from the Roche radius, computed as if the star was synchronized. Even for binaries with eccentric orbits, there is a saddle point between the two components. Because its exact place is depending on the orbital phase, a situation may arise where mass loss by critical lobe is not constant in one revolution, but shows on and off phases. On the other hand there are circumstances where synchronism cannot be attained. Evolving primaries expand. This occurs on a nuclear time scale

$$t_{nuc} = 10^{10} M/L \ (years) \tag{5.23}$$

during CHB, further on the Kelvin-Helmholtz time scale

$$t_{KH} = 3.110^{7} M^{2}/RL \ (years) \tag{5.24}$$

during hydrogen shell burning. Owing to this expansion the star spins down and rotation and orbital motion become asynchronized. Tidal interaction will force this star to absorb orbital angular momentum in order to spin up. For small mass ratios ($q \leq 0.2$), the available orbital angular momentum is not sufficient to meet the need of the primary. The low mass companion is then swallowed by the primary (Sparks and Stecher, 1974).

5.2.7 The Mass Loss Rate During RLOF

An evolving primary expands and its photosphere can reach L_1 from where mass loss will occur. For the computation of the mass loss rate one can proceed as follows:

1. Calculate the interior structure of the star with the set of eqs. 5.1-5.7 derived for single stars

2. Compare the radius of the star to the Roche radius determined by eq. 5.20

If $R > Rc$, calculate by iterations the mass loss rate, the mass loss and the remaining mass of the star, so that its radius equals the Roche radius. For the calculation of the Roche radius of a binary component it is assumed that the binary is circularized, and that the star rotates synchronously.

The assumption that the stellar radius equals the Roche radius derived from the binary period and mass ratio, gives very reasonable mass loss rates. It gives satisfactory evolutionary tracks of the two components.

This numerical procedure works well in situations where the mass loss leads to a decrease of the stellar radius. This is always the case when the massive star has a radiative envelope but not at all when the star has a deep convective envelope (red supergiants). Let us approximate the envelope convection in a massive star by adiabatic convection. Most of the envelope of a massive star is fully ionized so the equation of state satisfies the relation

$$P \propto \rho^{5/3} \tag{5.25}$$

Combined with the hydrostatic pressure this leads to the following relation between radius and mass

$$R \propto M^{1/3} \tag{5.26}$$

A decreasing mass leads to an increasing radius. Or, mass loss (on a dynamical timescale, shorter than the thermal time scale) leads to stellar expansion. The star will undergo a very violent mass loss phase on a dynamical time scale of the order of hours. The heavy increase of the stellar radius will make that the secondary star will be swallowed by the primary; this phase is known as the common envelope phase of a binary.

5.2.8 Mass Transfer

In order to have an idea of the process of mass transfer we consider the changes of the stellar radius and the Roche radius as the mass of the star decreases (Fig. 5.9). The evolution of the radius of an isolated star is shown by the straight line AB', vertically upward, at constant mass. In a close binary the radius R_1 increases until the critical Roche radius r_1 is reached, in point B, where the mass exchange process starts. During the beginning stages of the mass exchange the stellar radius remains more or less constant (BC) and then decreases as the mass loss continues (CD_1). When mass loss starts the Roche radiusr_1 decreases. The orbit shrinks. The change of r_1 as function of M_1 (decreasing) is shown by the curve r_1, from B to E. When the masses M_1 and M_2 of primary and secondary are equal, r_1 attains its minimum value (at E). For $M_1 < M_2$, r_1 will increase again, (ED), i.e. the orbit widens again. Since $R_1 > r_1$ between B and C, mass loss is required to reduce the mass of the primary. For radiative envelopes this occurs on a thermal time-scale. The curve for r_1 shows how the radius of a star in thermal equilibrium changes as its mass decreases by removal of the outer layers. Beyond point C the stellar radius of the primary is smaller than the Roche radius. Two possibilities may be distinguished:

- $R_1 < r_1$: the binary becomes detached
- If a new nuclear burning stage starts, the star fills its Roche lobe and again mass loss occurs, but now on a nuclear time-scale (CD).

Hence two mass exchange stages may be associated with the evolution of binaries: a rapid mass transfer phase, where the mass ratio of the two components is reversed from $q > 1$ to $q < 1$ (q = M_1/M_2), this phase is followed by a gradual increase of the mass M_2 of the secondary. The rapid phase can be very short, of the order of 10^4 to 10^5 years. Hence the succession of the events during mass transfer is as follows (see Fig. 5.9):

- A - B : star 1 expands on a nuclear timescale
- B - C : rapid mass transfer takes place, on a thermal timescale. The system consists of a subgiant (remnant of the primary, now the less massive star) and a main sequence star (the most massive star).
- Beyond C the subgiant evolves on a nuclear time-scale, expanding slowly, and transferring a small amount of matter.

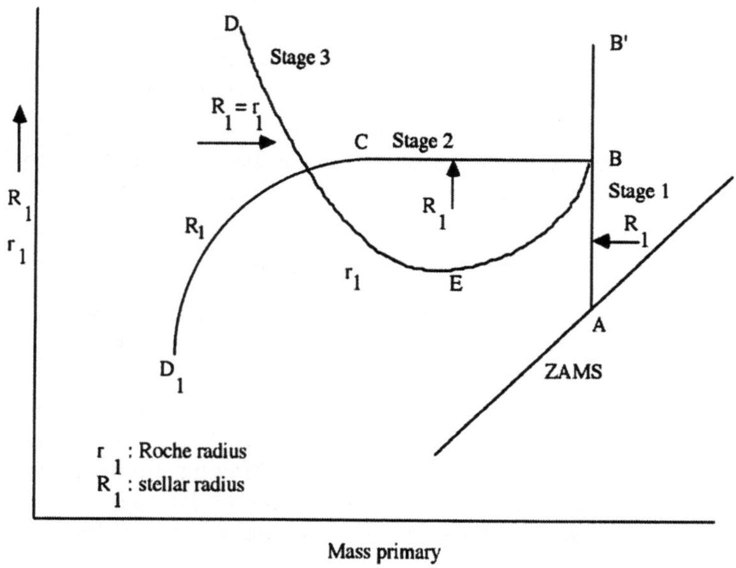

Fig. 5.9. The behaviour of the stellar radius and the Roche radius for a close binary system, with mass exchange. BCD_1 represents the radius of equilibrium models with decreasing mass

5.2.9 The Different Types of Unevolved Close Binaries

Depending on their initial mass, stars are characterized by one, two or three major expansion phases: the CHB phase, the hydrogen shell burning phase and the He shell-burning phase. Primaries of close binaries, depending on the initial orbital period, will fill their Roche lobe during one of these three expansion phases. Following the original idea of Kippenhahn and Weigert (1967) and Lauterborn (1969), we divide the unevolved close binaries into:

- Case A: the primary fills the Roche lobe during its CHB phase;
- Case B: the primary fills the Roche lobe during its H shell burning phase,
- Case C: the primary fills the Roche lobe during its He shell burning phase;

Case B binaries are by far the most frequent class of observed binaries. One may expect a different evolutionary behaviour in binaries where RLOF starts when the mass loser has a mainly radiative envelope (case Br) compared to binaries where RLOF starts when the mass loser has a deep convective envelope (case Bc). Among all case B types, case Br binaries are the most frequent ones.

5.2.10 Direct Accretion or Formation of a Keplerian Disc

For matter leaving the primary through L_1 where the stream is caught into the potential well of the secondary, we can consider two situations:

- the matter streams directly towards the secondary
- the gas-stream forms first a disc around the secondary.

This has been examined by Lubow and Shu (1975). For binary masses M_1 and M_2, considered as point masses and a given period P, a detailed calculation of the trajectories of the particles shows that a disc is produced, provided that the radius R_2 of the secondary is smaller than R_{disc}, the radius of the disc.

5.2.11 The Spin-Up Process of the Secondary

In the case the gas-stream forms a Keplerian disc around the secondary, it can be checked that a mass element dM hitting this secondary with mass M_2, has a specific angular momentum dJ (=dM$\sqrt{GM_2R}$). R is the radius of the secondary. This value has to be compared with the rotational angular momentum J_{rot} (=Iω_{rot}) of the star itself, (for rigid rotation). I is the moment of inertia, ω_{rot} the rotational angular velocity of the star. Making use of the evolutionary computations for single stars, we can calculate the total amount of mass that has to be accreted to force the whole star to rotate at the critical break-up velocity

$$\omega_{crit}=\sqrt{\frac{G(M_2+dM)}{R^3}}$$

Here we assume that during the accretion the radius R remains unchanged. Detailed computations of the accretion process reveal that in reality this is not exactly true, unless the amount of accreted matter is small. And this is precisely the case here: if a $25M_\odot$ mass gainer accretes about $3M_\odot$ through a Keplerian disc it starts to rotate at its break up velocity. The computations were made with the assumption that the complete star is spinning up. If the spin-up process affects only the envelope, a smaller amount of matter is required to force these outer layers to rotate at break-up velocity. We may conclude that when a Keplerian disc is formed around the secondary during the RLOF of the primary, a very efficient mechanism to remove angular momentum from the disc is required to produce efficient mass transfer.

5.2.12 Conservative and Non-conservative Evolution

For the computation of evolution sequences we may consider the following possibilities: a) conservative evolution: the total mass and angular momentum are considered as remaining constant. When the mass transfer stage starts, mass is removed from the primary and added to the companion. From the changing masses of the two components the variation of the period and the ensuing variation of the Roche radius can be derived. b) non-conservative evolution: mass and angular momentum losses are taken into account. This treatment allows the consideration of contact phases. In this case specifications are required to define the fraction of the lost mass, not accreted by the secondary, hence leaving the system, stored in a disc, or in a common envelope, and also for the momentum losses. It is possible for both cases to restrict the computations to the evolution of the primary, and to use the changing mass of the secondary only for the calculation of the changing period, or to calculate the evolution of the two components simultaneously. For the non-conservative treatment mass and angular momentum losses have to be taken into account. A quantity ΔM leaves the primary during mass exchange, determined in such a way that the star remains within its Roche lobe; since the mass loss time-scale and the accretion time-scale are not necessary similar, one may assume that only a fraction β, hence a quanti ty $\beta\Delta M$ is accreted by the secondary. For the angular momentum loss a similar procedure can be adopted: if one neglects the rotational angular momentum, and assuming that matter expelled by one or even two of the components is leaving the system, the orbital angular momentum transported by matter leaving the system may be calculated.

The fractional mass loss c can be expressed as

$$c = \frac{\Delta M}{M_{1i} + M_{2i}}$$

The orbital angular momentum transported by the material leaving the system may be evaluated using a Jeans' mode, hence as $\Delta J =$ cJ. Consequently the distance A and the period P vary, according to Kepler's third law as

$$\frac{A}{A_i} = \frac{(1-c)^2 (M_{1i} M_{2i})^2 (M_1 + M_2)}{(M_1 M_2)^2 (M_{1i} + M_{2i})} \tag{5.27}$$

$$\frac{P}{P_i} = \frac{(1-c)^3 (M_{1i} M_{2i})^2 (M_1 + M_2)}{(M_1 M_2)^3 (M_{1i} + M_{2i})} \tag{5.28}$$

5.2.13 Evolution of the Primary

The structure of both components can be computed by applying a code for simultaneous evolution. The radii of the two components and their respective Roche radii are calculated and compared to each other.In this way contact phases can be treated properly. The structure of the accretion star can be followed, according to the mass accretion formalism and the adopted treatment. Depending on the accretion process this star can be rejuvenated by the accreted matter and its lifetime can increase significantly. This rejuvenated secondary can also overflow its own Roche lobe, and possibly reversed phases of mass transfer can occur.Simultaneous evolution computations reveal that several mass ratio reversals can occur during the evolution of binary systems.

- Low mass case B - $1M_\odot < M_1 < 2.8M_\odot$
  As an example of the evolution of low mass stars, where mass transfer starts after core hydrogen exhaustion, hence case B-, we describe the evolution of a system of $2M_\odot + 1M_\odot$. Electron degeneracy sets in at the core, when the hydrogen shell is decreasing. The contraction of the core occurs not very fast, hence there is a rapid expansion of the envelope. The mass of the hydrogen-exhausted core is less than $0.35M_\odot$ for $M < 2.8M_\odot$; the temperature increase is not high enough to start He-burning. The mass exchange stops by extinction of the H-shell; the star becomes a white dwarf. Only the evolution of the primary was calculated in detail; the mass of the secondary was just used to compute the changing semi-major axis and the period. A Case B evolution of a low mass, close binary system, of $2M_\odot + 1M_\odot$, is depicted in Fig. 5.10. The evolutionary track in the HRD is given in Fig. 5.11. Details comcerning the evolution are given in Table 5.1.

- Case A
  As an example the simultaneous case A evolution of a system of $9M_\odot + 5M_\odot$, with an initial period of 1.45 days (Kippenhahn and Weigert, 1967) is described. The primary starts filling its Roche volume during core hydrogen burning. A phase of fast mass transfer sets in and a large fraction of the mass of the primary is transferred towards the companion. The mass ratio is reversed, i.e. the initially most massive star becomes the less massive one. The system becomes semi-detached. Hydrogen burning in the core of the mass increasing secondary is accelerated, hence the core hydrogen abundance of the secondary decreases faster than in the core of the primary. The H-abundance decreases from 0.602 to 0.259. During the mass transfer matter is removed from the outer layers. The lower layers expand in order to restore hydrostatic equilibrium. The energy for this expansion is stolen from the luminosity. So the luminosity decreases drastically. At the end of the RLOF phase the luminosity increases again. After this phase of fast mass transfer (in 12.5 million years) a phase of slow mass transfer starts. Hydrogen burning continues in the convective core. The secondary remains near the ZAMS for million years.The system shows the characteristics of Algol stars. The RLOF phase comes to an end when the mass of the primary is $3.02M_\odot$. The evolution is depicted in Fig. 5.12; evolutionary tracks are shown in Fig. 5.13.

5.2.14 The Evolution of the Two Components - Simultaneous Evolution

The constraints for the evolutionary computations are as follows: as long as the system does not overflow its outer critical surface (through L_2) the computations occur in the conservative way; when the outer critical surface is overfilled mass loss from the system is considered. To allow matter to escape from the system it needs some extra energy, at least sufficiently large to bridge the potential energy difference between L_2 and the Roche lobe. This energy per unit mass can

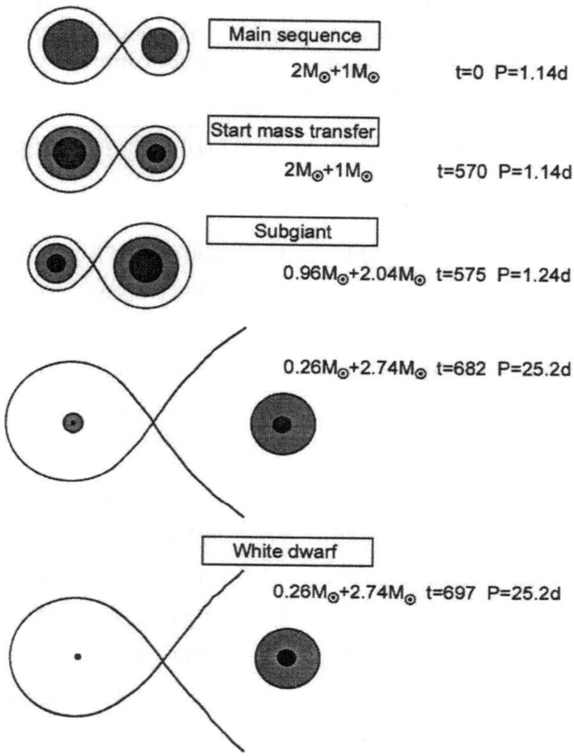

Fig. 5.10. Case B evolution of a $2M_{\odot} + 1M_{\odot}$ system. The evolutionary times t are in million years, the orbital periods in days. Black circlesrepresent hydrogen-rich, gray circles helium-rich regions

be estimated for q = 1 as 0.27 GM/A. This value is typical for a large range of mass ratios. The formula reveals that mass can more easily escape in wider systems than in closer ones; for extreme mass ratios the factor 0.27 becomes much smaller, hence for systems with large mass ratios matter can escape more easily. In order to check the value of simultaneous evolution, we compare evolutionary sequences for the two components of a system of $2M_{\odot} + 1M_{\odot}$, with an initial period of 1.14 days, with the track of the primary of the same system, with the same initial period, calculated by Kippenhahn, Kohl and Weigert (1967) in Fig. 5.14.

The system of $9M_{\odot} + 8.1M_{\odot}$ with an initial period of 3.13 days (Fig. 5.15) follows the conventional track of conservative evolution (Packet, 1988). The mass loss rate attains a maximum value of $1.610^{-4}M_{\odot}yr^{1}$ and the atmospheric hydrogen abundance by mass drops from 0.7 to 0.21 for the primary, from 0.7 to 0.67 for the secondary. The remnant of the primary evolves through helium burning

Table 5.1. Evolution of a system of $2M_{\odot} + 1M_{\odot}$ (Kippenhahn, Kohl, Weigert). The status, -detached (d), or semi- detached- (sd), L, T_{eff}, and the atmospheric hydrogen abundance of the primary, X_{at} are shown. The original composition is X = 0. 602, Y = 0.354, Z = 0.044 . The letters refer to Fig. 5.11

Age	M_1	M_2	*P*	*Status*	*logL*	$logT_{eff}$	X_{at}	*Model*
0.	2.	1.	1.14	d	1.307	3.972	0.60	A
4.8179	2.	1.	1.14	d	1.405	3.909	0.60	B
5.0591	2.	1.	1.14	d	1.500	3.947	0.60	C
5.6958	2.	1.	1.14	s.d	1.567	3.899	0.60	D
5.6989	1.55	1.45	0.80	s.d	0.715	3.781	0.60	E
5.7509	0.96	2.04	1.22	s.d	0.518	3.710	0.60	F
6.3538	0.28	2.72	21.40	s.d	1.444	3.634	0.57	G
6.4940	0.28	2.72	21.40	s.d	1.280	3.653	0.57	H
6.6808	0.28	2.72	23.74	s.d	1.489	3.642	0.57	I
6.8176	0.26	2.74	23.74	d	1.634	3.667	0.57	J
6.8710	0.26	2.74	23.74	d	1.698	3.795	0.57	K
6.9044	0.26	2.74	23.74	d	1.588	4.329	0.57	L
6.9328	0.26	2.74	23.74	d	1.024	4.492	0.57	M
6.9651	0.26	2.74	23.74	d	-0.572	4.297	0.57	N

and then expands again, so that a second phase of mass exchange starts. This second phase of mass transfer occurs for primaries in the mass range of 3 to 10 $M_{\odot}yr^1$ and the atmospheric H-abundance by mass drops from 0.7 to 0.21 for the primary, from 0.7 to 0.67 for the secondary, consequence of the expansion of the helium envelope resulting from the energy output of the small helium shell source.

5.2.15 Reversals of the Mass Ratio

Simultaneous evolution computations show mass ratio reversals during the evolution of binary systems. A case A evolution of a system of $9M_{\odot} + 8.1M_{\odot}$,initial period 0.77 days is examined (Packet, 1988). During a phase of fast mass transfer the mass ratio is reversed. Hydrogen burning in the core of the mass accreeting secondary is accelerated, hence X_{c2}, the core H abundance of the accretion star decreases faster than in the core of the primary. When X_{c2} is $\approx$ 0.4 the mass transfer is reversed, owing to the expansion of the secondary. This results in Roche lobe overflow. The mass ratio is reversed once more, and owing to

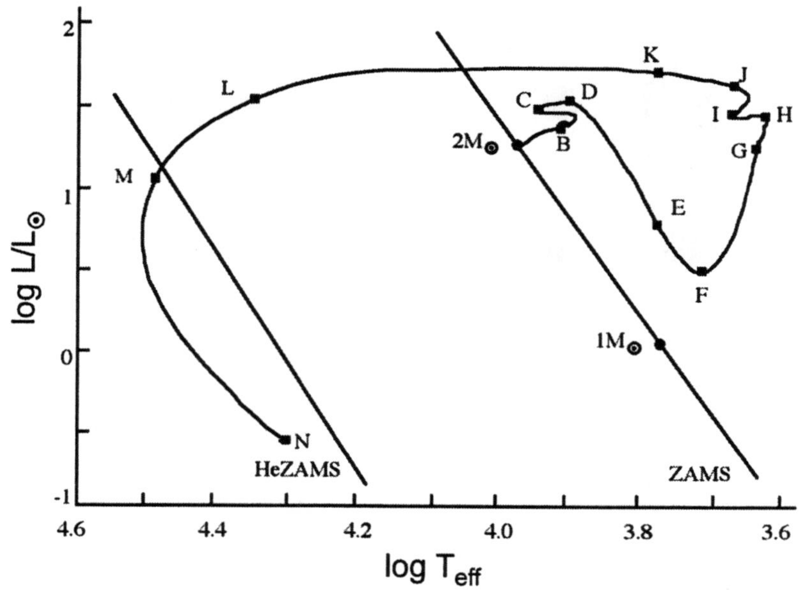

Fig. 5.11. The evolutionary track for the primary of a close binary system of $2M_\odot$ + $1M_\odot$. The zero age main sequences for normal hydrogen abundances(ZAMS), and for X = 0, (He ZAMS) are also shown. The letters near the track refer to Table 1

the mixing in, of new H in the core of the primary,X_{c1} increases. The system becomes semi-detached. The primary, which is now again the most massive component evolves faster, and after some time overflows its Roche lobe. The critical surface through L_2 is now rapidly attained ; mass and angular momentum leave the system ; the result will be a spiral-in, leading to to coalescence of the two components. Rather complicated situations may occur,with successive phases of mass transfer where the mass ratio is reversed twice. The history of a system of $9M_\odot$+$8.1M_\odot$ with an initial period of 3.13 days is shown in Fig. 5.15. Figure 5.16 shows the changing core H abundance and the mass ratio.

5.2.16 Simultaneous Evolution with Moderate Overshooting

If extended mixing is taken into account, the evolution of the stars is altered; the duration of the core H burning phase is increased, hence the main sequence lifetime of the stars becomes larger, and also the gradient in the hydrogen profile in the stellar interior, consequence of the shrinking of the convective core, is nearer to the surface. Evolutionary tracks, computed with and without overshooting, reveal that the radius at red point, i.e. the maximum radius reached during main sequence evolution is much larger for overshooting models. Consequently the relative occurrence of case B and case A-evolution for close binaries will be influenced drastically by overshooting.

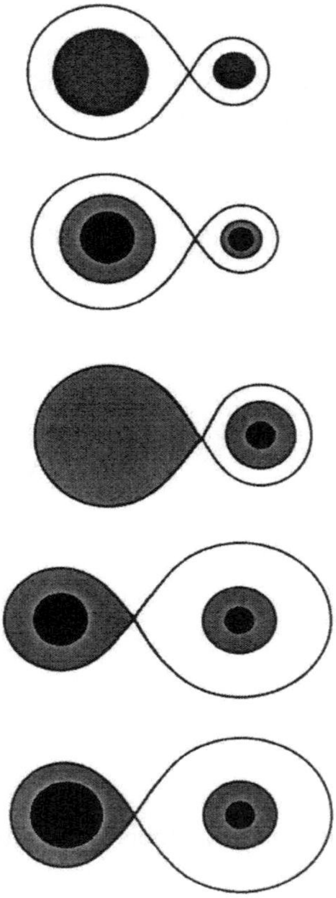

Fig. 5.12. The evolution of a medium mass close binary, case A. The time t is in million years, the orbital period P in days. The primary fills its Roche lobe during core hydrogen burning and transfers matter towards its companion. There are two phases of mass transfer, a rapid phase where somewhat more than $5.3M_{\odot}$ is transferred in 60000 years, followed by a slow phase. Black circles represent hydrogen rich-, gray circles helium-rich regions (Kippenhahn and Weigert, 1967)

5.2.17 The Evolution of Intermediate Mass Close Binaries (IMCB) for Galactic and MC-Abundances

A grid of evolutionary models was computed with the simultaneous code for intermediate mass binaries (masses between $3M_{\odot}$ and $15M_{\odot}$ and with periods small enough to allow mass transfer by Roche lobe overflow. This grid, together with grids for massive close binaries could be used as input for population synthesis, to investigate e.g. starbursts. Moreover it is interesting to examine the influence of different initial abundances on the evolution of IMCB, the physics

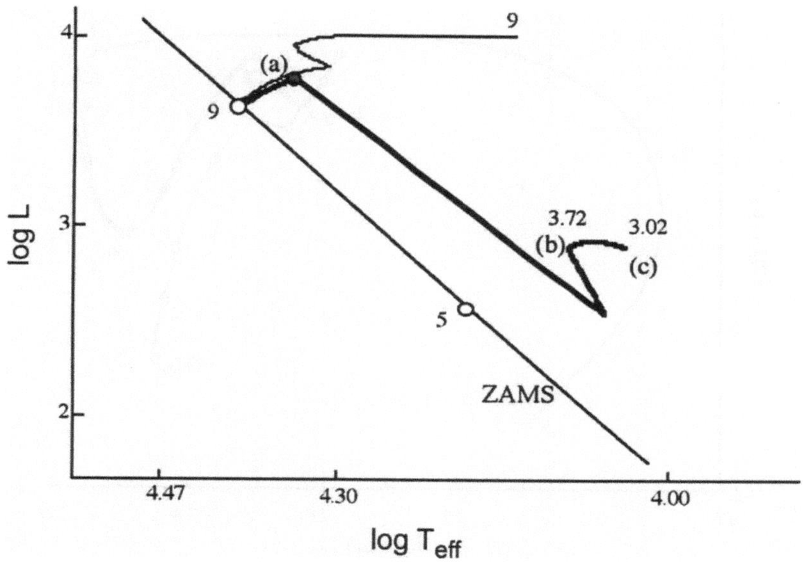

Fig. 5.13. Evolution of a $9M_\odot + 5M_\odot$ system. The evolution of the primary of the binary system is shown as a thick line. The mass transfer starts in (a), the rapid mass transfer ends in (b). From b to c slow mass transfer occurs. In fine line the evolution of a non-disturbed $9M_\odot$ is sketched

of accretion and evolutionary consequences and the evolution of the secondary before, during and after the mass transfer The computations were restricted to case Br systems, i.e to cases where the primary has a mostly radiative envelope. To examine the effect of accretion on the structure and evolution of the secondary two accretion models were used:

Soft mass transfer model: Matter whirls onto the surface of the secondary with zero velocity and is assumed to have the same specific entropy as the stellar surface layers (Neo et al., 1977; Flannery and Ulrich, 1977, Kippenhahn and Meyer-Hofmeister, 1977). This means that when the outer layers of the secondary star are in radiative equilibrium, this will not be destroyed by the accretion The gravitational energy of the accreted matter is converted into radiation which perhaps blows up the outer layers of the secondary, and so leading to the formation of a contact binary. The physics of this process are included in the code. If matter was transferred from primary to secondary during earlier nuclear processes, matter with a larger molecular weight might be deposited onto the surface layers of the secondary. This causes thermohaline convection, which is treated as an instantaneous process (Kippenhahn et al., 1980; Packet, 1988). Semiconvection was treated in the code by the Schwarzschild criterion (Stothers and Chin, 1976).

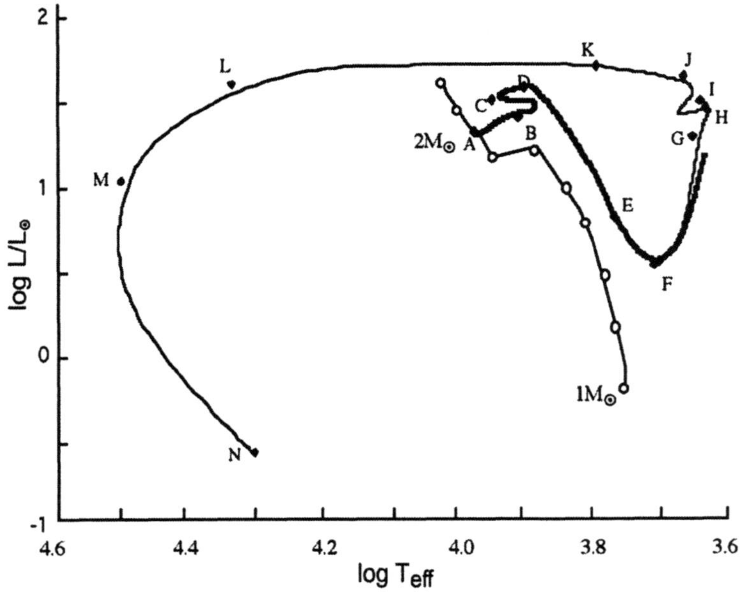

Fig. 5.14. Comparison of the evolution of a $2M_{\odot} + 1M_{\odot}$ system, with an initial period of 1.14 days, computed with a simultaneous evolution code, with earlier computations of Kippenhahn, Kohl and Weigert, 1967. The track of the observatio primary of the simultaneous computations is shown as thick line, the track of the accreting secondary is a fine line, marked with open circles. The track of Kippenhahn et al. is shown as a full fine line, marked with dots

Full mixing model: If at the start of the RLOF the distance between the two is sufficiently large to allow mass transfer through a (Keplerian) disc, the secondary spins up. Even for rigid rotation, only a few solar masses are needed to speed up this secondary to break up velocity. Rapid rotation causes very efficient turbulent diffusion or even complete mixing of the star (Zahn, 1994). Vanbeveren and De Loore (1994) have studied the effect of full mixing on the evolution of MCB's in general. The used code allows a choice between soft mass transfer and complete mixing during the mass transfer process. The adopted opacities influence only marginally the evolution of the primary in a close binary. Indeed in the major part of the star electron scattering is the main opacity source, and only in the very outer layers the temperature is sufficiently low to allow recombination of the chemical elements and only here processes different from electron scattering become important. Some small quantitative differences exist between the Galactic and the SMC results, but this has probably no observational consequences. The evolution during Roche lobe overflow of the primary and the final remnant mass after the RLOF phase do not significantly depend on the details of the RLOF. Given the initial mass of the primary of a case Br binary, we can replace the more or less complicated behaviour during the mass transfer in the

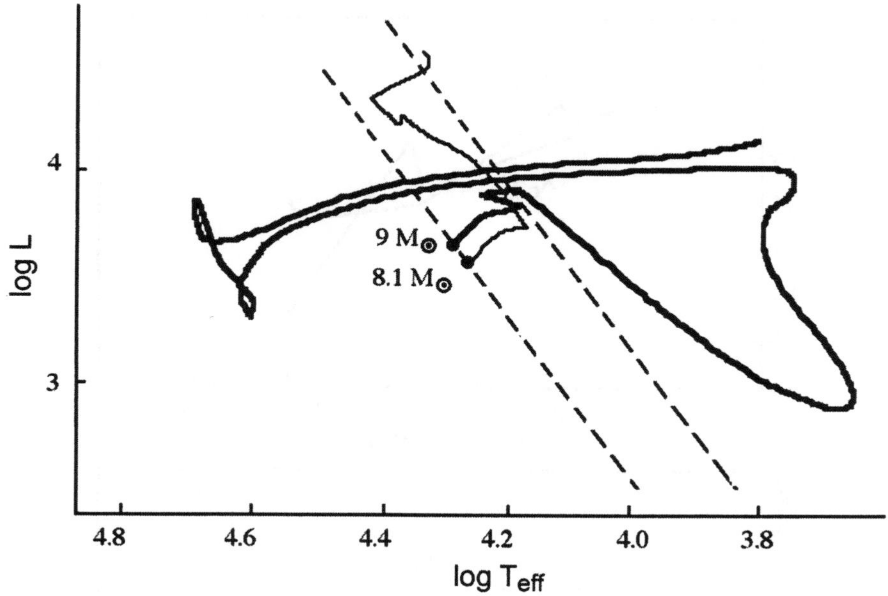

Fig. 5.15. The simultaneously computed evolution of an intermediate mass system of $9M_{\odot} + 8.1M_{\odot}$, with initial period of 3.13 days leading to an early case B of mass exchange

HR diagram by a box. What really happens in this box is not very important. In Fig. 5.20 we show the evolutionary tracks of all our primaries but we have replaced the RLOF track by a black box. The initial period has an effect on the mass loss rates during the RLOF. The larger the period the later the RLOF starts, the shorter the process and the more violent the mass loss rate.

Table 5.2 shows the best-fit relations between the remnant mass after RLOF, and the initial mass, for the Galaxy, the LMC and the SMC.

It is quite obvious that the evolution of the secondary depends critically on the details of the RLOF, i.e. on how much mass lost by the primary is accreted by the secondary (the parameter search) and on the adopted accretion model. However also for the secondary it is possible to propose a general scenario once the accretion model is chosen.

5.3 Evolution of Massive Close Binaries

5.3.1 Introduction

The qualitative evolutionary scenario (Fig. 5.17) of MCBs was introduced by Paczynski (1967) and was later completed by Van den Heuvel and Heise (1972). We consider a binary system where the primary is a massive star and possibly also the secondary. First both components evolve as single stars, independent

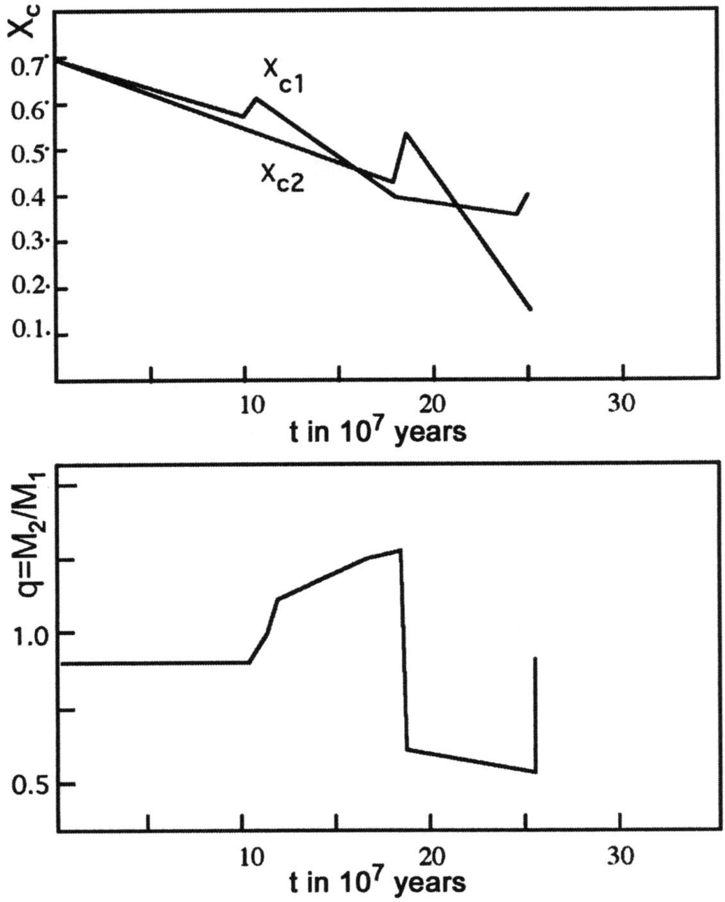

Fig. 5.16. The core H abundances X_{c1} and X_{c2} of the primary and secondary, and the mass ratio q, of a system of $9M_{\odot} + 2.7M_{\odot}$, with an initial period of 0.77 days, evolving according to case A. The mass ratio is reversed twice, connected with the changing speed of the variation of the core hydrogen

from each other. The most massive star expands faster than the companion does; hence it reaches first its critical radius and the RLOF starts. When the primary has lost most of its hydrogen rich envelope, it contracts; it shrinks within its Roche lobe and the RLOF stops. The system then consists of a hydrogen deficient CHeB star and an OB-type star that has accreted part or all the matter lost by the loser. Paczynski (1967) first suggested that these systems might represent the observed WR+OB binaries. After core He- exhaustion, the He star explodes. If the explosion does not disrupt the system, the post SN-binary resembles an OB+cc binary. Van den Heuvel and Heise (1972) suggested that these systems could represent the observed massive X-ray binaries. As the OB-type mass gainer

Table 5.2. Best-fit between the RLOF-mass and the initial mass, for binaries with galactic, LMC and SMC-composition

$3M_\odot \leq M \leq 7M_\odot$	
$M_f = 0.1M_i^{1.32}$	GAL
$M_f = 0.11M_i^{1.32}$	LMC
$M_f = 0.124M_i^{1.32}$	SMC
$M \geq 7M_\odot$	
$M_f = 0.093M_i^{1.44}$	Gal
$M_f = 0.085M_i^{1.52}$	LMC
$M_f = 0.048M_i^{1.7}$	SMC

further evolves, it expands and its radius may also reach a critical value. A second RLOF phase starts, but now, due to the extreme mass ratio of the binary, the low mass compact star will be swallowed into the envelope of the OB star and it begins spiraling in. The final outcome will be either a very close binary with a period of a few hours, consisting of a helium star and a compact companion, or the system can be disrupted and evolve into two neutron stars, i.e. two runaway pulsars. At the end of the evolution of the helium star, a second SN occurs. In most of the cases, the system will be disrupted leaving behind two single runaway pulsars. There remains also a very small probability that the binary remains bound, and evolves into a binary pulsar like PSR 1953+29.

5.3.2 The Evolution of the Primary During Mass Transfer

Before the start of the RLOF, both components of a MCB evolve as rotating single stars with the same mass. Just as for single stars, most of the evolutionary computations of MCBs use the structure equations established for non-rotating spherical symmetric plasma, i.e. the set of equations 5.1 to 5.7 together with the equation of state, expressions or tables for the opacity κ and for the nuclear energy ϵ. It turns out that MCB components have rotational velocities which are (on average) smaller than or equal to those of single stars.

1. Primaries with initial mass $\leq 40M_\odot$: case Br.
   Case Br means that the RLOF starts when the primary is a hydrogen shell burning star with an envelope which is mainly in radiative equilibrium (typical periods: between a couple of days and 1000 days; they are shown in Fig. 5.18). The evolution of the primary can be studied in a simplified way,

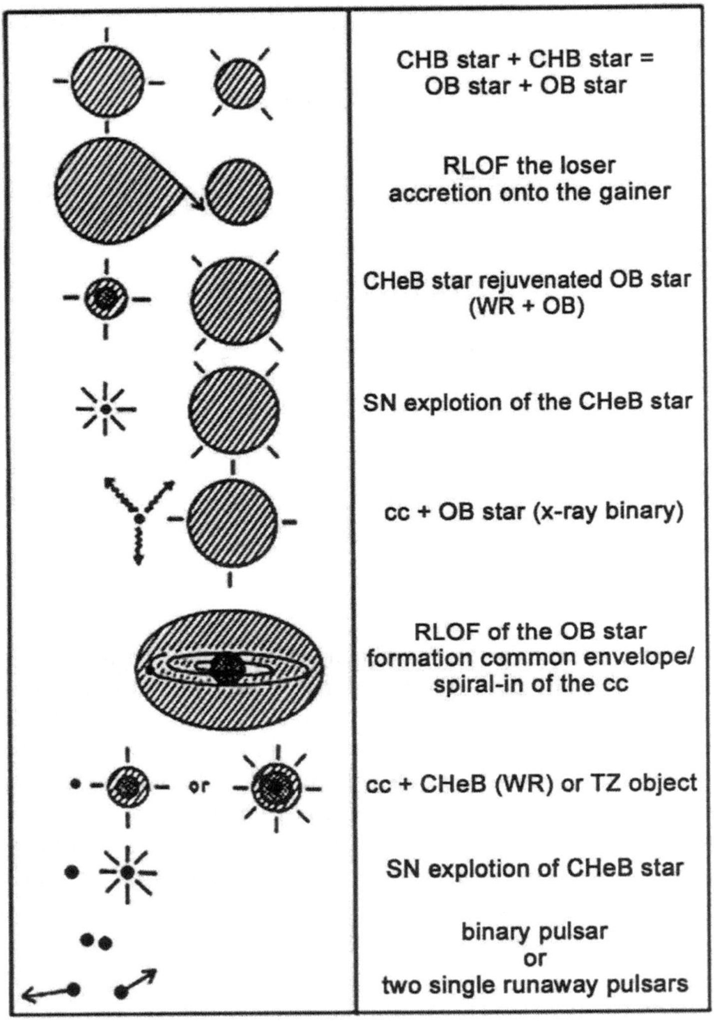

Fig. 5.17. The qualitative MCB scenario as it was introduced by Paczynski (1967) and completed by Van den Heuvel and Heise (1972)

without computing explicitly the evolution of the secondary, just using its mass for the determination of the critical Roche radius.

During the first 10 percent of the RLOF phase, the star evolves on the Kelvin-Helmholtz timescale and loses matter at very high rates. Roughly, these rates satisfy the relation

$$\dot{M} = \frac{M}{\tau_{th}} = 310^{-8} \frac{RL}{M}$$

(in $M_{\odot} yr^{-1}$, M, R and L in solar units).

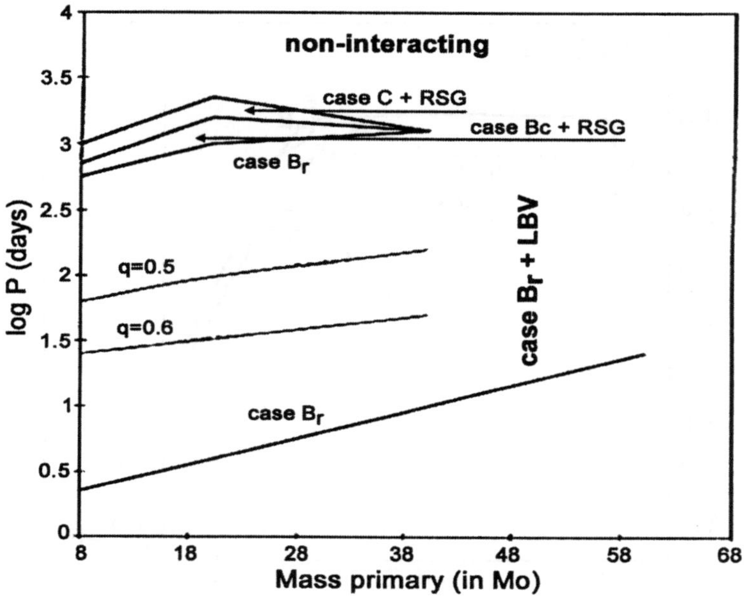

Fig. 5.18. Mass-period diagram showing the different RLOF cases

The maximum values for $\dot{M}$ are of the order of $10^{-3}M_{\odot}/\mathrm{yr}$: the star is then in the rapid phase of mass transfer. Due to the entropy term in the equation of energy this rapid mass loss leads to a rapid decrease of the luminosity. When the central temperature and density are sufficiently high to start the CHeB process, the expansion rate of an H-shell burning star decreases. This means that the mass loss required keeping the star within its Roche volume decreases: the star enters the slow RLOF-phase. During the slow phase the stellar luminosity increases again, and the star tries to restore thermal equilibrium. This equilibrium is reached when most of the H rich layers have been removed. Detailed evolutionary computations reveal that, a massive hydrogen shell-burning star tends to stop its expansion when He is burning in the core and when the atmospheric H- abundance X_{atm} drops below 0.2-0.3. The star is then in thermal equilibrium and contracts: the RLOF process is finished. The structure of an H- shell-burning star at the beginning of CHeB, in thermal equilibrium, does not depend on the details of the RLOF process0t. Consequently a rather unique relation exists between the mass just before RLOF (Mb) and at the end of the RLOF (Ma), practically not dependent on the initial mass ratio, the period of the binary and the treatment of the RLOF process, in a conservative or non-conservative way. Also the treatment of semi-convection has only a small effect on the computations of massive primaries in interacting binaries. Except for relatively small quantitative differences, the evolutionary properties of MCBs for the

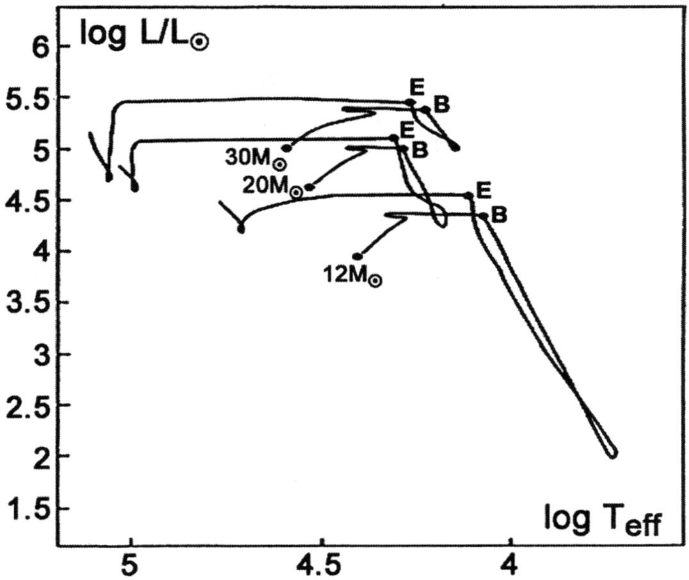

Fig. 5.19. Evolutionary tracks of massive primaries in a case Br binary assuming conservative RLOF; the points B and E indicate respectively the beginning and the end of the RLOF phase

Galaxy discussed above also apply to the Magellanic Clouds, which means they are independent of the initial metallicity.
Figure 5.20 shows evolutionary tracks of primaries of case Br MCBs with Galactic (full lines) and SMC (dashed lines) abundances. The evolution of primaries after RLOF is practically not dependent on the details of the RLOF phase; a box has replaced the RLOF stage. What occurs inside the box has practically no repercussion on the output.
The masses at the end of CHB and after RLOF are given in table 5.3. Best-fit relations between the pre-RLOF and post-RLOF mass are given in eq. 5.31, 5.32. The computations were performed with small convective core overshooting during CHB; SW mass loss with the following equation was included

$$log\dot{M} = 2.365logL - 3.710logT_{eff} - 9.639 \tag{5.29}$$

and compared with

$$log\dot{M} = 1.69logL - 15.4 \tag{5.30}$$

The results are very similar.
Best-fit relations for the remnant mass after RLOF:

$$Galaxy: \quad M_a = 0.093M_b^{1.44} \tag{5.31}$$

$$LMC: \quad M_a = 0.085M_b^{1.52} \tag{5.32}$$

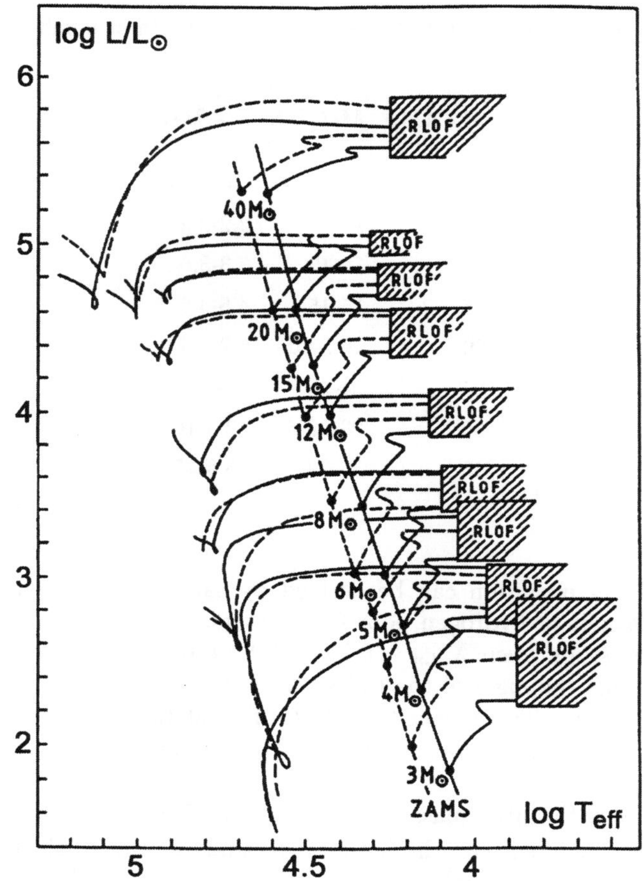

Fig. 5.20. The overall evolutionary behavior of primaries of case Br close binaries. Galactic tracks are in full lines, SMC tracks are dashed

$$SMC: \quad M_a = 0.048 M_b^{1.7} \tag{5.33}$$

With the same mechanism for the stellar wind, but when large core-overshooting is applied, we find:

$$Galaxy: \quad M_a = 0.055 M_b^{1.79} \tag{5.34}$$

A hydrogen stripped star with mass M starting CHeB is in thermal equilibrium and has an equilibrium radius R_e which can be approximated by

$$R_e = 0.62.10^{-4} M^3 - 0.49.10^{-2} M^2 + 0.18 M + 0.17 \tag{5.35}$$

R_e is in $R_\odot$, M in $M_\odot$. The relation is practically independent on the metallicity

Table 5.3. The mass of the primary for case Br evolution at the end of CHB (MeCHB) and at the end of the RLOF (MeRLOF) for the Galaxy (first number) and the SMC (second number).

MZAMS	*MeCHB*	*MeRLOF*
9	9	2.2-2.2
12	12-12	3.3-3.6
15	15-15	4.5-4.5
20	19.5-20	6.7-7.9
25	24-25	9.3-11.1
30	28-29.5	12.3-16.3
40	36-39	18.6-25.3

The further evolution can be derived by using the available evolutionary computations of hydrogen poor CHeB stars. RLOF in a MCB for case Br evolution stops when $X_{atm} \approx 0.2 - 0.3$. So the mass of the RLOF remnant must equals the mass of the CHB core of the corresponding progenitor for $X_c \approx 0.2 - 0.3$. This means that it is sufficient to calculate the CHB evolution to predict the overall evolution of the primary.

2. Primaries with initial mass 40 to $50 M_{\odot}$: case Bc and C
   Equation 5.24 tells us: the larger the radius of the H shell burning star, the smaller the thermal time scale. This means also: the larger the initial period, the more violent the rapid phase of the RLOF and the larger the . The mass loss rate becomes critical for periods sufficiently large enough to start RLOF when the primary has a convective envelope: this is case Bc. In this case the mass loss due to RLOF will be a very fast process (section 999). The star will lose matter on the dynamical time scale, of the order of hours. The secondary is unable to accept all this mass and soon after the beginning the RLOF a common envelope is produced. The further evolution depends then on the effect of viscosity when the secondary is moving into this common envelope. Detailed computations where this process was followed till the end have never been carried out. An attempt to predict the evolution of a case Bc binary was made Tutukov and Yungelson (1979). The scenario goes as follows: the primary in a case Bc binary, starts filling its Roche lobe, soon after the onset of the RLOF process. A common envelope phase starts. Deposition of orbital energy of the secondary by viscous drag will remove this envelope It is reasonable to assume that also for case Bc the mass loss process will stop when most of the H-rich layers have been removed and this means that relations (5.33) or (5.34) can also be applied

here. If we know the mass of the common envelope that has to be removed, one can calculate the period change. The removal of the common envelope implies a very large orbital period decrease. Detailed computations for case C binaries have never been performed. Probably the primary does not only lose all its H, but also a part of the He rich layers outside the He burning shell.

3. Primaries with initial mass 40-50$M_\odot$: case A
   Also for case A binaries, during the RLOF process a rapid and a slow phase occur. The rapid phase occurs on a thermal time scale (as in the case of Br systems) which is much longer for CHB stars than for H shell burning stars. Hence the mass loss rates during the rapid RLOF phase in case A binaries are much smaller than for corresponding case Br binaries. The rapid phase is followed by a slow phase and this lasts nearly the entire remaining CHB lifetime. At the end of this phase, layers, which were in the H burning core at the beginning of CHB, appear at the surface. These layers have CN equilibrium abundances, so we expect to observe these stars as OBN stars. The atmospheric H- abundance remains almost normal. At the end of CHB, the star contracts hence the RLOF ends. Later on H starts burning in a shell just outside the newly formed He-core, a new expansion phase starts, and the primary overflows its Roche again and a case Br type of evolution follows (this phase is called case ABr). Since the primary lost already a large amount of matter during the previous case A RLOF, the case ABr RLOF is less violent. The RLOF comes to an end when He burning in the core starts and when $X_{atm} \approx 0.2 - 0.3$, i.e. comparable to case Br, the remnant mass after case ABr is the mass of the H burning convective core of the progenitor when the central hydrogen content X_c = 0.2-0.3.
   So, the large mass loss during RLOF in an early case A binary, leads to a much smaller mass of the convective core during CHB is smaller than in a case Br binary. The remnant mass after case ABr RLOF in a binary after early case A evolution, is smaller than the remnant mass of the same primary of a pure case B binary. For periods sufficiently large to start case A RLOF when the central hydrogen content of the primary < 0.2-0.3, the final mass after the case ABr RLOF is equals to the final remnant mass after a pure case Br RLOF.
   As example of case A evolution of a MCB, Table 5.4 summarizes the general evolution of 15 $M_\odot$ and 30 $M_\odot$ primaries with companions of respectively 13.5 $M_\odot$ and 27 $M_\odot$ with an initial period of 2 days,evolving in a conservative way. The evolutionary tracks for a system of 15 $M_\odot$ +13.5 $M_\odot$ are depicted in Fig. 5.21.

5.3.3 The Variation of the Period During RLOF

The period of a massive close binary (MCB) varies as a consequence of the changing mass of the components. Two main processes affect these masses: stellar wind mass losses, and mass transfer- or loss by Roche lobe overflow. The effects

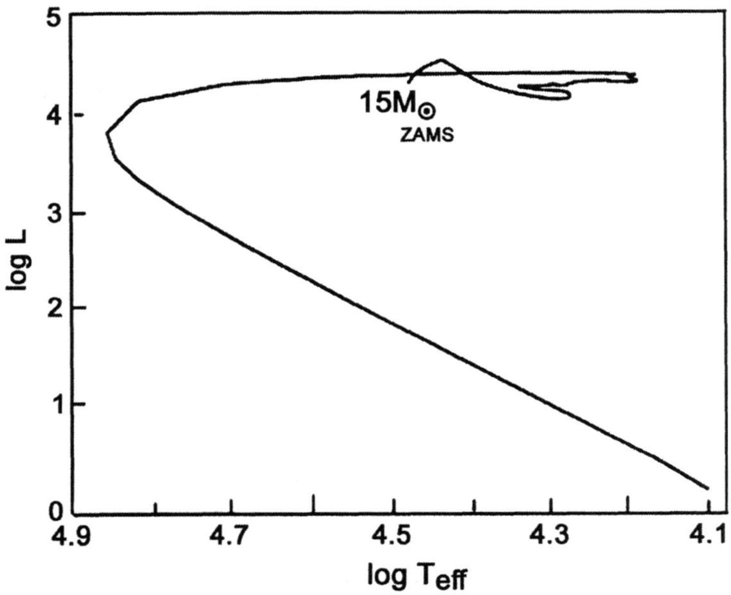

Fig. 5.21. Evolutionary track of the primary of a $15 M_\odot$ + $13.5 M_\odot$ galactic early case A binary; the initial period is 2 days

of these two are different; in some cases the wind losses are important, in other cases the effect of the stellar wind is only marginal compared to the RLOF mass loss. Stellar winds are driven by radiation pressure. This radiation pressure is the main driver of a more or less spherical symmetric stellar wind but it does not affect considerably the gas-stream caused by the RLOF process. We consider two cases:

The RLOF mass loss is much larger than a possible stellar wind (SW) mass loss. We consider binaries where RLOF occurs and where effects of SW mass loss on the variation of the binary period during RLOF can be neglected. It turns out that only the most luminous LBVs have SW mass loss rates that may be comparable to RLOF mass losses. So, we exclude the VMCBs. We consider the case of conservative RLOF, - all mass leaving the primary due to RLOF is accreted by the secondary, and the case of non-conservative RLOF- at least a part of the mass lost by the primary during its RLOF leaves the binary system. In the case of non-conservative RLOF one defines a parameter β, this is the fraction of the mass lost by the primary accreted by the secondary. So if M_1 (resp. M_2) is the mass loss (resp. mass gain) rate of the primary (secondary), one has

$$\dot{M}_2 = -\beta \dot{M}_1 \quad ; \quad 0 \leq \beta \leq 1 \qquad (5.36)$$

Table 5.4. The evolution of the early case A binaries $15M_\odot + 13.5M_\odot$ and $30M_\odot + 27M_\odot$. Period 2 days. The phase marked as Max is the phase where the mass loss rate is largest

Phase	*Time*	$M_1(M_\odot)$	$\dot{M}$	$logT_{eff}$	$logL/L_\odot$	X_c
ZAMS	0	15	0	4.47	4.26	0.7
Begin RLOF Case A	8.75	15	0.086	4.43	4.48	0.39
Max	8.88	13	92.2	4.39	4.3	0.38
End RLOF Case A	16.6	6.6	0	4.27	4.16	0.01
Begin RLOF Case B	16.8	6.6	2.42	4.29	4.24	0
Max	16.9	4.3	86.8	4.19	4.24	0
End RLOF Case B	16.9	3.1	0	4.2	4.32	0

Phase	*Time*	$M_1(M_\odot)$	$\dot{M}$	$logT_{eff}$	$logL/L_\odot$	X_c
ZAMS	0	30	0	4.58	5.03	0.7
Begin RLOF Case A	3.27	29.5	1	4.55	5.17	0.46
Max	3.34	26.6	211	4.53	5.07	0.45
End RLOF Case A	7.42	14.1	0	4.46	5.02	0.03
Begin RLOF Case B	7.67	14	10	4.48	5.12	0
Max	7.67	12.1	825	4.41	5.06	0
End RLOF Case B	7.68	9.52	0	4.39	5.16	0

Conservative RLOF.
The rotational angular momentum of a binary component is always much smaller than the orbital angular momentum. This means that efficient mass transfer

requires an efficient mechanism to move the surplus of angular momentum of the transferred matter back into orbital angular momentum of the components. Conservation of mass implies conservation of total orbital angular momentum. Combination of eq. 5.13 and 5.14 gives

$$\frac{M_1^2 M_2^2}{M_1 + M_2} A = const \tag{5.37}$$

and, using Kepler's law

$$\frac{P}{P_s} = [\frac{M_{1s} M_{2s}}{M_1 M_2}]^3 \tag{5.38}$$

Values at the start of the RLOF are denoted by the subscript 's'. The results of close binary evolutionary computations reveal that the period of the binary reaches a minimum during RLOF, and, after RLOF is always larger than the period before RLOF.

Non-conservative RLOF.

-a. Non-conservative RLOF in case A/case Br binaries.

In most of the MCBs (except the VMCBs), SW mass loss is not sufficiently large to allow the primary to escape RLOF. At the beginning of the RLOF phase, the primary suffers mass losses at very high rates ($10^{-3} M_\odot$/yr). If all this matter is accreted by the secondary, it expands and both stars may enter a contact phase. So the system may enter a situation as sketched in Fig. 5.22. Only one possibility is known by which matter can leave a case A/case Br MCB in an efficient way at rates comparable with the RLOF mass loss rate of the primary, is mass transport through the second point of Lagrange, L_2. In this way a ring forms around the binary with a radius L_2C, with C the center of mass. We assume as simplification that the orbits are circular and that a circular ring revolves in the same sense as the binary at a distance A_{ring} from C, the mass center. The angular momentum per unit mass in the ring is then

$$\jmath_{ring} = \sqrt{G(M_1 M_2) A_{ring}} \tag{5.39}$$

The variation of orbital angular momentum when $-(1-\beta)dM_1$ is lost through L2, is given by

$$dJ = \jmath_{ring}(1-\beta)dM_1 \tag{5.40}$$

$$\frac{dP}{P} = (1+3\alpha)\frac{(1-\beta)dM_1}{(M_1+M_2)} - \frac{3dM_1}{M_1} - \frac{3dM_2}{M_2} \tag{5.41}$$

with

$$\alpha = \frac{(M_1+M_2)^2}{M_1 M_2}\sqrt{\frac{A_{ring}}{A}} \tag{5.42}$$

The ring formed by matter passing across L_2 has as radius $L_r \approx L_2 C$. For a large range of binary mass ratios, $L_2 C = 1.3$A. Consequently, considering the ratio $A_r/overA = \eta$ as constant, equation 5.41 can be integrated.

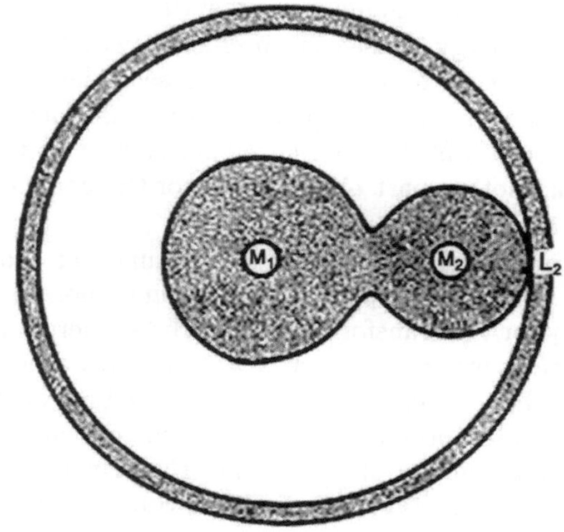

Fig. 5.22. Mass loss from a case A/case Br binary across L2 and the formation of a ring around the system, during RLOF

For $\beta \neq 0$

$$\frac{P}{P_s} = [\frac{M_1 + M_2}{M_{1s} + M_{2s}}][\frac{M_1}{M_{1s}}]^{3[\sqrt{\eta}(1-\beta)-1]}[\frac{M_2}{M_{2s}}]^{-3[\sqrt{\eta}\frac{1-\beta}{\beta}+1]} \tag{5.43}$$

For $\beta = 0$, and $\eta = 1.3$, eq. 5.41 becomes

$$\frac{P}{P_s} = [\frac{M_1 + M_2}{M_{1s} + M_{2s}}][\frac{M_1}{M_{1s}}]^{0.42} e^{3.42\frac{M_1 - M_{1s}}{M_2}} \tag{5.44}$$

As usual, "s"refers to values at the start of RLOF. This formalism leads to an important reduction of the orbital period.

-b. Common envelope phases for case binaries evolving according to case Bc/C.

The RLOF in MCB for case Bc/case C mass exchange is very vehement. For primaries with a convective envelope, mass loss leads to a rapid expansion soon after the beginning of the RLOF. As a result the secondary will be dragged in by the primary: this is the common envelope phase of a case Bc/case C binary. Binaries consisting of a massive primary and a low mass secondary ($q \leq 0.2$) will not enter a RLOF stage. Instead, this secondary will be swallowed into the envelope of the primary. Also here we may call this a common envelope stage. As energy sources in a massive binary that can remove matter from a common envelope we know stellar nuclear radiation energy and orbital energy. Stellar nuclear radiation can drive a stellar wind from the common envelope; the rate is comparable to that of supergiants, i.e. a few times$10^{-6} - 10^{-5} M_{\odot}/\mathrm{yr}$.

When the orbital energy is used in an efficient way this can lead to larger mass loss rates. Perhaps, orbital energy of the secondary is transformed into thermal energy of the common envelope (viscosity could play an important role) and the secondary spirals-in into the envelope. A part of the thermal energy goes into radiation; another part ejects matter of the common envelope out of the system.

As illustration of the consequences of this common envelope /spiral-in process, we consider a 20 $M_{\odot}$ OB type star with a companion of 2 $M_{\odot}$. We assume maximum efficiency for the transformation of orbital energy into escape energy. From our computations we know that $M_{1e} = 7\ M_{\odot}$, so we can estimate that the removal of $13 M_{\odot}$implies a reduction of the period by a factor 2000. We therefore conclude that the spiral-in process in a binary with very small mass ratio is able to remove a significant fraction of the mass of the primary star combined with a severe reduction of the orbital separation and consequently of the period.

The variation of the period by stellar wind mass loss.

We consider the orbit as circular during the mass loss phase, and the stellar wind mass loss as spherically symmetric. ω is the orbital angular velocity, and r_1 and r_2 the orbital radii (with respect to the center of mass of the system) of the stasr with mass M_1 and M_2 , dM_1 and dM_2 the mass of infinitesimal shells that leave the stars. The angular momentum loss is then

$$dJ = \omega {r_1}^2 dM_1 + \omega {r_2}^2 dM_2 \tag{5.45}$$

$$= \frac{2\pi}{P} \frac{{M_2}^2 A^2}{(M_1 + M_2)^2} dM_1 + \frac{2\pi}{P} \frac{{M_1}^2 A^2}{(M_1 + M_2)^2} dM_2 \tag{5.46}$$

$$= J[\frac{M_2}{M_1 + M_2} \frac{dM_1}{M_1} + \frac{M_2}{M_1 + M_2} \frac{dM_2}{M_2}] \tag{5.47}$$

We use now eq. 5.14 with e=0 and find

$$\frac{dJ}{J} = \frac{dM_1}{M_1} + \frac{dM_2}{M_2} - \frac{1}{2} \frac{d(M_1 + M_2)}{M_1 + M_2} + \frac{1}{2} \frac{dA}{A} \tag{5.48}$$

Combination of the last two (5.47 and 5.48) gives:

$$\frac{dA}{A} = \frac{d(M_1 + M_2)}{M_1 + M_2} = \frac{1}{2} \frac{dP}{P} \tag{5.49}$$

hence

$$\frac{P}{P_o} = [\frac{M_{1o} + M_{2o}}{M_1 + M_2}]^2 \tag{5.50}$$

Mergers.
For mass ratios q below 0.2, the low mass companion is swallowed by the massive component. This spiral-in process leads to drastic period reduction. So, when the low mass component is a normal CHB star, both stars will probably merge. The low mass secondary is practically a homogeneous star at the beginning of the spiral-in phase. Therefore the evolution of the massive component after merging is probably similar to the evolution of an accretion star with as mass the sum of its own mass and the mass of the low mass star.

This can also happen in binaries with initial q > 0.2, for non-conservative RLOF, leading to a significant reduction of the orbital period as predicted by equations 5.43 or 5.44. The further evolution is uncertain. Possibly the evolution of the merger could be similar to the evolution of a star that accreted an amount of mass equal to the mass of the secondary. However there remains the difficulty that the chemical composition of the secondary can be far from homogeneous, so that the effect of merging is not predictable.

The case when the loss rates by RLOF and by SW are comparable.
If the mass of the primary in VMCBs exceeds 40-50 $M_\odot$, the RLOF competes with the LBV stellar wind mass loss process. When the SW is large enough, the RLOF process can be avoided. In this case the period variation is entirely determined by equation 5.50, implying a significant binary period increase. When the SW and the RLOF mass loss rates are comparable, one can obtain a first order estimate of the period variation by combining equation 5.50 (to determine the period variation caused by the SW mass loss) and either equation 5.38 (conservative RLOF) or equation 5.43 or 5.44 (non-conservative RLOF). If the SW mass loss and the common envelope/spiral-in act togetherone has to take into account the binding energy of the common envelope.

5.3.4 The Evolution of the Primary Before and During its RLOF

Before the start of the RLOF, the two components of a MCB evolve as rotating single stars of the same mass. Most of the evolutionary computations of MCBs are performed using the structure equations established for a non-rotating spherical symmetric plasma (equations 5.1 to 5.7). Comparison of the observed rotational properties of MCB components with the v_e-distribution of massive single stars reveals that MCB components have rotational velocities (on the average) smaller than or equal to those of single stars.

In primaries with masses beyond 40-50 $M_\odot$, large LBV like SW mass loss occurs. Consequently the amount of matter lost by a primary during its RLOF is largely reduced. For very large SW, the RLOF will not even begin. After the removal of most of the H-rich layers by SW and possibly by RLOF the primary evolves into a WR star; SW mass loss continues at high rates (equation 5.50). In table 5.5 we summarize some evolutionary properties of the orbital elements of galactic binaries where this type of SW dominated evolution applies. We call ΔM_{SW} the total mass lost by the primary due to SW, and ΔM_{RLOF} the mass lost due to RLOF ($\Delta M_{SW} + \Delta M_{RLOF}$ is then the total mass of the H-rich envelope of the primary). The period variation caused by the spherical symmetric SW is given by equation 5.50.During thr RLOF the loss/transfer of ΔM_{RLOF} implies a period variation; for the 60 $M_\odot$ + 5 $M_\odot$ binary, the spiral-in formalism was adopted . Note that, even when the spiral-in process is responsible for the removal from the 60 $M_\odot$ star of only a small amount of mass, the orbital period reduction is very significant. Later on it will become clear that this reduction is necessary for the explanation of the low mass X-ray binaries with a BH component, i.e. low mass X-ray binaries with a BH component provide indirect evidence that LBV mass loss is not large enough for a very massive primary to escape the RLOF.

Table 5.5. The change of the period evolution in a binary with a 60 $M_\odot$ primary; during an LBV SW phase the star loses 10 $M_\odot$; during an ensuing RLOF phase the starloses an additional 6 $M_\odot$; the results are compared to the case where the whole 16 $M_\odot$ are lost by LBV SW; P is the period, P_o the initial period, SWR is the start of the WR phase, SWC is the beginning of the WC phase, ECHeB is the end of CHe-burning. CHeB; DMSW-O DMSW-LBV, DMRLOF are resp. the total mass lost by the primary during the O-type phase, the LBV phase and during the RLOF/spiral-in phase

System	ΔM_{SW-O}	ΔM_{SW-LBV}	ΔM_{RLOF}		$(\frac{P}{Po})_{SWR}$	$(\frac{P}{Po})_{SWC}$	$(\frac{P}{Po})_{ECHeB}$
60+30	12	16	0	-	2.2	3	5.2
-	12	10	6	b=0.5	1.1	1.5	2.5
-	-	-	-	b=1	1.8	2.4	4.1
60+10	12	16	0	-	2.8	4.5	11.5
-	12	10	6	b=0	0.2	0.3	0.8
60+5	12	16	0	-	3.1	5.4	17.4
-	12	10	6	spiral-in	0.03	0.05	0.2

Best-fit relations between the initial (M_i) and final mass (M_f) for initial mass between 3 and 7 $M_\odot$:

$$Galaxy: \quad M_f = 0.1 M_i^{1.32} \tag{5.51}$$

$$LMC: \quad M_f = 0.11 M_i^{1.32} \tag{5.52}$$

$$SMC: \quad M_f = 0.124 M_i^{1.32} \tag{5.53}$$

for initial mass between 7 and 15 $M_\odot$:

$$Galaxy: \quad M_f = 0.093 M_i^{1.44} \tag{5.54}$$

$$LMC: \quad M_f = 0.085 M_i^{1.52} \tag{5.55}$$

$$SMC: \quad M_f = 0.048 M_i^{1.7} \tag{5.56}$$

5.4 The Evolution of the Secondary

5.4.1 Accretion Effects

Mass transfer occurs either by direct accretion by the secondary, or by formation of a Keplerian disc. Matter falling onto a star has a velocity close to the free-fall velocity v_{f-f}

$$v_{f-f} = \sqrt{\frac{2GM}{R}} \tag{5.57}$$

This can be used to calculate the dynamical pressure P_{dyn} exerted by this matter:

$$P_{dyn} = \rho v_{f-f}^2 = \frac{\dot{M}}{4\pi R^2} v_{f-f} \tag{5.58}$$

$\dot{M}$ is the mass accretion rate.

For MCBs this dynamical pressure is always much smaller than the hydrostatic pressure and may be neglected. So to examine the reaction of a star accreting matter we may assume hydrostatic equilibrium. The gas stream produces a shock front near the stellar surface (Ulrich and Burger, 1976). The gravitational energy of this instreaming matter is converted into thermal energy and is radiated in the form of UV radiation and X-rays. Assuming that this process occurs with 100 percent efficiency, the radiated luminosity produced by accretion, hence called L_{acc} is

$$L_{acc} = \frac{GM}{R} \dot{M} \tag{5.59}$$

In MCBs, this luminosity only exceeds the critical Eddington luminosity when $\dot{M}$ is larger than $10^{-2} M_{\odot}/\mathrm{yr}$, which is one or two orders of magnitude larger than the mass transfer rates during a typical RLOF. We therefore conclude that the accretion luminosity does not alter the hydrodynamics of the in-falling matter. Although accretion of matter in a binary is local, thus not spherical we may assume that this matter is very rapidly redistributed over the stellar surface so we can model the effect of accretion by a spherical symmetric formalism. So the main effect of mass accretion is the compression of all mass layers in the stellar interior due to the increase of gravity and consequently the release of gravitational energy of these layers. This release is represented by the entropy term in the energy equation 5.10. Adopting the mass fraction m = $\frac{M_r}{M}$ as independent variable (see also Neo. et al., 1977), the energy equation reduces to

$$\frac{dS}{dt} = \frac{\partial S}{\partial t} - \frac{\dot{M}}{M}[\frac{\partial S}{\delta \ln m}] \tag{5.60}$$

The second term is called the homologous term, the first one the non-homologous term.

We can distinguish two cases, the standard accretion model, and the accretion induced mixing model (full mixing model).

Model for standard accretion. If we assume that the radiative equilibrium of the outer layers is not disturbed by accretion, in other words, if the entropy of the infalling matter is equal or larger than the entropy of the envelope of the mass gainer we are dealing with the 'standard accretion model'. The entropy gradient remains positive during the accretion, the homologous term in equation 5.60 leads to an enhancement of the luminosity: the star becomes overluminous with respect to its mass. This overluminosity becomes significant if the star does not have sufficient time to get rid of this excess energy (accretion time scale ($\approx \frac{M}{\dot{M}}$) shorter than the thermal time scale (equation 5.24) thus

$$\dot{M} \geq 3.210^{-8} \frac{RL}{M} \tag{5.61}$$

(in $M_{\odot}$/yr, R, L and M in solar units). As an example, a mass gainer of $15 M_{\odot}$ star at the end of CHB becomes significantly overluminous for an accretion rate $> 10^{-3} M_{\odot}$/yr. A significant overluminosity may cause the outer layers of the star to expand and so a situation can arise where also the mass gainer fills its critical equipotential surface: the binary is then a contact binary. The reaction of the convective core of an accreting star in the case of the standard model is very important. Secondaries in interacting binaries may have already performed a large part core hydrogen burning when RLOF of the primary starts. This means that a molecular weight gradient in its interior has been built up. An increase of the stellar mass leads to an increase of the stellar convective core, but this leads on its turn to the formation of a molecular weight barrier at its border, prohibiting a fast increase of the convective core. The treatment of convection and semi-convection is extremely important. The timescale of mixing depends on the diffusion coefficient.

We can consider two limiting cases:

1: diffusion in the semi-convective layers on top of the convective core occurs very fast. The boundary of the convective core is determined by the Schwarzschild criterion; the chemical profile of the elements in the semi-convective region satisfies the condition $\nabla_{rad} = \nabla_{ad}$.

2: the diffusion coefficient is small, so the diffusion process in semi-convective regions is slow (small diffusion coefficient). We may expect a much slower increase of the convective core when the star is accreting matter and a situation can even occur where the convective core does not increase.

During the RLOF, a primary may transfer to its companion mass layers, which have been nuclearly processed during an earlier evolutionary phase. This mass with a molecular weight larger than the molecular weight of the outer layers of the mass gainer, when accreted, causes an inverted μ-gradient. This situation is very unstable and initiates mixing. The process, commonly known as 'thermohaline convection', has been studied by e.g. Zahn (1983) and by Kippenhahn et al. (1980). They conclude that it is a very fast process, which may be treated as an instantaneous process. The numerical procedure for this fast mixing is then straightforward: the accreted mass can be mixed with the outer layers of the gainer until a situation occurs where the molecular weight of the mixed layers is equal to the molecular weight of the layer just below the mixed region. It is obvious that this process will produce stars with increased N and depleted C and O at their surface.

Model for accretion induced full mixing. Accretion of mass implies accretion of angular momentum; hence a mass gainer will spin up, which will induce mixing and thus destruction of the radiative equilibrium. If moreover, the entropy of the accreted matter is much lower than the entropy of the envelope of the gainer, convection will smear out the entropy profile. This convection zone will grow inwards. The homologous term in equation 5.60 vanishes in this convection zone; therefore the overluminosity will not be as large as in the standard accretion model. The limiting situation is of course the complete mixing of the whole star. We will call this the "model for accretion induced full mixing ".

5.4.2 Evolutionary Computations

1. Accretion case 1a.
   Diffusion in the semi-convective layers on top of the convective core occurs very fast.
   For accretion time scales, smaller than the thermal time scale of the CHB star, the mass gainer has a position in the HR diagram very close to that of a normal CHB single star of the same mass and composition. After the accretion phase, the further evolution of the star is practically similar to that of a normal single star of the same mass. If the accretion time scale exceeds the thermal time scale, the mass gainer becomes overluminous and its radius increases significantly. When this occurs in a binary, the mass gainer as well may fill its critical Roche volume and a contact system is formed. At the end of the rapid accretion phase, thermal equilibrium is restored very fast and the star moves in the HRD to the position of a normal single star of the same mass and composition. Also in this case the further evolution is normal, After accretion, the stars are rejuvenated. In the HRD, they are on a time-isochrone with a time scale lower than the real lifetime. For binaries with a mass ratio close to one, the RLOF starts when the secondary is also in its hydrogen shell burning stage. In Fig. 5.23 we show the evolution of a 12 $M_\odot$ hydrogen shell burning star accreting 8 $M_\odot$ and we compare the track with that of a 12 $M_\odot$ CHB mass gainer.
   The He core is fixed and somehow protected by the hydrogen burning shell. Accretion does not increase the core but instead produces an extended fully convective region on top of the H- burning shell. After the accretion phase the star is (slightly) underluminous for its mass,
   The most important feature is the further evolution of the gainer after the accretion process. The star has a He core, small for its mass, and, as consequence of a large convection region on top of the burning shell; this shell possesses a large amount of fuel. So a rapid expansion of the star does not occur during the H shell burning and CHeB phase. The star remains in the blue part of the HRD and does not make evolve towards the RSG region.

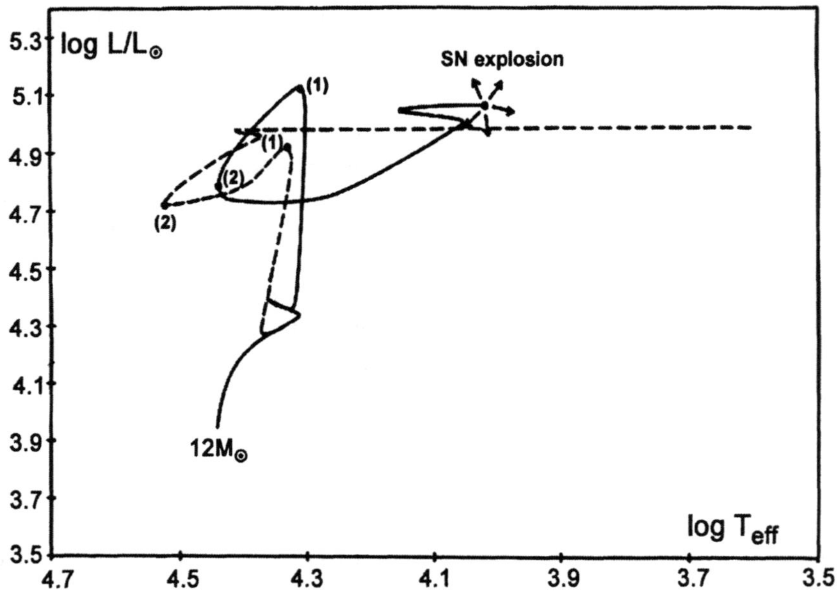

Fig. 5.23. The evolution of a 12 $M_\odot$ star accreting 8 $M_\odot$. A dashed track shows the evolution for accretion onto a CHB star, a full track for accretion onto a hydrogen shell burning star. Some evolutionary stages are indicated: (1) : end of the accretion phase: the star contracts and restores rapidly its thermal equilibrium; (2) : end of the contraction phase; the star has attained thermal equilibrium and has a position in the HR diagram normal for its mass and chemical composition. The evolution is continued till the end of CHB.

We did not continue our evolutionary computations after CHeB up to the SN explosion, but we can reasonably believe that the star remains a blue star till the end.

2. Accretion case 1b.

In this case the diffusion in semi-convective layers is very slow. The formation of a μ-barrier on top of the convective core right at the start of the accretion phase, inhibits a rapid increase of the core mass. Even when the gainer is a CHB star, a situation can arise which is comparable to the one for an accretion case 1a, and when the gainer is an H shell-burning star. Braun and Langer (1995) carried out detailed computations. After accretion, the stars are underluminous; rejuvenation is less pronounced, compared to case 1a. Even gainers, which were CHB stars at the beginning of the mass transfer, can remain blue stars during their entire further evolution after accretion. Furthermore, although the stellar mass was considerably increased, their core mass remained almost the same as before the accretion. Their further lifetime (and certainly their CHeB lifetime) may therefore be significantly larger than that of a normal single star of the same mass. This scenario explains (at least partly) the large number of AB-type supergiants observed in the blue Hertzsprung gap in the HR diagram.

3. Accretion case 2.
   Even for large accretion rates, the large expansion of the star, - typical for accretion case 1-, does not occur. When the star is fully mixed, the treatment of semi-convection is of no importance. Figure 5.24 shows evolutionary tracks of a 20 $M_\odot$ star accreting 5 $M_\odot$ at the end of CHB. The three accretion cases described above, are shown; and the evolution track of a normal 25 $M_\odot$ single star as well. The accretion tracks of the various models are replaced by a dashed line joining the HRD positions before and after the mass accretion. The rejuvenation of a mass gainer is very pronounced with the accretion induced mixing model. When the accretion process starts during the second half of the CHB phase of the gainer, the star after accretion is significantly overluminous andremains so during the further evolution. For a given spectral type and luminosity class of the gainer, the mass can be 20 to 30 percent lower than that of normal stars of the same class.

5.4.3 The Variation of the Chemical Abundances in the Surface of a Mass Gainer

During the second part of the RLOF, the primary transfers CNO processed matter. In this way, RLOF and thermohaline mixing produce stars with surface layers significantly N enhanced and CO depleted. If, owing to accretion, the star is fully mixed the effect is obviously much more pronounced. The surface H abundance in mass gainers is not very much affected in the standard case. With the accretion induced mixing process, on the contrary, the hydrogen abundance in the outer layers of the gainer can significantly be reduced.

5.4.4 The Further Evolution of a MCB after the Core-Collapse of the Primary

The evolution of the primary is finished when the core is composed mainly of iron and nickel, and nuclear processing ceases. When the mass of the FeNi core is larger than a given critical value, the whole star will probably collapse and will form a massive black hole (BH). This critical value depends on the equation of state of matter where electrons and nuclei are degenerate, and this is not well known. Also the post-CHeB evolution of massive stars is not well understood, hence linking the critical core mass value to a critical value for the initial mass of a star is uncertain. Possibly above a minimum initial mass of 40-50 $M_\odot$ binary components might form BHs. Observations of massive X-ray binaries reveal that neutron stars are remnants of binary components with initial mass up to at least 40 $M_\odot$, and this is not in contradiction with the value quoted above. SN explosions do not occur when BHs are formed, hence it can be expected that the majority of binaries with primaries beyond $\approx 40 - 50 M_\odot$ will pass an LBV mass loss phase, possibly followed by a RLOF but, due to

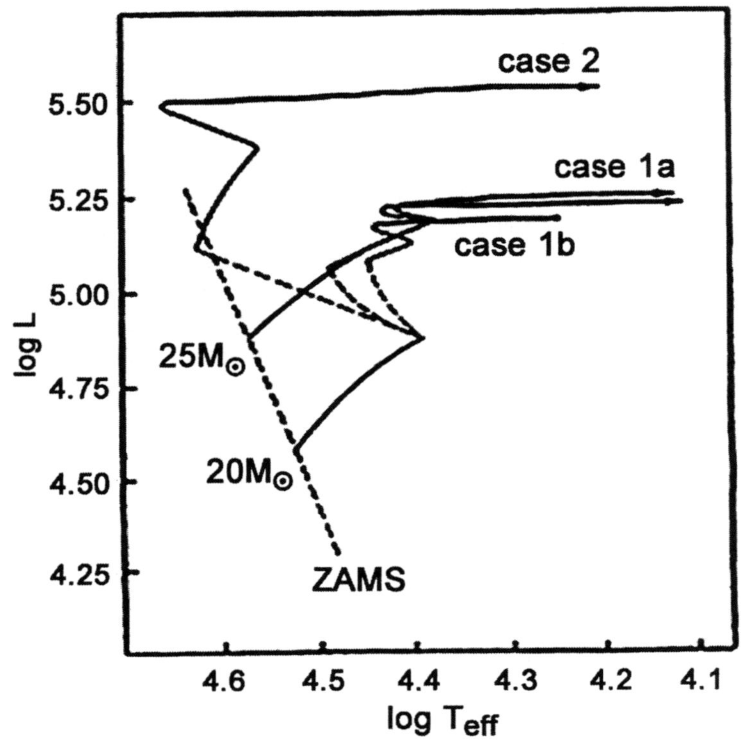

Fig. 5.24. Evolutionary tracks of a 20 $M_\odot$ star, accreting 5 $M_\odot$ at the end of its CHB phase, with the three accretion models 1a, 1b, 2, discussed above. For comparison also the track of 25 $M_\odot$ single star

the LBV stellar wind, with a reduced mass loss and mass transfer, and then form an OB+BH binary. For primaries with initial mass below 40-50 $M_\odot$, the final FeNi core will explode and finish as a NS. The mass layers outside the core are ejected. The event can possibly disrupt the binary. In this case, the further evolution of the OB component is that of a single star. Since there was accretion, the chemical composition may be different from that of a normal single star. We can consider the following possibilities:

- The mass of the OB type mass gainer is larger than 40-50 $M_\odot$. In this case we expect the further evolution of the gainer is similar that of a single star of the same mass and hence will evolve according to the LBV/RSG mass loss processes.
- The mass of the OB type mass gainer is smaller than 40-50 $M_\odot$. Primaries in binaries with a mass ratio not to close to one will evolve after the mass transfer phase like single stars of the same mass. When the initial mass ratio is close to one, the star can remain a blue star during its whole life. These stars will then afterwards explode as blue supergiants, producing events like

SN 1987A. When the SN explosion does not disrupt the binary, an OB+NS binary remains. The evolution of an OB+ cc binary (cc, a compact object, is a NS or a BH) depends on the orbital period, the mass of the OB component and the initial mass ratio q of the system. The q-value determines the evolutionary phase of the mass gainer at the start of the mass transfer.

When the mass of the OB type component exceeds the minimum mass of LBVs (40-50 $M_{\odot}$ for LBV's of the Galaxy), the OB star will lose a significant amount of its H rich layers by an LBV type SW, possibly followed by a spiral-in stage. The total amount of matter lost by the OB star by this spiral-in is strongly reduced by the preceeding SW mass loss. Such binaries will probably evolve into a core helium-burning star accompanied by a cc, (CHeB+cc); these are (WR+cc) binaries. Such systems are rare. When the mass of the OB type component is smaller than 40-50 $M_{\odot}$, we can consider three subclasses:

- The period is large enough to let the OB star evolve into a RSG before it fills its critical Roche volume; the OB star can then lose most of its H- rich layers by a RSG stellar wind and the binary evolves into a WR+cc, with a very large orbital period. Such large period systems are probably very rare.

- The mass gainer in the progenitor binary started late enough with accretion, so that afterwards it remains in the blue part of the HR diagram during its remaining lifetime. Probably the OB+cc binary will continue as OB+cc binary till the SN explosion of the OB star. This explosion will be of type II SN but from a blue progenitor, like SN1987A. The class of OB+cc binaries are candidates for massive X-ray binaries, so the X-ray lifetime of this class could be very long, possibly the entire CHeB lifetime of the OB type star.

- After accretion the OB star evolves more or less as a normal star; the period of the OB+cc binary is small enough to allow the OB star to fill the critical Roche volume before it evolves into a RSG. Later on the binary enters a spiral-in stage.

5.5 Evolutionary Computations for Massive Close Binaries and Observations

5.5.1 Introduction

We have now, for a large sample of binaries, with various masses, periods and mass ratios computed extended series of evolutionary tracks. The aim is to find out if we can match the characteristics of observed binaries by these tracks. Some groups of stars are in this respect very interesting, a.o. the Wolf-Rayet stars, the massive X-ray binaries, OB runaways, ...

5.5.2 The WR Stars

WR stars show in their spectrum emission lines of helium and nitrogen, -WN stars-, helium and carbon, -WC stars-, or helium and oxygen, -WO stars. The majority of the WR stars are hot hydrogen deficient core helium burning stars. WN stars show CNO equilibrium abundances, the WC and WO types show the products of the 3α-process. Some WR stars show WC and WN characteristics together: these are called WNC stars. The relative strengths of NIII, NIV, NV in WN stars and CIII, CIV in WC stars, allow a further subdivision of WR stars from WN2 to WN9, WC4 to WC9 (Smith et al., 1996). In most of the late type WN stars we see hydrogen in their spectrum; the majority of the early types are hydrogen deficient.

WN stars without hydrogen (mostly early types) are called WNE types; WN stars with hydrogen (mostly late types) are classified as WNL types (Vanbeveren and Conti, 1980). Figure 5.25 shows the WR area in the HR diagram. This has to be considered with much care, since the determination of T_{eff} in the case of WR stars, is very difficult. Owing to the very large stellar wind mass loss rates, the outflowing mantle is very dense and all photospheric information is hidden. The WR spectrum is the spectrum of the mantle; hence the derived temperature is the temperature of this mantle. Hence the temperature resulting from the NLTE analyses is too small, the real Teff is larger, perhaps 100000K. The WR region in the HR diagram is in the upper-left corner, $LogT_{eff} \geq 4.4$ and Log L/ $L_{\odot} \geq 4.5$.

For a discussion of their relation to other massive stars, one has to consider volume-limited samples and therefore, good distance determinations are required. If we are sure the WR star is a member of a well-populated stellar aggregate, the distance can be determined. For WR stars not belonging to a cluster or association, the problem is more difficult and one has to rely on WR subtype-absolute visual magnitude calibrations. Stellar wind mass loss rates for galactic WR stars were derived in a semi-empirical way. A linear best fit for stellar wind mass loss rates for galactic WR stars (but with very poor statistical significance (Hamann, 1994)): is given by

$$log(-\dot{M}) = 1.5logL - 12 \tag{5.62}$$

with M in $M_{\odot}$/yr, L in $L_{\odot}$.

WR stellar winds are not homogeneous, but they are clumpy; the use of homogeneous models could overestimate the mass loss rates for WR stars perhaps by a factor 3. For more details, and tables with calibration, we refer to Vanbeveren, Van Rensbergen and De Loore, 1998. The dependence of the WR mass loss rates on metallicity is also not clear.

5.5.3 WR+OB Binaries

For a number of WR+OB binaries, the WR mass loss may be driven from the orbital properties and the inclination. In the case there is not enough information, the mass of the OB-type component may be estimated from a mass-spectral

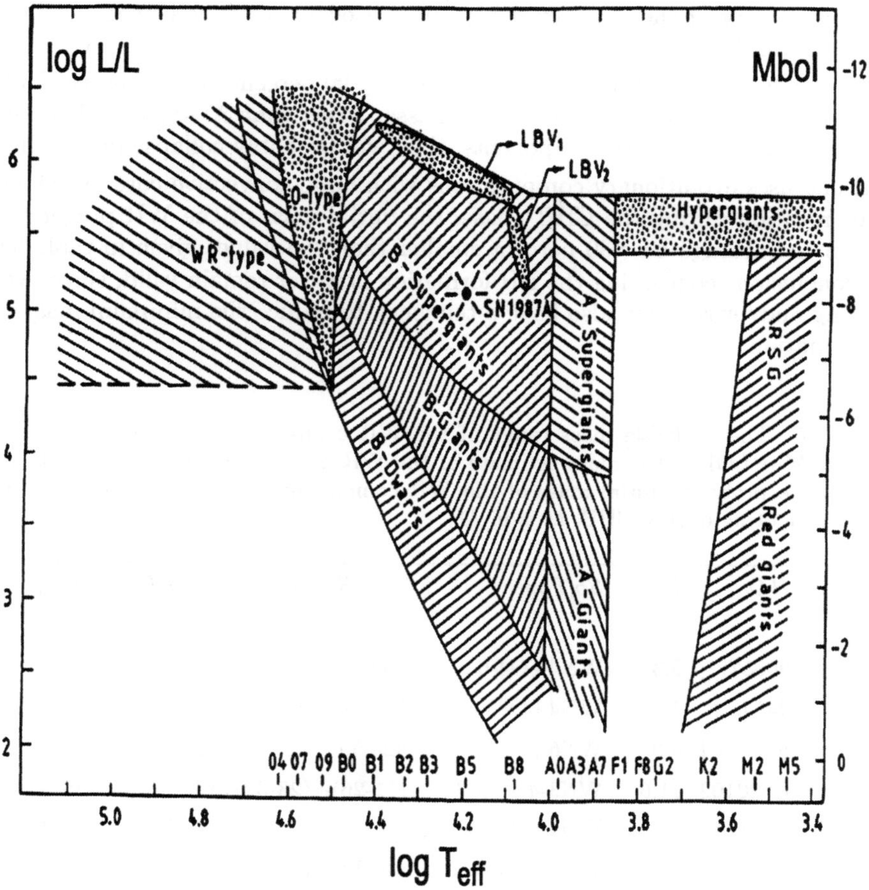

Fig. 5.25. The HRD area occupied by WR stars

type-luminosity class calibration. We can consider the mass gainers as normal stars calculated with the standard accretion model, or we can use the results for gainers calculated with the accretion induced mixing model during the mass transfer. From the evolutionary computations of hydrogen poor CHeB remnants after RLOF and the mass relation (5.31,5.32) (small overshooting during CHB), one can estimate the initial mass of these systems. For WR stars of the WNL or WNE- type, and assuming the star at the beginning of its WNL (resp. WNE) phase gives a minimum progenitor mass; assuming the star at the end of its WNL (resp. WNE) phase gives a maximum progenitor mass. When the WR star is a WC type, we get only results when it is assumed that the WC star is at the beginning of the WC phase. We may conclude that all the observed WC binary components seem to have an initial mass larger than 35 $M_{\odot}$. So probably for the majority of them no SN explosion will occur; they will finally collapse and form a BH.

A large fraction (at least 60 percent, possibly 90 percent) of the observed WN binary components seem to have initial masses larger than 30 $M_\odot$ and therefore, just as for the WC binary components, they will probably collapse into a BH.

If one uses evolutionary computations with large convective core overshooting or rotation during the CHB phase, the initial mass of the WR progenitors may on the average be ≈30 percent smaller than the values given in table 5.6. The conclusions concerning the BH formation however remain valid, since overshooting and/or rotation also have a tendency to reduce the minimum mass for BH formation.

Table 5.6. The probable mass range of the WR components in galactic WR+OB binaries; $M_\odot$RLOF is the mass estimate of the WR progenitor just after RLOF and M_{init} is the probable initial mass of the WR component assuming small convective core overshooting during CHB. M is in $M_\odot$

WRnr	*HD*	*P*	*Sp.Type*	$MWR + MOB$	$M_\odot RLOF$	M_{init}
21	90657	8.3	WN4+O4-6	15+29	17-21	35-45
31	94546	4.8	WN4+O8V	(9-13)+(20-30)	11-18	27-40
47	E311884	6.3	WN6+O5V	42+50	42-44	>70
97	E320102	11.6	WN3-4+O5-7	(11-26)+(23-52)	14-29	35-55
127	186943	8.6	WN4+O9/B0V	9+22	11-15	27-35
133	190918	111.8	WN4+O9Ib	(14-18)+(33-42)	16-23	35-50
139	193576	4.2	WN5+O6	9+26 12-15	30-35	
151	CX Cep	2.1	WN5+O8	11+21	13-17	33-38
153	211853	6.7	WN6+O6	(13-17)+(26-35)	16-23	35-50
155	214419	1.6	WN7+O	31+26	31-33	57-65
9	63099	14.7	WC5+O7	(16-25)+(25-40)	>22	>45
11	68273	78.5	WC8+O	9+23	>16	>38
30	94305	18.8	WC6+O6/8	(11-16)+(23-35)	>17	>37
42	97152	7.9	WC7+O7V	(12-18)+(20-30)	>18	>39
79	152270	8.9	WC7+O5-8	(8-19)+(22-52)	>15	>35
113	168206	28.7	WC8+O8-9III-V	12+24	>18	>39
140	193793	7.9yrs	WC7+O4-5	23+62	>29	>56
	V Sag	0.51	WN5?b+F8V	1+3.8	>1	>10(?)

5.5.4 WR+OB Systems without RLOF

1. LBV-Wind-made systems
   For primaries of MCBs larger than 40-50 $M_{\odot}$ the binary may attain the WR+OB phase without mass transfer. A spherically symmetric SW increases the orbital period of the binary, so these WR+OB binaries should have periods larger than 15 days for WN types, and larger than 20 days for WC types. Another point is that, since there was no accretion, the stars were not rejuvenated, hence the WR star and the OB type companion need to be on the same time-isochrone, i.e. the OB companions in these binaries should be preferentially late O-type giants or supergiants or early B type supergiants.

   The LBV scenario leads to good values for the spectral type of the OB type component of the WC binaries HD 94305 and HD 168206, and is probably a good evolutionary scenario. The O type components in E311884 and HD 193793 have an early spectral type; they cannot be explained if rejuvenation has not played a role. HD 68273 ($\gamma^2 Vel$) and HD 190918 are good candidates for the LBV wind scenario. Figure 5.26 summarises the evolution of both systems with the assumption that both systems passed an LBV phase rather than RLOF.
   In the figure is also sketched their further evolution. Primaries with initial mass larger than 40-50 $M_{\odot}$ will probably finally collapse and form a BH (no SN!). Both systems will evolve into a BH+OB binary, with a period of the order of 100 days. When the OB type component starts filling its Roche lobe, the BH will spiral-in into the atmosphere of the OB star. If the conversion of orbital energy into potential energy of the outer layers of the OB star the system after the spiral-in will be a WR+BH binary with a period of 5-8 hours. Around many single WR stars shell like structures seem to be present, possibly produced during previous efficient LBV- or RSG-SW mass loss.

2. RSG-Wind-made systems
   Primaries smaller than 40 $M_{\odot}$ may lose their hydrogen rich layers by a large SW during the RSG if the original binary period was large (> 600 days). The binary period increases with SW mass loss, so it can be expected that the period of post-RSG WR+OB binaries is larger than 1000 days. Since accretion (hence rejuvenation) did not occur in these binaries, just as for binaries evolving according to the LBV scenario, the OB type component should be a late O type giant/supergiant or an early B type supergiant. Three good candidates are HD 137603, HD 192641 and HD 193077.

5.5.5 WR+OB Systems Made by RLOF

In MCBs where the SW of the primary is sufficiently small to allow RLOF, the removal of an important fraction of the matter lost by the primary due to

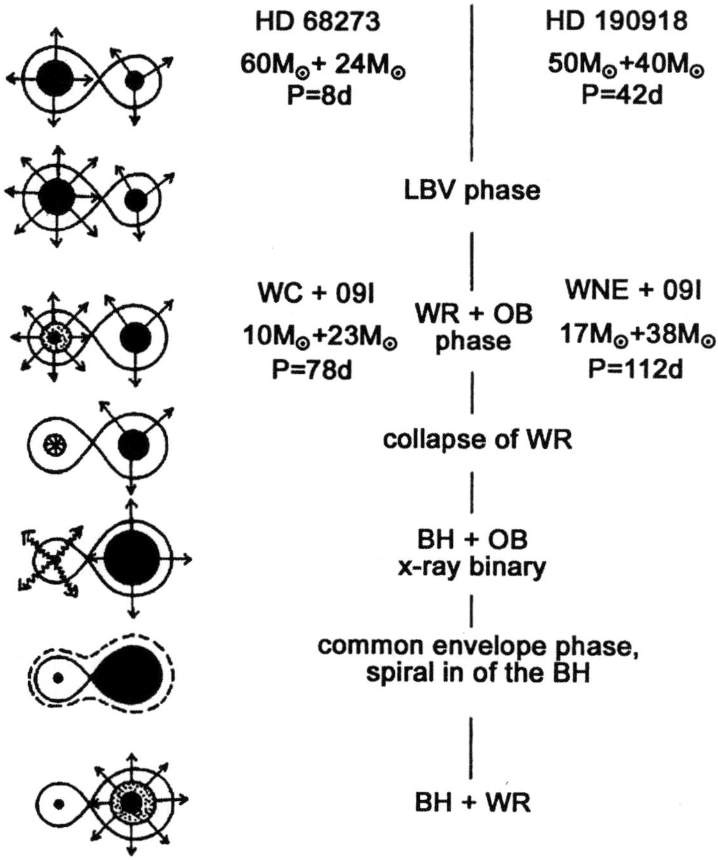

Fig. 5.26. Evolution scenario for the WR binaries $\gamma^2 Vel$ and HD 190918, LBV wind scenario candidates

RLOF and flowing out of the system leads to a very large reduction of the orbital period.

Mass transfer during RLOF rejuvenates the mass gainer, so when the OB component in a WR+OB binary looks younger than the WR star, this is a proof that mass transfer occurred. Inspection of the spectral types of the OB stars in observed WR+OB binaries, reveals that mass transfer must have taken place in the binaries HD 90657, HD 94546, E320102, HD 186943, HD 193576 (V444 Cyg), HD 63099, HD 97152, HD 193793(?). All these systems are practically circularized, except HD 193793, and this indicates that probably tidal effects played a significant role. An example of the class of binaries where mass transfer must have played an important role, is V444 Cyg. The observed mass of the WNE component ≈ 9 $M_{\odot}$, so from our evolutionary catalogue we derive as mass

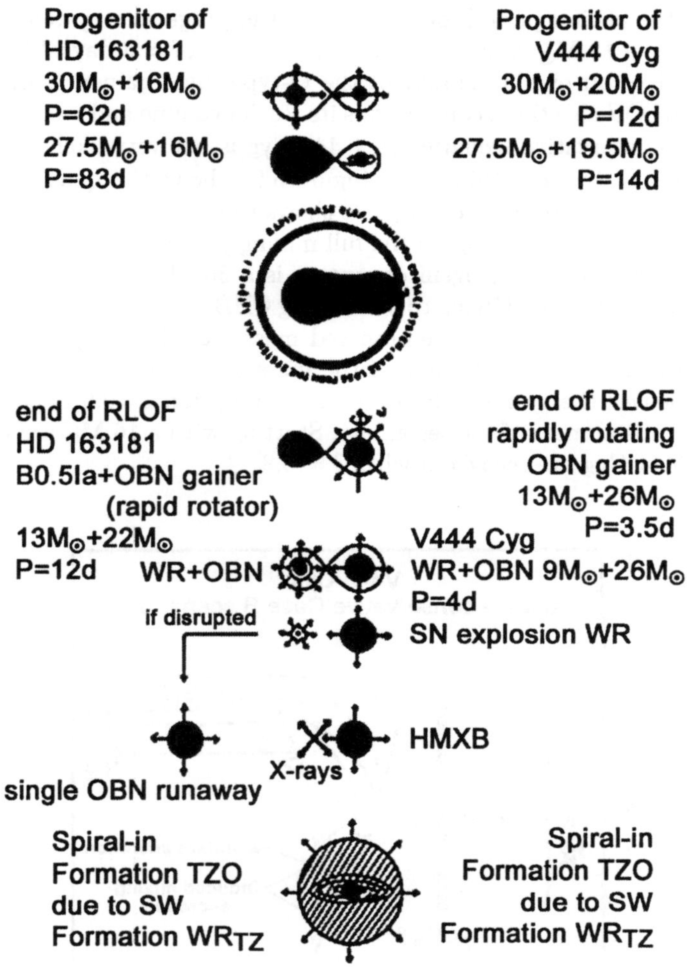

Fig. 5.27. Evolution scenarios for the OBN+OB binary HD163181 and the WR+OB binary V444 Cyg

of the progenitor ≈ 30 M$_\odot$ and an age of the binary of 7 million years. The OB companion is an O6 star. The age of a normal O6 star is 1-2 million years, hence the OB star must have been rejuvenated, and this means that mass accretion must have occurred. The observed mass of the O6 component is $\approx 25 - 26M_\odot$. A normal O6 star has a mass of 37 M$_\odot$ (when it is a class V) and $\approx 48M_\odot$ (when it is a class III). Therefore we can conclude that either the O-type component is not an O6-type star but rather an O7-O8, or the O6 component is significantly undermassive for its spectral type, in other words, overluminous for its mass. If one adopts a conservative RLOF, to obtain the observed post-RLOF component masses and the orbital period, one finds that the progenitor system must have been a 30 $M_\odot$ + 12 $M_\odot$ case A binary with a period of $\approx$ 3 days. This scenario does not lead to an overluminosity of the O-type mass gainer. Soon after the start of the RLOF also the secondary fills its Roche volume and a contact system is formed. A conservative scenario for V 444 Cyg is not very likely.

So one has to try something else. Figure 5.27 shows the quasi-conservative MCB evolutionary scenario and Fig. 5.28 shows the evolutionary tracks. For the evolution of the gainer we adopted the full mixing process and treated accretion in the standard way. The progenitor system is a 30 $M_\odot$ + 20 $M_\odot$ MCB with initial period P = 46 days. Owing to SW during CHB, the orbital period increases slightly to 54 days. To match the observed masses of the components of V444 Cyg we had to assume β=0.5 i.e. half the mass lost by the primary during its RLOF also leaves the system. However the assumption β=0.5 depends on the choice of the initial mass of the secondary. Starting with a 16 $M_\odot$ secondary, the masses of V444 Cyg are recovered with β = 0.8.

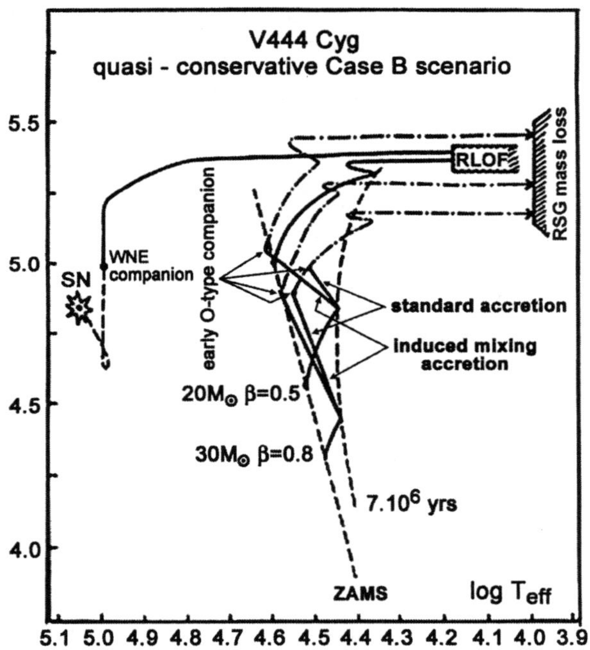

Fig. 5.28. Case B evolution for the WR+OB binary V444 Cyg. It is assumed that the RLOF process was quasi-conservative

The period at the end of the RLOF is 3.5 days: the system matches a WNL + OB binary. Further mass loss by SW removes the remaining hydrogen rich layers. When the WR star has evolved into a WNE star, its mass is 9 $M_\odot$, the period is 4.2 days. At the end of the mass transfer, when the secondary has been fully mixed, the O6-type component may be an overluminous star with mass M= 26 $M_\odot$ and a homogeneous composition (X, Y, Z) = (0.5, 0.48, 0.02).

The further evolution of V444 Cyg.:

The WR star will explode leaving as remnant a NS. If the binary is disrupted, the single OB-type runaway will become a RSG. Mass loss by SW will remove the hydrogen rich layers and the star will evolve into a WR star with mass between 10-14 $M_\odot$, depending on the previous RLOF, conservative or quasi-conservative, and on the adopted accretion model, the standard accretion model or the accretion induced mixing model. If the system is not disrupted, the binary will evolve into a spiral-in phase. The observed period is 4.2 days, so the NS will spiral-in completely and a Thorne-Zytkow-object (TZO) will be formed. Further SW mass loss during the RSG phase of this TZO will remove its H-rich layers and a WRTZ is formed. Figure 5.27 shows the evolutionary similarity, -the present situation, the component masses -, of the binary HD 163181 and V444 Cyg. HD 163181 is on its way to evolve into a WR binary like V444 Cyg.

5.5.6 WR Stars with an OB Type Companion of Lower Luminosity

Important mass transfer will not occur occur in OB+OB binaries where the LBV SW mass loss was very efficient, or if the period is sufficiently large so that the primary suffers a RSG stellar wind mass loss, before starting RLOF, or if the binary is a case Bc type evolving through a common envelope, Since there was no mass transfer, the mass ratio of the binary (MOB/MWR) may be significantly smaller than in those WR+OB binaries where mass transfer has occurred, and it will be more difficult to detect the OB companion. The WR stars WR19, WR104 and WR119 (catalogue of Van der Hucht et al., 1981) are possibly binaries with a fainter OB type; their progenitors might be MCBs that evolved according to one of these scenarios. We can expect the presence of shell or ring structures as observed around some single WR stars.

5.5.7 WR Stars with a Normal Low Mass Companion

Her X-1-like binaries, and LMXBs with a BH component, are an indirect proof of the existence of MCBs with extremely small mass ratio with a rather high frequency. When the primary in such binaries is sufficiently massive and when the LBV SW mass loss was very efficient, a WR accompanied by a normal low mass companion is formed with a period of the order of some days to a few decades. For primary masses below 40-50 $M_\odot$ and for sufficiently large binary periods, the binary will evolve according to the RSG scenario, leaving as remnant a WR star and a normal low mass companion with a very large period (order of years). Also a case Bc binary with an initial mass ratio below 0.2, may evolve into a WR accompanied by a normal low mass companion after a SW/common envelope phase. Figure 5.29 illustrates the MCB evolution leading to the formation of binaries with a CHeB star (WR) + low mass companion.

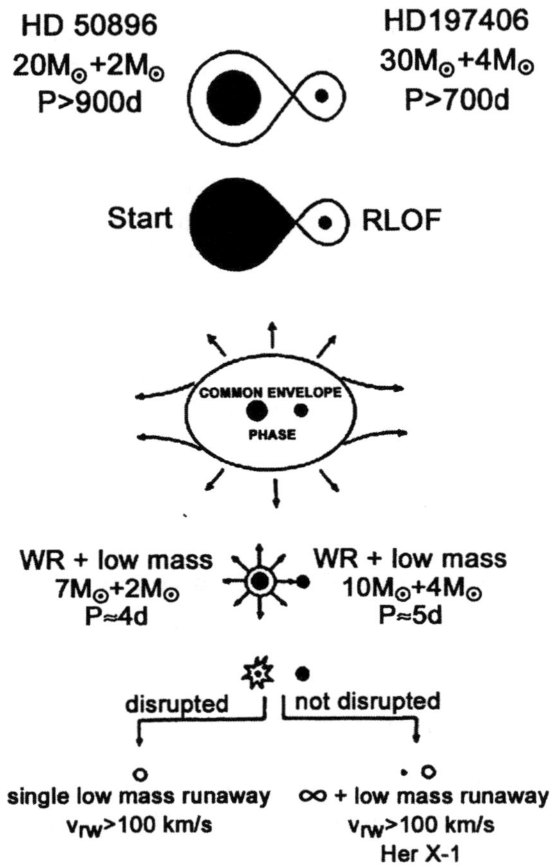

Fig. 5.29. Scenario for MCBs with small mass ratio, evolving into WR stars with a low mass solar type component (possible evolutionary scenario for HD 50896 and HD 197406)

CHeB (WR) stars with an initial mass below $\approx 40M_\odot$ will end their evolution as NS accompanied by a SN explosion; if their mass is larger than 40-50M$_\odot$ they will collapse into a BH without a SN. If the SN explosion does not disrupt the system, the cc+low mass star system may further evolve into a LMXB, like Her X-1 for the lower masses, or an LMXB with a BH candidate for larger masses. Binaries with primaries below 40 M$_\odot$, hence where a SN occurs, will in most cases be disrupted. and will leave a single NS and a low mass runaway with a large peculiar space velocity, exceeding 100 km/s. HD 50896 and HD 197406 are two WR candidates with a low mass companion orbiting with a period of a few days,

5.5.8 The Descendants of CHeB+OB Binaries

The CHeB remnant after RLOF in MCBs evolves into an NS or a low mass BH preceded by a SN explosion, or the CHeB star can collapse completely (without a SN explosion) and form then a massive BH. The SN explosion can lead to the disruption of the binary. The evolution of the OB-type star determines the further history of the binary (or remaining single star). OB stars with a compact companion may become HMXB.

5.5.9 X-ray Binaries

About 200 Galactic X-ray sources are known, with fluxes in the energy range 1-10 keV (hard X-rays). Some strong sources were found in the Magellanic Clouds. From the observed fluxes X-ray luminosities $L_X \approx 10^{33} - 10^{38}$ erg s^{-1} were derived. Zeldovich and Guseynov (1966) suggested that at least some strong X-ray sources are rotating neutron stars, accreting matter expelled by a companion star. This was confirmed by Schreier et al. (1972); they showed that the source Cen X-3 pulses with a period of 4.84 s; such a period is characteristic for the rotation period of a neutron star with a strong magnetic field, member of an eclipsing binary with a 2.087 day period.

1. High mass X-ray binaries (HMXB)
   HMXBs are X-ray binaries with as optical component a massive star. Some 70 HMXBs have been found. For 22 systems an orbital period could be derived. They are listed in Table 5.7. For pulsating X-ray sources, the pulse period is given. One can distinguish two subclasses:
   (a) HMXBs with an OBe type optical component
   OBe/X-ray binaries have orbital periods between 15 days and a few hundred days. OBe/X-ray binaries have spectral types earlier than B2. The X-ray component is pulsating hence indicating it is a rotating, magnetic NS. OBe optical components lay well within their Roche volumes. They lose matter by stellar wind. As a consequence of their high rotation velocities a large part of this stellar wind matter may be confined within a disc. X rays are produced by accretion of this disc matter by the neutron star.
   (b) Standard HMXBs
   The optical components of standard massive X-ray binaries are giants or supergiants. They have very small eccentricities and their orbital periods are in most cases smaller than 10 days. They are persistent X-ray sources. There are two groups of HMXBs:
   - a large group where the X-ray source is pulsating and where most likely a rapidly rotating NS is operating (the spin period equals the pulse period) with strong magnetic fields
   - a smaller group with non-pulsating X-ray sources (ex. Cyg X-1, LMC X-3)

1. Pulsating standard HMXBs.

If the optical star in an X-ray binary is identified, the system is eclipsing and the X-ray source pulsates, accurate mass estimates of the two components can be obtained. From the Doppler shifts in the spectra at various orbital phases and Doppler tracking of the pulse period of the pulsar the radial velocities can be found.

When the optical star nearly fills its Roche lobe, it has the shape of a pear, which causes photometric variability. Comparing the theoretical light curves of nearly Roche lobe-filling stars and the observed light curve allows a determination of the inclination angle i. This has been performed for six standard pulsating HMXBs It turned out that the observed average mass value of the six pulsars is about 1.4 $M_{\odot}$, the Chandrasekhar limit of a neutron star.

2. Non-pulsating HMXBs.

From the mass- spectral type- luminosity class relation for the optical component the mass of the compact object can be derived. For LMC X-3, was found $M_X \approx 6$ $M_{\odot}$, a value beyond the upper mass limit of a stable neutron star; therefore the X-ray source is classified as a black hole. The most probable BH candidate is the X-ray source in Cyg X-1. In 1966 (Giacconi et al., 1967) discovered the X-ray binary Cyg X-3. The X-ray flux shows a 4.8 hour period modulation and Van Kerkwijk et al. (1992) classified it as a WR+cc system. Two other WR+cc candidates are HD 50896 (variability period: 3.76 days) and HD 197406 (variability period: 4.32 days). Both candidates sit far above the galactic plane. For HD 50896 the radial component of the runaway velocity is about100 km/s (Schmutz, 1997). If HD 197406 is a WR+cc, it might be a WR+BH binary (Drissen et al., 1986).

2. Low mass X-ray binaries (LMXB) with a MCB history

LMXBs are X-ray binaries with as optical component a low mass star. Many LMXBs are old binaries belonging to Globular Clusters. They are probably formed by close encounters, i.e. tidal capture or by exchange collisions of old neutron stars (Verbunt, 1990; Bhattacharya and Van den Heuvel, 1991). Some LMXBs are found in the galactic disc. A formation mechanism as for globular cluster sources is not very likely, since the stellar density in the disc is much smaller. More plausible is the 'accretion-induced collapse' process (Canal et al., 1990). A Roche lobe filling low mass blue star sends matter to a white dwarf companion (an O-Ne-Mg white dwarf) with a mass near the Chandrasekhar limit. When the mass is sufficiently large, the WD collapses, leaving an NS and a low mass companion. A few LMXBs are important for a better understanding of MCB evolution, the LMXB BH candidates and the X-ray pulsar Her X-1. Their characteristics are given in Table 5.8.

Table 5.7. Orbital data of known HMXBs. M_{cc} is the mass of the compact companion and $M_{o.s}$ that of the optical star ($M_\odot$)

Source	Sp.Type	$P_{orb}(d)$	e	M_{cc}	$M_{o,s}$	$P_{pulse}(s)$
HMXBs with OBe component						
A 0538-66	B2 IIIe	16.7	≥0.4			104
4U 0115+63	Be	24.3	0.34			3.61
E0236+610	B0e	26.5				
V 0332+53	Be	34.2	0.31			4.38
EXO 2030+375	Be	46	0.4			41.8
A 0535+26	O8.7 IIIe	111	0.2-0.4	104		
GX 304-1	B2Vne	133				272
4U 1145-679	B1Vne	188				292
Standard HMXBs						
A 0114+65	B0.5Ib	11.6				850(?)
X Per	O8.5III-V	580?				835
4U 1700-37	O6.5f	3.4	0.01	1.8±0.4	52±2	
Vela X-1	B0.5Ib	9	0.09	1.9±0.6	24±2.5	283
1E 1145.1-614	B2Iae	9.8				297
LMC X-1	O7-9III	4.22				
4U 1538-52	B0I	3.7		1.1±0.4	21±9	529
Wray 977	B1.5Ia	41.5	0.47	48		696
SMC X-1	B0I	3.9	0	1.2±0.3	17±4.5	0.78
Cen X-3	O6-8(f)p	2.1	0	1.1±0.6	21±4	4.84
LMC X-4	O7III-V	1.4	≤0.1	1.5±0.4	13.5	
Cyg X-1	O8.7Iab	5.6	0			
LMC X-3	B3V	1.7	0.13			
HMXBs with a WR type component						
Cyg X-3	WR?	0.2				

5.5.10 HMXBs with OB-Type Components

Massive X-ray binaries (HMXBs) are the end products of the evolution of MCBs, after explosion of the evolved mass gainer, leaving a compact companion. For spherically symmetric SN explosions, the OBe/HMXBs should have eccentricities ≤ 0.1; the observed values however are much higher (Table 7). This can be interpreted as indirect (but strong) evidence that the SN explosion was asymmetric. The available information derived from the masses of the OB components in standard HMXBs is more reliable. For evolutionary purposes, the three HMXBs Vela X-1, Wray 977 and Cyg X-1 are particularly interesting.

Table 5.8. LMXB candidates with an MCB history (Van Kerkwijk et al, 1995 (Her X-1); McClintock, 1992, (A0620-00, GS 2023+338, GS 1124-68)

System	LX (erg/s)	Sp.type type	Period (days)	f(m) (Mo)	pulse period (sec)
Her X-1	2.10^{37}	A	1.7	0.9	1.24
LMXB black hole candidates					
A0620-00	10^{38}	K5V	0.	2.91 0.08	-
GS 2023+338	2.10^{38}	K0III-G9V	6.47	6.26 0.31	-
GS 1124-68	10^{38}	K0-4V	0.433	3.07 0.4	-

- Vela X-1.
  In the atmosphere of the supergiant optical companion we see an overabundance of He; one expects also an overabundance of N and CO. From the orbit and X-ray eclipse one can derive the mass and radius of the supergiant: R = 28 - $35_{\odot}$, M = 21.5 - $26.5 M_{\odot}$ From a combination of NLTE results, orbital and X-ray eclipse observations, follows: $\frac{logL}{L_{\odot}} = 5.5 - 5.$ and a distance d of 1.8 - 2 kpc.
  Using the annual proper motions and the radial velocity, Van Rensbergen, Vanbeveren and De Loore (1996) found that Vela X-1 left the association Vela OB1 some 2$\pm$1 million years ago. The distance of Vela OB1 is$\approx$ 1.9 kpc (Blaha and Humphreys, 1989), close to the value for Vela X-1 and thus confirming the proposition that the progenitor of Vela X-1 was a member of the association.
  A first attempt to match the observations is to compute evolutionary models, treating the accretion in the standard way. After the accretion, the evolution of the star is continued up to the beginning of hydrogen shell burning. The have a final mass roughly equal to the observed mass of the optical star in Vela X-1. The tracks are depicted in Fig. 5.30. The accretion process is treated in the standard way. The numbers along the tracks are mass values. The position of the tracks can be compared with the observed HRD position (and masses) of the optical component of Vela X-1. Comparison of the HRD position of the B0.5 supergiant with these tracks, reveals that evolutionary models of mass gainers computed with the standard accretion model are unable to match the characteristics of the optical component of Vela X-1. The H and He abundance remain close to the solar values, and this is different of the ϵ value derived with actual NLTE atmosphere codes.

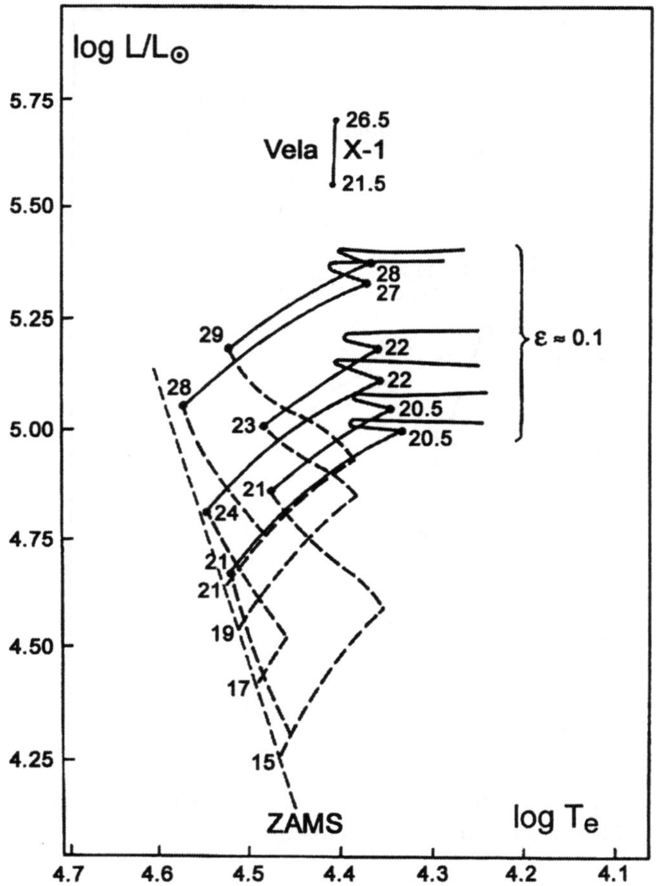

Fig. 5.30. Possible models for Vela X-1. Evolutionary tracks of mass gainers with masses of 15 to 21 $M_{\odot}$. Accretion in the standard way

The accretion process is treated in the standard way. The numbers along the tracks are mass values. The position of the tracks can be compared with the observed HRD position (and masses) of the optical component of Vela X-1. Comparison of the HRD position of the B0.5 supergiant with these tracks, reveals that evolutionary models of mass gainers computed with the standard accretion model are unable to match the characteristics of the optical component of Vela X-1. The H and He abundance remain close to the solar values, and this is different of the ϵ value derived with actual NLTE atmosphere codes. Since the standard accretion model does not work, we tried the accretion induced mixing model. Figure 5.31 shows the evolutionary tracks, calculated with full mixing. As mass gainers we selected secondaries where the accretion process begins at the end of CHB,

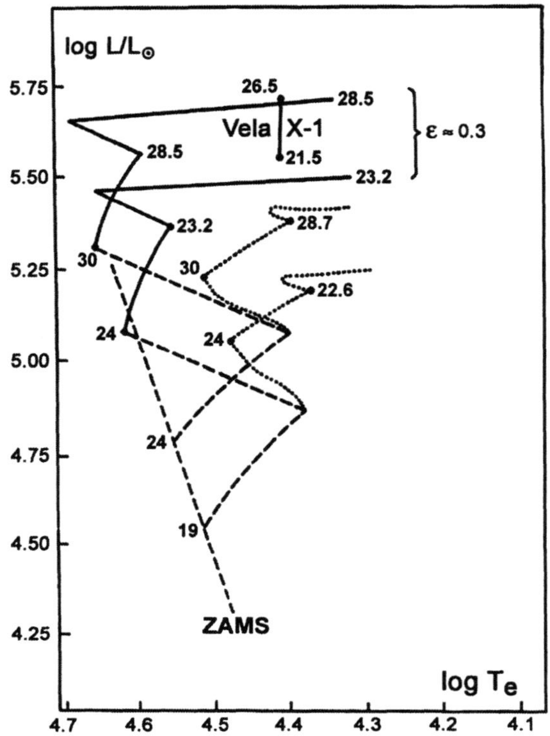

Fig. 5.31. Possible models for Vela X-1. Evolutionary tracks of mass gainers with masses of 15 to 21 $M_\odot$. Accretion by full mixing

corresponding to MCBs with initial mass ratio near one. The figure also gives the predicted H to He ratio.

The accretion induced mixing model is used. The position of the tracks can be compared with the observed HRD position (and masses) of the optical component of Vela X-1. These tracks are able to reproduce the optical component of Vela X-1 The accretion induced mixing model is able to reproduce the observed HRD position of the B0.5 supergiant of Vela X-1, and explains also explain the ϵ value. Evolutionary tracks for the progenitor of the binary Vela X-1, matching all observational characteristics, are depicted in Fig. 5.32. Figure 5.33 shows the evolutionary scenario. The model predicts that the SN explosion kicked the X-ray binary out of the association Vela OB1 less than 3 million years ago.

The binary Vela X-1 will evolve further by a spiral-in process of the neutron star into the B supergiant. A TZO will be formed. When the SW has removed the hydrogen rich layers, the star will have evolved into a Wolf-Rayet Thorne Zytkow object (WRTZ).

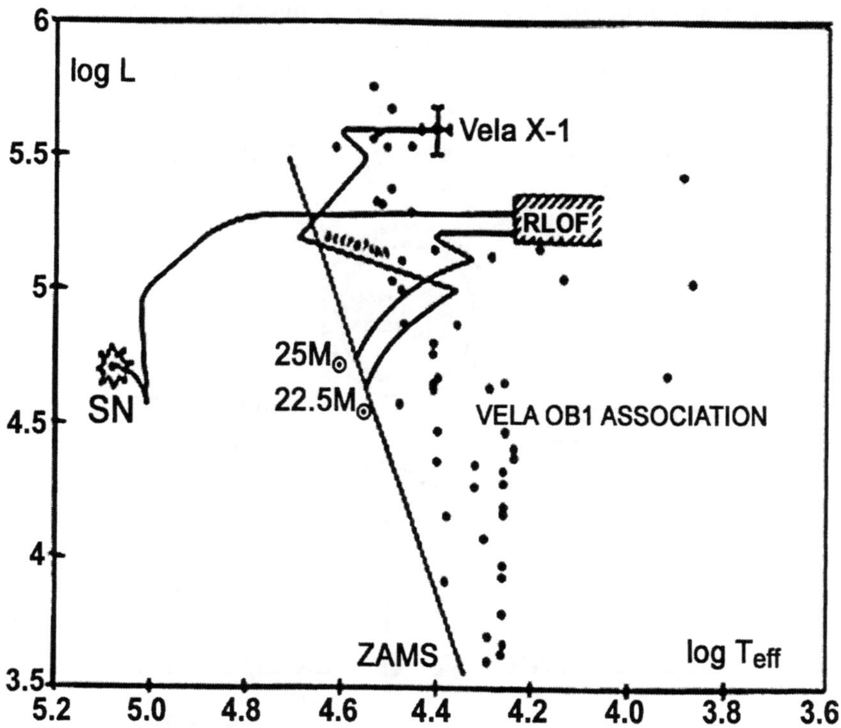

Fig. 5.32. Evolutionary tracks of a MCB explaining the observed properties of the HMXB Vela X-1. We started with a MCB of 25 $M_{\odot}$+22.5$M_{\odot}$. Check the HRD position of the optical component to the HRD of the members of the association Vela OB

- Wray 977

  The X-ray source Wray 977 is a pulsar, hence the compact object is an NS. The mass of the optical companion is very large, between 38 and 48 $M_{\odot}$ (Sato et al. (1986), Kaper et al. (1995).

  The evolutionary scenario is shown in Fig. 5.33. We tried out various scenarios, and found that that a non-conservative scenario, with β=0.5, and starting with initial masses between 39 and 46 $M_{\odot}$ can match the observations; if the RLOF is assumed to be conservative, these initial masses are reduced to 33 $M_{\odot}$ and 40 $M_{\odot}$ respectively. We can generalize and conclude that primaries in interacting binaries with initial mass up to 30 $M_{\odot}$ (perhaps up to 40 $M_{\odot}$) end their life as neutron stars, accompanied by a SN explosion). The further evolution of Wray 977 is depending on the SW mass loss of the supergiant and the effect on a possible spiral-in phase. If the mass loss rate is small, Wray 977 will evolve into a TZO (like Vela X-1). If the supergiant evolves into an LBV, the SW can reduce the spiral-in process and a WR+cc binary could be formed.

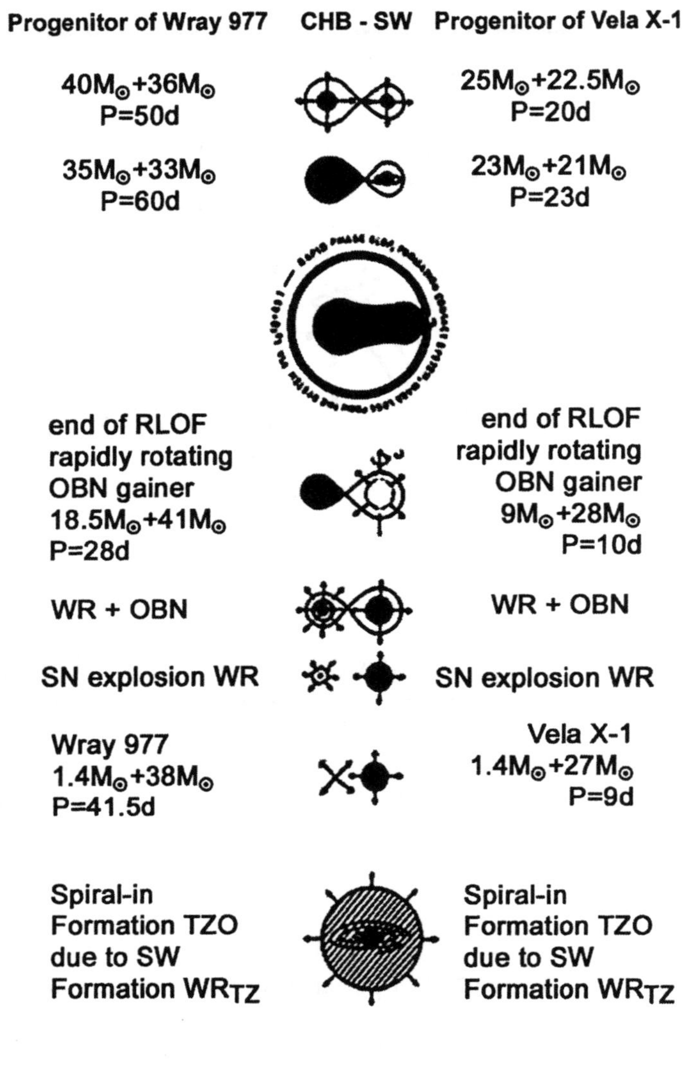

Fig. 5.33. Evolutionary scenario for Wray 977 and Vela X-1

If the compact object in 4U1700-37 turns out to be a neutron star, its evolution is probably similar to that of Wray 977.

- Cyg X-1
  A lot of uncertainties exist so the characteristics of this X-ray source should be handled with care. The optical star is an O9.7 Iab type star; 28000 K $\leq$ Teff *le* 33000 K; log $(L/L_\odot) \approx 5.4$ The mass of the optical component M $\geq$ 17 $M_\odot$; most probably (accounting for its luminosity), the mass $\approx$ 30 $M_\odot$,

He might be overabundant. $M_{cc} \geq 7\ M_\odot$; the most probable mass ranges between 10 $M_\odot$ and 16 $M_\odot$.
Since the mass M_{cc} of the compact star exceeds by far the maximum mass of a stable NS; it may be considered as a BH candidate. Observations of the HMXB Wray 977 seem to indicate that primaries with initial mass as high as 40 $M_\odot$ evolve into neutron stars. So the BH in Cyg X-1 originated most probably from a star with initial mass > 40$M_\odot$. If we may generalize we can conclude that some massive stars with initial mass $\geq$ 40 $M_\odot$ end their life with a mass $\geq$ 10$M_\odot$. Since black holes most probably result from stars with initial mass larger than 40-50 $M_\odot$, the progenitor binary probably followed the LBV scenario, without (or with very small) mass transfer.

5.5.11 The OB Type Runaways ζ Oph and ζ Pup

Supernova explosions occur in many cases in an asymmetric way. Hence a large fraction of the OB type stars that became runaway as a consequence of such explosions in binary systems, are expected to be in reality single.

1. ζ Oph

   ζ Oph is an O 8.5 V star. It has a peculiar velocity of about 46 km/s, so it is a runaway, and a very high rotational velocity ($v_e$sini = 365 km/s). The H-abundance $X_{atm} \approx 0.577$ (Herrero et al., 1992) is much lower than the solar value ($X_{atm} = 0.7$). The star is in the vicinity of the Sco-Cen OB association; this association consists of three subgroups, Upper Scorpius, Upper Centaurus and Lower Centaurus Crux. Van Rensbergen, Vanbeveren and De Loore (1996) showed that single star evolutionary computations cannot match the abundance constraint, the past trajectory of the star, the position of the three clusters and the assumption that the star was kicked out of one of these clusters.

2. ζ Pup

   This is a rapidly rotating O4If star ($v_e sini = 210$ km/s) with a runaway velocity of 65 km/s, /s and $X_{atm} \approx 0.546$ (Herrero et al. 1992). The distance $d \approx 400$ pc, so $log(L/L_\odot) \approx 5.8$. Van Rensbergen, Vanbeveren and De Loore (1996) examined the possibility that ζ Pup was ejected from one of the two clusters in the direction of this star, from Vela 2 or Vela R2, as single star as a consequence of cluster ejection less than 2.10^6 years ago and found that this is practically impossible.

3. Evolutionary scenarios

   We examine a massive close binary scenario for both runaways. Neither in ζ Oph nor in ζ Pup are indications for the presence of a compact object.

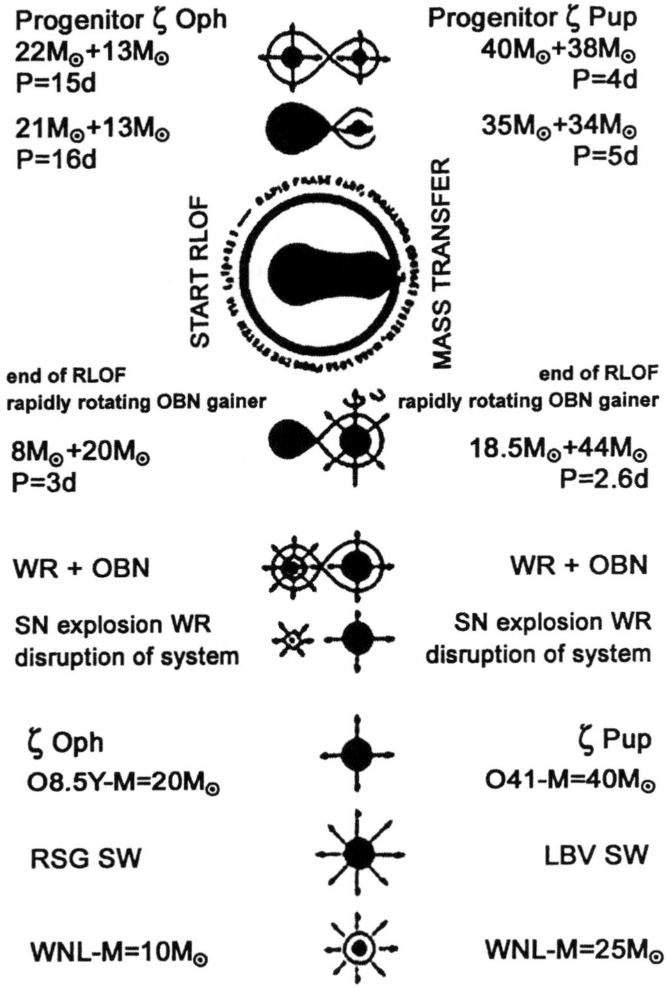

Fig. 5.34. A possible evolution scenario for ζ Oph and for ζ Pup

Hence we start from the assumption that the supernova explosion disrupted the binary. So the space velocity of both runaways is related to the masses of the components of the pre-SN binary ζ Oph.

The observed value of $X_{atm} \approx 0.577$ can be used as evidence for the accretion induced mixing model. A possible evolutionary scenario is shown in Fig. 5.34. We started from an initial system of 22 $M_\odot$ + 13$M_\odot$ and calculated the RLOF phase with β=0.5. After some 10 million years the primary exploded, and about 3 million years later the mass gainer hits the observed HRD position of ζ Oph. The binary originally belonged to Upp Cen Lup (a

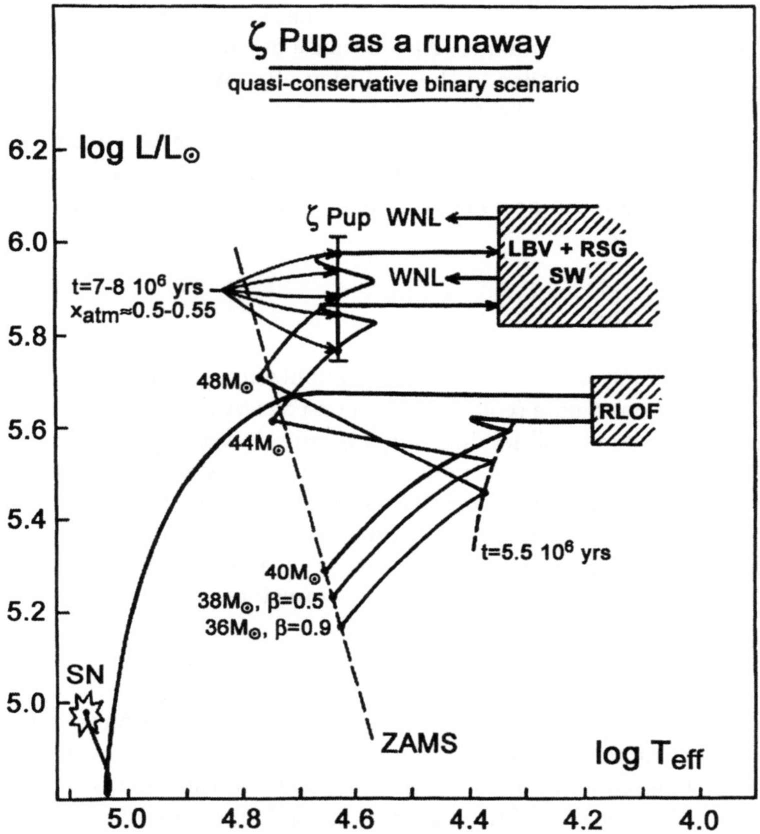

Fig. 5.35. The computed evolution of a MCB matching the observational characteristics of the runaway ζ Pup. The track track leading to the SN depicts the evolution of the 40 $M_\odot$ primary. Two mass gainers of 38 and 36 $M_\odot$ match the observed HRD position and the atmospheric chemical abundance of ζ Pup

subgroup of the Sco-Cen-OB association).

Also in ζ Pup the observed H abundance is very low, so just like forζ Oph we may assume that extended mixing occurred during the accretion process. Figure 5.35 depicts a possible evolutionary scenario of the runaway. As initial system we used a $40M_\odot$ + 38 $M_\odot$ massive close binary. After the mass transfer phase the primary leaves as remnant an $18M_\odot$ WNL star, the secondary is a 44 $M_\odot$ early O-type dwarf. The remaining H-rich layers are removed by SW mass loss, and the binary evolves into a WC star and an early O-type star. The SN explosion of the WC star disrupts the system. From the observed space velocity of 70 km/s, we can estimate a pre-SN period of ≈ 4 days; this limits the initial period of the binary. The mixing in the gainer during mass transfer explains the overluminosity of ζ Pup and the high surface He-abundance. Also because of this mixing the further evolution of the star is similar to the evolution of a single star but with initial (X, Y) abundances significantly different from normal i.e. (X, Y,) = (0.5-0.55, 0.48-0.43). The LBV and/or RSG type SW removes $\approx 15M_\odot$ more. It evolves into a 25$M_\odot$ WNL star. ζ Pup will probably end its life as a massive BH. The evolution can be compared to that of the optical component of the HMXB Wray 977.

5.6 Conclusions

Using detailed evolutionary simultaneous computations for massive close binaries with various constraints (conservative, non-conservative evolution, standard mixing, induced mixing, full mixing, starting from initial distribution functions for the period, the masses, the mass ratios, the binary frequency, taking into account the repercussion on the orbital parameters of supernova explosions, we succeded in explaining the observed features of a number of evolvced systems. Comparison of the computed evolution of these systems with observations allows to restrict possible values on a number of parameters. accounting properly for the effect of the supernova explosion.

Although the observations of massive stars are only complete for a restricted domain of our Galaxy, it is possible to find a massive star evolutionary model explaining most of the observations. The question if all this may be extended to the whole Galaxy or even to other galaxies is at the moment not solved. Improvement of the quality of the observations and extending the number of observations to more remote regions of our galaxy and to other galaxies is therefore very important. For more details on observations and evolution of massive single stars and binaries we refer to an extended review of Van den Heuvel (1993), a monograph of De Jager 'The Brightest Stars ', a monograph by Vanbeveren, Van Rensbergen and De Loore, 'The Brightest Binaries ', 'Stellar Structure and Evolution 'by Kippenhahn and Weigert, a review 'Massive Stars 'by Vanbeveren, De Loore and Van Rensbergen, 'Structure and Evolution of Single and Binary Stars ', by De Loore and Doom.

References

1. Bhattacharya, D., Van den Heuvel, E.P.J.: 1991, Phys. Rev. 203, 1.
2. Braun, H., Langer, N.: 1995, A.A. 297, 483.
3. Canal, R., Isern, J., Labay, J.: 1990, Ann. Rev. Astron. Astrophys. 28, 183.
4. De Jager,C.,1980, The Brightest Stars, D.Reidel Publ. Co.,Dordrecht
5. De Loore, C., Doom, C.: 1992, 'Structure and Evolution of Single and Binary Stars', Kluwer Academic Publishers, Dordrecht, The Netherlands.
6. Drissen, L., Lamontagne, R., Moffat, A.F.J., Bastein, P., Seguin, M.: 1986, Ap. J. 343, 426.

7. Eggleton, P.P., 1983, ApJ 268, 368.
8. Flannery, B.P., Ulrich, R.K., 1977, ApJ 212, 533.
9. Giacconi, R., Gorenstein, P., Gursky, H., Waters, J.R.: 1967, Ap. J. 148, Letters 118.
10. Hamman, W.-R.,1994, Space Sci. Rev. 66, 237.
11. Herrero, A., Kudritzki, R.P., Vilchez, J.M., Kunze, D., Butler, K., Haser, S.: 1992, A.A. 261, 209.
12. Kaper, L., Lamers; H.J.G.L.M., Ruymaekers, E., Van den Heuvel, E.P.J., Zuiderwijk, E.J.: 1995, A.A. 300, 446.
13. Kippenhahn, R., Weigert, A.: 1967, 'Stellar Structure and Evolution', Springer, Heidelberg.
14. Kippenhahn, R., Kohl, K., Weigert, A., 1967, Zeitschrift für Astrophysik, 66,58
15. Kippenhahn, R., Meyer-Hofmeister, E., 1977, A.A. 54,539
16. Kippenhahn, R., Weigert, A.: 1967, Zeitschrift für Astrophysik 66, 58.
17. Kippenhahn, R., Ruschenplatt, G., Thomas, H.C.: 1980, A.A. 91, 175.
18. Kruszewski, A., 1963, Acta Astron. 13, 106.
19. Lauterborn, D., 1969, in ' Mass Loss from Stars', ed. M. Hack, D. Reidel Publ. Com. Dordrecht, Holland, p.262 .
20. Levato, H., 1976, ApJ. 203, 680.
21. Limber, D.N. , 1958, Ap.J. **127**, 387
22. Lubow, S.H., Shu, F.H. 1975, Ap.J. 198, 383.
23. Lynden-Bell, D., Pringle, J.E.,1974, MNRAS 168,603.
24. McClintock, J.E.: 1992, in 'X-ray Binaries and Recycled Pulsars', eds. E.P.J. Van den Heuvel and S.A. Rappaport, Kluwer Acad. Publish. Dordrecht, p. 27.
25. S. Nakamura, M. Senoh, N. Iwasa, S. Nagahama: Jpn. J. Appl. Phys. **34**, L797 (1995)
26. Neo, S., Miyaji, S., Nomoto, K., Sugimoto, D.: 1977, Publ. Astron. Soc. Japan 29, 249.
27. Packet, W.: 1988, Ph. D. Thesis, Vrije Universiteit Brussels.
28. Paczynski, B.: 1967, Acta Astronomica 17, 355.
29. Penny,L.R., 1996, Ph. D. Thesis, Georgia State University
30. Plavec, M., 1970, IAU Coll. on Stellar Rotation.
31. Sato, N., Nagase, F., Kawai, N. et al.: 1986, Ap. J; 304, 241.
32. Schmutz, W.: 1997, A.A. 321, 268.
33. Schreier, E., Levinson, R., Gursky, H., Kellog, E., Tannanbaum, H., Giacconi, R.: 1972, Ap. J. 172, L79.
34. Schwarzschild, M., 1958,' Structure and Evolution of the Stars', Dover, Publ. Inc.,N.Y.
35. Smith, L.F., Shara, M.M., Moffat, A.F.J.: 1996, MNRAS 281, 163.
36. Sparks, N.M., Stecher, T.P. 1974, Ap. J. 188, 148.
37. Stothers, R., Chin, C,-W., 1976, Ap. J. 204, 472.
38. Tassoul, J.- L.: 1987, Ap. J. 322, 856.
39. Tassoul, J.- L.: 1988, Ap. J. 324, L71.
40. Tassoul, J.- L.: 1990, Ap. J. 358, 196.
41. Tassoul, J.- L., Tassoul, M.: 1992, Ap. J. 395, 258.
42. Ulrich, R.K.,Burger, H.L.: 1976, ApJ 206, 509.
43. Vanbeveren, D., Conti, P.S. 1980, A.A. 88, 230.
44. Vanbeveren,D., De Loore,C., 1994, A.A. 290 129.
45. Vanbeveren,D., Van Rensbergen,W., De Loore,C., 1982, Astron. Astrophys., 115, 68.

46. Vanbeveren,D., Van Rensbergen,W., De Loore,C., 1998, 'The Brightest Binaries', Kluwer, The Netherlands
47. Vanbeveren,D., De Loore,C., and Van Rensbergen,W., 'Massive Stars', The Astronomy and Astrophysics Review, 1998,9,63.
48. Van den Heuvel, E., Heise, J., : 1972 ,Nat. Phys. Sc. 239, 67.
49. Van den Heuvel, E., : 1993, in 'Interacting Binaries', Saas-Fee Advanced Course 22, eds. H. Nussbaumer and A. Orr, Springer Verlag.
50. Van der Hucht, K.A., Conti, P.S., Lundsrom, I., Stenholm, B.: 1981, Space Sci. Rev. 28, 227.
51. Vanderlinden, F.: 1982, Ph. D. Thesis, University of Amsterdam.
52. Van Kerkwijk, M.H., Charles, P.A., Geballe, T.R., King, D.L., Miley, G.K., Molnar, L.A., Van den Heuvel, E.P.J., Van der Klis, M., Van Paradijs, J.: 1993, Nature 355, 703.
53. Van Kerkwijk, A.H., Van Paradijs, J., et al.: 1995, A.A. 303, 483.
54. Van Rensbergen, W., Vanbeveren, D., De Loore, C.: 1996, A.A. 305, 825.
55. Verbunt, F.: 1990, in 'Neutron Stars and Their Birth Events', ed. W. Kundt, Kluwer, Dordrecht, p. 179.
56. Zahn, J.P.: 1977, A.A. 41,328.
57. Zahn, J.P.: 1983, in Astrophysical Processes in Upper MS Stars, 13th Saas-Fee Course, eds. B. Hauck and A. Maeder, p.253.
58. Zahn, J.P.: 1994, Space Science Rev. 66, 285.
59. Zeldovich, Y., B., Guseynov, O.H.: 1966, Ap. J. 144, 840.

6 X-Ray Binaries and Black Hole Candidates: A Review of Optical Properties

Jorge Casares

Instituto de Astrofisica de Canarias, 38200 – La Laguna, Tenerife, Spain

Abstract. This chapter summarizes the optical properties of X-Ray Binaries, with special emphasis on the class of Low Mass X-ray Binaries and Soft X-Ray Transients. The latter provide the most compelling evidence for the existence of black holes in the Universe, with nine well-established dynamical studies. We review the techniques employed to extract the component masses and discuss the importance of systematic effects. Despite the growing number of black hole cases, the uncertainties involved are still too large to draw statistical conclusions on the mass distribution of collapsed objects. We also present new observational techniques which may help to improve the mass determinations and set constraints on the theory of Supernovae and black hole formation.

6.1 X-ray Binaries

Our Galaxy is populated with about 10^3 powerful X-ray sources with typical luminosities $L_X \simeq 10^{36} - 10^{39}$ erg s^{-1}. This energy output is most easily explained invoking a model of mass-exchanging binaries with a compact star (i.e. neutron star or black hole) [164],[152]. This scenario is solidly supported by many empirical results obtained since the dawn of X-ray astronomy . Before highlighting the most significant contributions let me summarize the basis of the canonical model.

6.1.1 Introduction

X-ray binaries are close binaries consisting of a compact object and an optical (*companion*) star. The optical star fills its Roche lobe[1] and transfers gas onto the compact object. Material shares the angular momentum of the companion star and this needs to be reduced for accretion to take place. This is done by the action of viscous processes which take place around the compact object, in the so-called *accretion disc*. The net effect of the accretion disc is both the removal of angular momentum and the extraction of gravitational potential energy ($\Delta E_{\rm acc} = \frac{GMm}{R}$ with m the accreted mass, M and R the mass and radius of the accretting object) from the accreted gas which is the ultimate source of X-rays. Matter being accreted at a rate $\dot{m}$ will power an accretion luminosity

$$L_{\rm acc} = \frac{GM\dot{m}}{R} \tag{6.1}$$

[1] Roche lobes are the binary equipotential surfaces intersecting in a single point, the so-called *inner Lagrangian Point* L_1.

which is the maximum gravitational energy available. As a matter of fact, only a fraction $L_{\rm rad} = \eta \dot{m} c^2$ is radiated, where η stands for the efficiency to convert the rest mass at infinity into radiation. Note that, for a given $\dot{m}$, the accretion efficiency depends on the ratio M/R or "compactness" of the central object. This can also be defined as R_s/R, where $R_s = 2GM/c^2$ is the *Schwarzschild radius.* Only in black holes we do reach "compactness" ≥ 1 but, also in the less compact neutron stars we reach $\eta \sim 0.1$, far greater than the efficiency of nuclear reactions[2] $\eta \sim 0.007$. Thus, the radiation energy delivered by accretion onto neutron stars and black holes exceeds nuclear combustion energy $\Delta E_{\rm nuc} \simeq 0.007mc^2 = 6 \times 10^{18}$ erg g^{-1}, which is the most common energy resource in the Universe. For example, a typical neutron star with $M \simeq 1.4 M_{\odot}$ and $R \simeq 10$ km is able to extract $\Delta E_{\rm acc} \simeq 10^{20}$ erg g^{-1} from its gravitational potential well. This is a powerful engine which, when accreting at a characteristic rate $\dot{m} \simeq 10^{16}$ g s^{-1} would yield a luminosity $L_{\rm acc} \sim 10^{36}$ erg s^{-1}. Assuming that this luminosity is emitted as a black body we can obtain a rough estimate of the mean radiation energy $L_{\rm acc} = 4\pi R^2 \sigma T_b^4$ which for $R \simeq 10$ km yields a black body temperature $T_b \simeq 10^7$ K. For a typical photon energy $h\overline{\nu} = kT \simeq 1$ keV and hence most of the accreted energy is radiated in the X-ray band.

A maximum luminosity will be reached when radiation pressure balances the gravitational force of accreted matter. This is known as the *Eddington luminosity* which, under the assumption of spherically symmetric accretion of hydrogen plasma, is given by

$$L_{\rm Edd} = \frac{4\pi G M m_p c}{\sigma_T} \simeq 1.3 \times 10^{38} \left(\frac{M}{M_{\odot}}\right) {\rm ergs}^{-1} \tag{6.2}$$

where m_p is the proton mass and σ_T the Thomson scattering cross-section. With $L_{\rm X} \simeq 10^{36} - 10^{38}$ erg s^{-1} the galactic X-ray sources are consistent with neutron stars accreting matter in X-ray binaries up to the Eddington luminosity. More detailed information on accretion physics onto compact binaries can be found in [53].

In the late 60's the advent of X-ray astronomy opened a new window on the Universe with the discovery of a large population of galactic X-ray souces. The first optical identifications were Sco X-1 [147] and Cyg X-2 [57] and their optical spectra showed marked similarities with Cataclysmic Variables i.e. interacting binaries containing an accretting white dwarf. In particular, evidence for broad H and He emission lines suggested the presence of an accretion disc whereas the detection of late-type photospheric absorptions in Cyg X-2 was tentatively atributed to the mass donor companion. Intensive efforts to discover periodicities in the lightcurves and radial velocities failed but the model of accreting compact binaries survived based only on the spectroscopic properties.

[2] This assumes a hydrogen rich fuel.

The first X-ray source shown to be a true binary was HDE 226868, the 9th magnitude optical counterpart of Cyg X-1. Optical spectroscopy [193] unveiled a 5.6d radial velocity curve with an amplitude of 64 km s^{-1}. The optical companion is an OB supergiant with $M \geq 15M_{\odot}$ which, combined with the radial velocity parameters, yield a compact object of $M_X \geq 3M_{\odot}$. This limit exceeds the maximum mass of a neutron star (see Sects. 2 and 4) strongly suggesting that Cyg X-1 contains a black hole! The canonical model for X-ray binaries was widely accepted shortlt thereafter when X-ray observations of Her X-1 revealed the presence of a 4.8s pulsed signal. The X-ray pulse showed an accumulated delay of arrival times, following a 2d sine-wave, and regular eclipses at the times of maximum delay [152] (see Fig. 6.1). The simplest explanation requires a 4.8s pulsing neutron star orbiting around a giant star every 2d at high inclination: the X-ray signal experiences a periodic phase delay caused by the finite light travel time across the system and is eclipsed at every passage through the superior conjunction with the non-collapsed star.

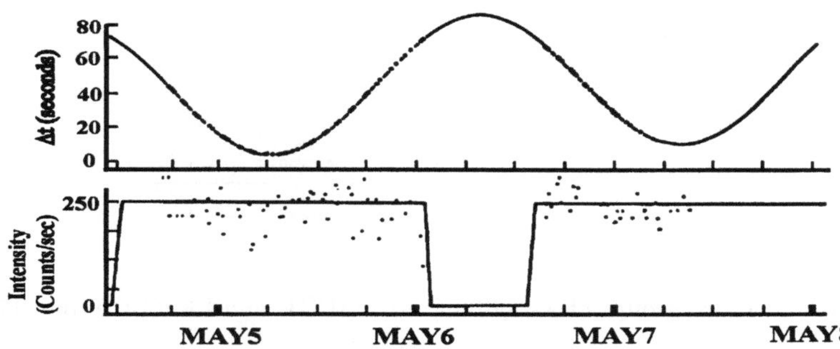

Fig. 6.1. Her X-1 observed in X-rays. Top panel shows the arrival time of the 4.8s pulsations. Bottom panel displays the X-ray light curve with evidence for a total eclipse. Adapted from [152].

Since the early days of X-ray astronomy, significant progress has been achieved in the number of orbital period detections. This has been possible through the discovery of X-ray/optical eclipses and modulations, with timescales of $\sim$ hr-d and radial velocity modulations of emission/absorption lines, with amplitudes of $\sim 10 - 10^2$ km s^{-1}. This information has been efficiently exploited using sophisticated techniques, such as *eclipse mapping* and *Doppler tomography* which have rendered the first brightness distribution maps of accretion discs and companion stars both in the continuum and emission lines. They have provided unprecedented insights into the physics of these systems (e.g. radial temperature distributions, irradiation effects). A comprehensive review of these techniques and their results can be found in [108].

6.1.2 Classification: HMXBs and LMXBs

X-ray binaries can broadly be divided into two main groups according to the properties of the optical companions: High Mass X-Ray Binaries (HMXBs) and Low Mass X-Ray Binaries (LMXBs).

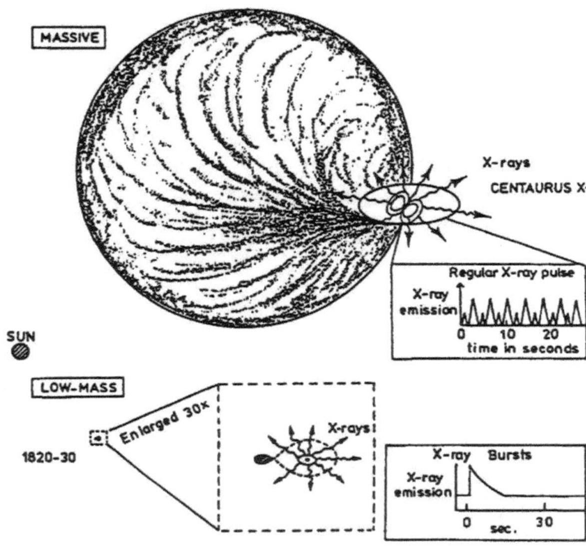

Fig. 6.2. The two groups of X-ray binaries: HMXBs and LMXBs. From [179].

Companion stars in HMXBs are OB supergiants transfering mass *via* stellar wind (with typical $\dot{m} \sim 10^{-6}$ M$_\odot$ yr^{-1}) or possibly with incipient Roche lobe overflow. The luminous supergiant dominates completely the optical emission, whereas the X-rays frequently show regular pulses, indicative of strong magnetic fields in the accreting neutron stars. They have characteristic $L_{\rm opt}/L_{\rm X} > 1$ and, due to the large size of the companion's Roche lobe, the orbital periods are always longer than 1d and a significant fraction exhibit eclipses.

LMXBs, on the other hand, have late-type companion stars which are normally overwhelmed by the optical flux caused by reprocessing of the X-rays in different parts of the binary (mainly the accretion disc). They have typical $L_{\rm opt}/L_{\rm X} < 0.1$ and the X-rays can exhibit fast bursts whose profiles have exponential decays on timescales $\sim$ 30s. These are interpreted as thermonuclear explosions on the surface of weakly magnetized neutron stars [102]. There are more than 175 LMXBs known but only a modest fraction ($\sim$ 4 %) show X-ray eclipses, far less than expected for a random distribution of binary orientations. This puzzling result was noted in the early days of X-ray astronomy and used to cast some doubts on the binary model. However, Milgrom [121] proposed that the reason behind this was a simple selection effect: LMXBs have *thick* accretion discs which efficiently screen the companion star from the X-rays produced in the vicinity of the compact object; in other words potentially eclipsing LMXBs are not seen in X-rays because the accretion disc blocks the compact object's line of sight at all orbital phases. Milgrom's theorem provides a basic geometrical constraint on LMXBs: the opening angle of the accretion disc γ is related to the system mass ratio $q(= M_2/M_1)$ *via* Paczynski's formula [138] for the size of the Roche lobe

$$\frac{R_2}{a} \simeq 0.462 \left(\frac{q}{1+q}\right)^{1/3} \simeq \tan\gamma \tag{6.3}$$

where R_2 is the *equivalent* radius [3] of the companion's Roche lobe and a is the binary separation. LMXBs typically have $q \sim 0.1 - 0.7$ and hence $\gamma \simeq 15°$. Therefore, the distribution of inclinations in LMXBs has a maximum at $i \sim 75°$ (note that i is defined as the angle between the line of sight and the perpendicular to the binary orbit).

The spread of HMXBs in the galactic plane, with a dispersion in latitude $\sigma \langle b^{II} \rangle = \pm 1.9°$, is consistent with Pop I objects, as expected from the early spectral types of their companion stars. On the other hand, LMXBs cluster towards the galactic center and show a much wider distribution of distances above the galactic plane, with a dispersion $\sigma \langle b^{II} \rangle = \pm 9.1°$ corresponding to 1 kpc. This difference is naturally explained in the context of neutron star formation models where supernovae explosions can provide $\sim 10^2$ km s^{-1} kick velocities to the binary from an asymmetric collapse which will spread out the distribution of parent binaries [186]-[144]. Obviously neutron stars in HMXBs also receive kick velocities although their high mass companions will efficiently "absorb" the system's recoil velocity (e.g. see [72])[4].

In addition, kick velocities are also needed to accommodate the observed number of LMXBs ($\sim 100 - 200$) with birthrates predicted by current evolutionary models [172],[182]. For instance, we can assume that most LMXBs are formed from parent binaries consisting of a massive primary (the progenitor of the neutron star) with $M_1 \sim 10 - 15 M_\odot$ and a low mass secondary with $M_2 \leq 1 M_\odot$[5]. The Common–Envelope phase will remove the hydrogen envelope of the massive star, leaving a naked He core with $M_1(He) \sim 2.5 - 4 M_\odot$ which will undergo a Type Ib supernova explosion, forming a collapsed neutron star

[3] Roche lobes are, by definition, not spherical and hence this represents the circular radius of a sphere whose volume is equivalent to that of the Roche lobe.

[4] The longest period ($\geq$ 15 d.) HMXBs have Be companions in highly eccentric orbits which can only be explained by kick velocities at the formation of the neutron star [183]. They constitute a well-defined subclass of HMXBs called "Be X-ray Binaries" [180].

[5] The contribution to the final population of LMXBs expected from the $M_1 > 15 M_\odot$ tail of the distribution can be neglected, assuming a Salpeter IMF.

of $M_{ns} \sim 1.45 M_\odot$ [188]. The condition for the binary to survive a *symmetric* supernova, and hence become an LMXB, is that the ejected mass cannot exceed half the mass of the binary prior to the explosion [11], i.e. $M_{ns} + M_2 \geq 1/2 \times (M_{He} + M_2)$ or $M_{He} \leq 2 \times M_{ns} + M_2 = 3.9 M_\odot$ and consequently almost every system will survive the supernova and become LMXBs. The birthrate implied by these numbers would be $\leq 10^{-4}$ yr^{-1}, in obvious contradiction with the birthrate required by the observed number of LMXBs coupled with their estimated lifetime ($10^7 - 10^8$ yr), that is $10^{-6} - 10^{-5}$ yr^{-1} [83]. van den Heuvel & van Paradijs [182] (see also [144]) have shown that kick velocities of $\simeq 10^2$ km s^{-1} would disrupt the required fraction of parent binaries so as to reconcile both birthrates.

6.1.3 A Subclass of LMXBs: the Soft X-ray Transients

There is a subgroup of LMXBs called Soft X-ray Transients (SXTs), sometimes also referred to as *X-ray novae.* These are characterised by massive X-ray outbursts (usually lasting for several months) during which they are discovered by X-ray or γ-ray all-sky monitors such as those on Ginga, GRO and XTE (Fig. 6.3). SXTs are being discovered at a rate of 1-2 per year and, although the sample is still limited, it has been noted that their galactic distribution (unlike neutron star LMXBs) is consistent with that of Population I objects, i.e. they do not cluster towards the galactic center [197]. This suggests that the progenitors of the galactic black holes were massive Population I OB stars. Furthermore, the distribution of galactic latitudes is more concentrated towards the galactic plane, indicating that compact primaries are somehow formed through a symmetric supernova collapse. Table 6.1 presents an updated list of the currently known X-ray transients.

The outburst lightcurve usually exhibits a fast rise (a few days) followed by exponential decay with typical e-folding times of ~ 45 d [39] (see Fig. 6.3). They have long recurrence times of typically several decades, much longer than dwarf novae outbursts. During outburst $L_X \sim 10^{38} - 10^{39}$ erg s^{-1} and they are indistinguishable from (X-ray) *persistent* LMXBs. The X-ray spectra exhibit a hard power-law (with associated flickering) extending to several hundred keV; superposed on this there is a softer X-ray component [6], with the exception of a few remarkable cases like GS2023+338 or J0422+32 (more details on X-ray properties of SXTs can be found in [170]). At quiescence, the X-ray luminosity drops to $L_X \leq 10^{33}$ erg s^{-1} and the optical brightness decreases by $\sim 3-7$ mag. The companion star is almost entirely responsible for the quiescent optical flux, opening the possibility to perform dynamical studies and constrain the nature of the compact stars. Following this strategy we now know that most SXTs ($\geq 70\%$) contain black holes, whereas almost all *persistent* LMXBs contain neutron stars, as implied by the detection of X-ray bursts. The high incidence of black holes

[6] hence their name, coined so as to distinguish them from the harder Be X-ray transients, a sub-class of HMXBs where X-ray activity is driven by variable mass accretion in highly eccentric orbits [106].

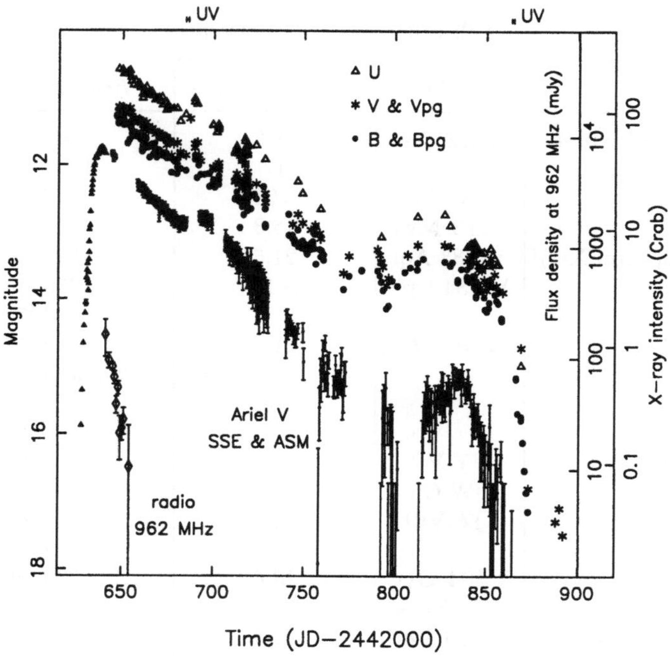

Fig. 6.3. Example of multiwavelength lightcurves of the prototype SXT A0620-00 during the 1975 outburst. Note the presence of secondary maxima or "reflares" in the lightcurve (best seen in X-rays) at ~ 50 days and ~ 170 days after the outburst peak. From [101].

among SXTs is a natural consequence of the low accretion rates implied by their *evolved* secondaries [94] and this will be discussed in more detail in Sect 4. Prior to this, we need to investigate the influence of irradiation effects in accretion discs to understand the reason behind this behaviour.

6.1.4 Accretion Discs in LMXBs: Evidence for X-ray Heating

The importance of X-ray irradiation in LMXBs is overwhelming. The amplitude of the optical outburst $(F_{\rm outb.}/F_{\rm quie.})_{\rm opt} \sim 10^3$, for instance, is roughly the square root of the amplitude of the X-ray outburst $(F_{\rm outb.}/F_{\rm quie.})_{\rm X} \sim 10^6$. This strongly suggests that the whole optical emission is powered by the huge X-ray flux available in the system. In addition, some key features of SXT lightcurves, such as the exponential decay and the secondary maxima (see Fig. 6.3) are nicely reproduced by including X-ray irradiation [122],[98],[161]. Two main reprocessing sites can be identified in the system: the companion star and the accretion disc. X-ray reprocessing in the companion star is almost negligible because of the modest solid angle subtended from the X-ray source, i.e. $(R_2/a)^2 \sim 0.1$. Also, according to Milgrom's theorem, the companion star is largely shielded by the accretion disc and this leaves the accretion disc as the main reprocessing site.

Table 6.1. *X-Ray Transients*

Source	*Alter. Name*	*Quiescent V mag.*	*Reported Outbursts*
A1455-314(=Cen X-4)†	V822 Cen	18.3	1969, 79
4U1543-47*		17	1971, 83, 92
A1524-62	KY Tra	>21	1974, 90
A0620-00*	**V616 Mon**	18.3	1917,75
X1908+005(=Aql X-1)†	V1333 Aql	21.6	1975 and every 200-300 d.
X1658-298†	V2134 Oph	> 23	1976, 78, 99
H1743-32			1977
H1705-25*	**V2107 Oph**	21.5	1977
4U1630-47			1978 and every $\simeq$ 600 d.
X1608-522†		>20	1978
EXO 1846-031			1985
EXO 0748-676†	UY Vol	>23	1985
GS1354-645	BW Cir	22	1987, 97
GS2000+25*	**QZ Vul**	22.5	1988
GS1826-24			1988
GS2023+338*	**V404 Cyg**	18.4	1938, 56, 89
GRS1124-68*	**GU Mus**	20.4	1991
GRO J0422+32*	**V518 Per**	22.4	1992
GRS1009-45*	**N Vel 1993**	22	1993
GRS1716-249	V2293 Oph	>22	1993
GRO J1655-40*	**N Sco 1994**	17.3	1994
XTE J1755-324			1997
XTE J1748-288			1998
XTE J1550-564		B~22	1998, 99
XTE J2012+381		21.3	1998
XTE J2123-058†		22.5	1998
XTE J1859+226		~ 23	1999
XTE J1118+48			1999

*Firm black holes supported by dynamical evidence.
†Neutron star binaries.

There are two solid pieces of observational evidence supporting the important role of X-ray reprocessing as the source of optical emission:

a) Emission at $\lambda\lambda4640 - 4650$ or "Bowen blend": The spectra of LMXBs are characterised by a blue spectral distribution and broad emission lines of HI, HeI and high excitation lines of FeII, HeII $\lambda4686$ and the Bowen blend at $\sim \lambda4645$. Unlike Balmer and HeI lines, the EWs of high excitation HeII$\lambda4686$ and the Bowen blend are correlated with L_X, disappearing when the source reaches quiescence. The Bowen blend is a combination of high excitation lines (mainly CIII, OII and NIII at $\lambda\lambda4634 - 4642$) produced by the fluorescence resonance mechanism which initially requires seed photons of HeII Lyα at $\lambda303.78$ [115]. Several intermediate resonance lines are also produced (e.g. $\lambda3133$, $\lambda3444$) and line intensity ratios can be compared to models in order to derive the physical conditions of the reprocessing region [150].

b) Time delay between optical and X-ray bursts: Sometimes, LMXBs exhibit optical bursts delayed by a few seconds with respect to the X-ray bursts. The delay times are consistent with the light-travel across the binary (typically $\sim$ 2 secs) and suggest that X-ray reprocessing takes place in the accretion disc and (occasionally) in the companion star. The optical signal also appears smeared out with respect to the X-ray pulse, an indication that the reprocessing site is extended (e.g. [177]). Correlated rapid variability between optical/UV and X-ray lightcurves has been detected in the SXT J1655-40 during its 1996 outburst[82]. The optical and UV variability lag the X-rays by 10-20 secs, pointing to the accretion disc as the reprocessing site in this long period binary. For a localized reprocessing site, the time-delay distribution is a function of binary phase. The shape and amplitude depend on the inclination and the separation between the X-ray source and the reprocessing site, and can be used to determine both by employing "echo mapping" techniques[80].

The optical properties of LMXBs contain statistical information of irradiated discs. For example, the histograms of dereddened colours $(B-V)_0 = -0.09 \pm 0.14$, $(U-B)_0 = -0.97 \pm 0.17$ are consistent with a flat optical distribution $F_\nu \simeq$constant rather than the standard $F_\nu \propto \nu^{1/3}$ law expected in viscously heated discs. These colours agree well with irradiated model predictions, where most of the reprocessed energy is radiated in the UV [103]. The average ratio of optical to X-ray luminosity yields $L_{\rm opt}(300-700\ nm)/L_{\rm X}(2-11\ keV) \simeq 0.002$ which can also be expressed as a magnitude difference $\overline{\xi} = \overline{B_0 - m_x} = 21.8 \pm 1.0$, with the X-ray magnitude defined as $m_x = -2.5 \log F_{\rm X}(\mu Jy)$ [185]. If we apply the bolometric correction[7] to the optical flux one obtains $L_{\rm opt}/L_{\rm X}(2-11 keV) = \epsilon \simeq 0.02$. ϵ stands for the efficiency of reprocessing and depends on geometry parameters i.e. inclination and disc solid angle, as seen from the X-ray source, and albedo (i.e. fraction of X-rays reflected).

A crude estimate of the averaged disc temperature is provided through simple energy balance in a black-body approximation $R_d^2 T_d^4 \simeq \epsilon R_x^2 T_x^4$, where R_d and R_x are the radius of the disc and the X-ray source (neutron star) and T_d and T_x are the temperatures of the disc and the X-ray source respectively. Assuming typical values $R_x \simeq 10$ km, $T_x \simeq 10^7$ K and $R_d \simeq 2$ light-secs we obtain $T_d \simeq 30000$ K. This *averaged* temperature is impossible to attain in accretion discs heated by viscous processes such as in Cataclysmic Variables (CVs hereafter). Furthermore, the histogram of absolute V magnitude of LMXBs peaks at $M_V \simeq 1.0$. For comparison, CVs have $M_V \simeq 5-7$ and, therefore, accretion discs in LMXBs are, on average, $\simeq 10^2$ more luminous than in CVs. More information about the optical properties of LMXBs can be found in [187].

[7] i.e. by including the UV flux, as determined in the prototype system Sco X-1 [190].

The ionization parameter ξ provides a rough guide to the ionization state of the gas inside accretion discs in LMXBs. This is defined as $\xi = L_X/nR_d^2$, where n is the number density and ξ is therefore proportional to the ratio of ionizing energy available divided by the total mass stored in the disc. LMXBs (and SXTs in outburst) typically have $L_X \simeq 10^{38}$ erg s^{-1} and $n \simeq 10^{15}$ cm^{-3}. In the inner disc regions (*i.e.*$R_d \simeq 10^{10}$ cm) $\xi \simeq 10^3$ and hence the gas will be highly ionized. This is comparable to the values observed in the BLRs of AGNs, and so the same physical processes are at work (mainly photoionization, recombination and fluorescence). The spectra will thus be dominated by FeII permitted lines, HeII λ4686 and Bowen resonance lines, in addition to the absence of forbidden lines

To summarize, accretion discs in LMXBs are strongly affected by irradiation. With $L_X \simeq 10^{36} - 10^{38}$ erg s^{-1} the optical radiation is generated by X-ray reprocessing, and accretion discs appear much hotter and brighter than in CVs, where viscous heating becomes the dominant mechanism. In the next section we will see how irradiation effects can transform the temperature profile and geometry of the disc. This has a crucial impact on the stabilization of accretion discs and the development of outbursts cycles.

6.1.5 Irradiated Disc Models

Model by Vrtilek et al. *A&A* 235, 162 (1990): A simple model to investigate the effects of strong X-ray heating on the structure of accretion discs was proposed by Vrtilek et al. [189] (VR model hereafter). The model starts by assuming a standard Shakura & Sunyaev disc [163], i.e. geometrically thin, optically thick and stationary ($\dot{M} = constant$) with extra-heating input by X-ray illumination. By combining the eq. of hydrostatic equilibrium

$$h = \frac{c_s}{\Omega_K} = \sqrt{\frac{P}{\rho}\frac{1}{\Omega_K}}$$

and the perfect gas equation

$$P = \frac{\rho kT}{\mu m_H}$$

one obtains

$$\frac{kT}{\mu m_H} = \Omega_K^2 h^2 \tag{6.4}$$

where $\Omega_K = (\frac{GM}{r^2})^{1/2}$ is the Keplerian angular velocity and h the disc height. When irradiation is negligible, the thermal equilibrium condition in an optically thick disc yields [163]

$$\sigma T_0^4 = \frac{3GM\dot{M}}{8\pi r^3}\left(1 - \sqrt{\frac{r_{\rm int}}{r}}\right), \tag{6.5}$$

i.e. the standard temperature profile of stationary optically thick discs is $T_0 \propto r^{-3/4}$ (for $r \gg r_{\rm int}$). However, if irradiation becomes important the accretion disc will absorb a fraction $(1-\eta)$ of the incident X-rays (with η the albedo) which will be thermalised and will raise the local temperature. The optical radiation emitted by a disc surface element ds will, therefore, have two contributions due to viscous heating and X-ray irradiation

$$\sigma T_{eff}^4 = \sigma T_0^4 + (1-\eta) F_X^{irr} \tag{6.6}$$

where F_X^{irr} is the X-ray flux received by ds. For LMXBs we have $F_X^{irr} \gg F_{vis}$ and the first term on the right-hand side of eq 6.6 can be neglected. In addition, under the assumption that the disc is geometrically thin we can approximate

$$F_X^{irr} \simeq \frac{L_X}{4\pi r} \frac{\delta(h/r)}{\delta r}$$

so

$$T_{eff}^4 \simeq (1-\eta) \frac{L_X}{4\pi\sigma r} \frac{\delta(h/r)}{\delta r} . \tag{6.7}$$

For the disc to be irradiated we have to assume a concave geometry, i.e.

$$h \propto r^n \tag{6.8}$$

And now we can solve eqs 6.4, 6.7 and 6.8 if we assume that the vertical structure of the disc is isothermal, i.e. $T_{eff} = T(r)$. This is the crucial assumption of this model and is expected to hold when $F_X^{irr} \gg F_{vis}$. By solving the system of equations we get

$$\begin{aligned} h &\propto r^{9/7} \\ T &\propto r^{-3/7} \end{aligned} \tag{6.9}$$

which can be compared to the standard (viscously heated) Shakura & Sunyaev discs [163]

$$\begin{aligned} h &\propto r^{9/8} \\ T &\propto r^{-3/4} . \end{aligned} \tag{6.10}$$

Therefore, irradiated discs are geometrically thicker and have a flatter temperature profile than non-irradiated discs. In the next section we will show that this result is fundamental to understanding the outburst cycles in interacting binaries.

As an example of applying the VR model we have analyzed the lightcurves of J2123-058, an X-ray transient discovered by XTE in June 98. The lightcurves exhibit regular eclipses with a period of 6hr and occasional optical (and X-ray) bursts, an indication that the compact object is a neutron star. We followed the evolution of J2123-058 in the optical from the peak of the outburst down to

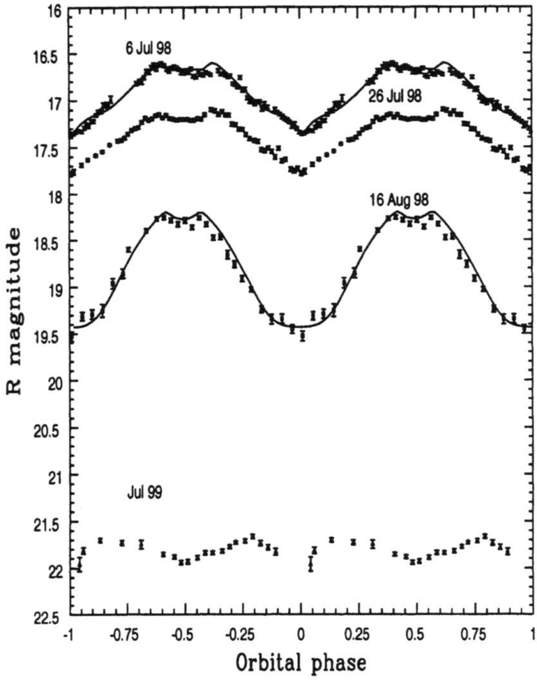

Fig. 6.4. Lightcurves and best model fits at different stages of the outburst. X-ray luminosity was fixed at $L_X = 1.3 \times 10^{37}$ erg s^{-1}, $L_X = 1.3 \times 10^{36}$ erg s^{-1} and $L_X = 0$, from top to bottom. The best fits provide $i = 76°$, $R_d = 0.75\ R_{L_1}$, $\gamma = 7.6°$ and $i = 72°$, $R_d = 0.56\ R_{L_1}$, $\gamma = 5.7°$ on 6 July and 16 August respectively. An ellipsoidal model fixing $i = 73°$ is overplotted to the quiescence data on September.

quiescence using the IAC-80 and the Optical Ground Station (OGS) telescopes at the Observatorio de Izaña in Tenerife[8].

Figure 6.4 depicts the evolution of the R-band orbital lightcurve as the outburst progresses. It shows dramatic variations of shape and amplitude, from irradiation-dominated to a pure ellipsoidal modulation when the X-rays switch off and represents an ideal case to study lightcurve variations as a function of L_X. The lightcurves were fitted using synthetic models which include eclipses, X-ray heating of the companion and the accretion disc (according to VR model) and shadowing of the companion by the accretion disc. Model fits yield consistent values for the inclination at two different epochs *independently* ($i = 76 \pm 1°$ and $i = 72 \pm 3°$) and succesfully reproduce the shape and amplitude of the lightcurves by requiring that both the accretion disc radius R_d and flaring angle γ shrink by a few percent (see Fig. 6.4). Note that L_X has dropped by a factor 10 between

[8] The OGS and the IAC-80 telescopes are operated on the island of Tenerife by the European Space Agency (ESA) and the Instituto de Astrofísica de Canarias, respectively.

the two fits and yet X-ray irradiation in the disc is the dominant contribution to the optical flux. More details of this analysis can be found in [206].

Model by Dubus et al. *MNRAS* 303, 139 (1999): An alternative "self-consistent" model (i.e. with no *a priori* assumptions on disc geometry or vertical structure of temperature) of irradiated discs has been recently proposed by Dubus *et al.* [45] (DU model hereafter). They start by showing that the key assumption of VR model (i.e. isothermal profile in the vertical direction when $F_X^{irr} \gg F_{vis}$) does not necessarily hold. The temperature at the disc mid-plane T_0 is given by

$$T_0^4 = T^4(\tau_{tot}) = \frac{3}{8}\tau_{tot} T_{eff}^4 + T_{irr}^4$$

where T_{eff} is the non-irradiated effective temperature (produced by viscous heating), T_{irr} the temperature generated by irradiation and

$$\tau_{tot} = \int_0^{\infty} \kappa \rho dz$$

the total disc optical depth integrated in the vertical direction z, with κ the opacity and ρ the disc density. For the disc to be isothermal (i.e. $T_0 \simeq T_{irr}$) $T_{irr}^4 \gg \tau_{tot} T_{eff}^4$ or, equivalently (in flux units)

$$\frac{\sigma T_{irr}^4}{\tau_{tot}} = \frac{F_X^{irr}}{\tau_{tot}} \gg F_{vis}.$$

Therefore, VR model's assumption of vertically isothermal discs depends critically on τ_{tot} and this assumption may break down in the outer disc, where τ_{tot} can reach $\sim 10^2 - 10^3$.

DU integrate numerically the equations of disc structure, i.e. conservation of mass, angular momentum and energy, including the radiative transfer equation in the vertical direction. The results of their computations indicate that, as opposed to VR model, irradiated discs are *convex* and the temperature profile is not affected unless $\frac{F_X^{irr}}{\tau_{tot}} \gg F_{vis}$. Due to the convex geometry the outer disc would be shielded from X-rays by the inner disc, unlike the VR model. However, as we will see in the next section, there is compelling observational evidence that *outer* discs in LMXBs are strongly affected by irradiation [184]. To overcome this contradiction DU suggests that either accretion discs in LMXBs are *warped* or the X-ray source is extended e.g. accretion discs have a corona which scatter the high energy photons arising from the vicinity of the compact object. Warped discs (i.e. each ring is tilted to the orbital plane and precessing) could be triggered and maintained by radiation pressure [142]-[143], a very likely mechanism to work in the strong X-ray irradiation fields characteristic of LMXBs.

6.1.6 Disc Instabilities: Persistent and Transient LMXBs

Although the standard Shakura & Sunyaev model [163] assumes accretion discs are stationary (i.e. $\dot{M} = constant$) the very existence of SXTs indicates that this approximation is no longer valid. In order to trigger the outbursts we need to invoke a physical process responsible for modulating the (steady) accretion flow from the companion star onto the compact object. Two competting mechanisms, inspired by previous models for dwarf novae outbursts, have been proposed to account for outbursts in LMXBs:

Mass-Transfer Instability: Developed by Hameury, King and Lasota in a set of papers [64]-[66], this model needs a companion star underfilling its Roche lobe by a few percent. Steady hard X-ray (≥ 7 keV) irradiation during quiescence is required to expand the star's atmosphere and initiate the mass transfer phase. The star tries to adjust the radius expansion (and $\dot{M}$) to be in equilibrium under the irradiation effect. However, for a certain range of $\dot{M}$ the process can be unstable and an outburst will be triggered. This requires a substantial fraction of the accretion luminosity during quiescence to be emitted as hard X-rays and a lower limit has been proposed:

$$L_X(> 7keV) > 4 \times 10^{34} \left(\frac{M_1}{3M_\odot}\right)^2 \left(\frac{M_2}{M_\odot}\right) ergs^{-1}$$

Thanks to the increase in sensitivity of X-ray observatories (e.g. ROSAT, ASCA and XTE) it is now possible to study the quiescent X-ray spectra of several SXTs in the 0.5-10 keV band and they are found to be very soft ($T_{bb} \sim 0.2 - 0.3$ keV) [171]. The observed luminosities are much lower than expected for standard disc models (e.g. [119]) and has inspired a new scenario for accretion during quiescence, the *Advection Dominated Accretion Flow or ADAF model.* In the ADAF model the inner disc is replaced by a very hot, optically thin accretion flow which has extremely low radiative efficiency and carries most of the accretion energy directly onto the compact object [123],[124]. ADAFs are the only scenario which can succesfully fit the whole spectral distribution of quiescent black hole X-ray transients from IR to soft X-rays [204],[125],[47]. ADAF models predicts hard X-ray luminosities below the limit required by mass-transfer instability models although detections above $\simeq 10$ keV are not yet possible (we have to wait for the better sensitivity of XMM and INTEGRAL).

Additional complications for mass-transfer instability models come from the first Doppler tomograms of SXTs obtained in quiescence [109],[29],[21],[69]. Accretion disc images consistently show strong Balmer emission from the *hot spot* or impact region between the accretion stream and the outer disc rim. These observations imply that mass transfer (and hence Roche lobe overflow) takes place during quiescence. The mass-transfer instability model has been finally excluded by the lack of a physical mechanism to trigger the outburst (i.e. how the companion is heated at the low quiescent L_X) and also the success of the rival model.

Disc Instability: This model was initially proposed by Huang & Wheeler [81] and subsequently refined by Cannizo *et al.* in a sequence of papers [18]-[20]. The outbursts are likely triggered by thermal-viscous instabilities in the accretion disc, caused by a sudden change in the disc opacity κ_R at $T \leq 10^4$ K, when HII starts to recombine. The dependence of opacity on temperature can be parameterised as $\kappa_R \propto \rho T^a$ (see Fig. 6.5) with

$$\begin{aligned} a &\simeq -3.5 \qquad if\ T \geq 10^4 K\ , \\ a &\simeq 5-10 \qquad if\ T \leq 10^4 K\ , \\ a &\leq -5 \qquad if\ T \leq 10^3 K\ . \end{aligned} \tag{6.11}$$

The condition for thermal instability to occur in a disc annulus Σ is given by [53]

$$\left(\frac{\delta \ln Q^+}{\delta \ln T_0}\right)_\Sigma > \left(\frac{\delta \ln Q^-}{\delta \ln T_0}\right)_\Sigma , \tag{6.12}$$

where Q^+ and Q^- represent the heating and cooling rate in the disc annulus Σ. If $Q^+ > Q^-$ the disc mid-plane temperature T_0 will rise and, due to eq. 6.12, the heating gradient will be higher than the cooling gradient so T_0 will continue to increase until an instability occurs. The cooling rate depends on the opacity [163] and it is easy to show that $(\delta Q^-/\delta \ln T)_\Sigma = 4.5 - a$. Therefore, according to eq. 11, if $T \leq 10^4$ K then $(\delta Q^-/\delta \ln T)_\Sigma < 0$ whereas $(\delta Q^+/\delta \ln T)_\Sigma > 0$ always and a thermal instability will take place. A more detailed description of this mechanism is provided in [166].

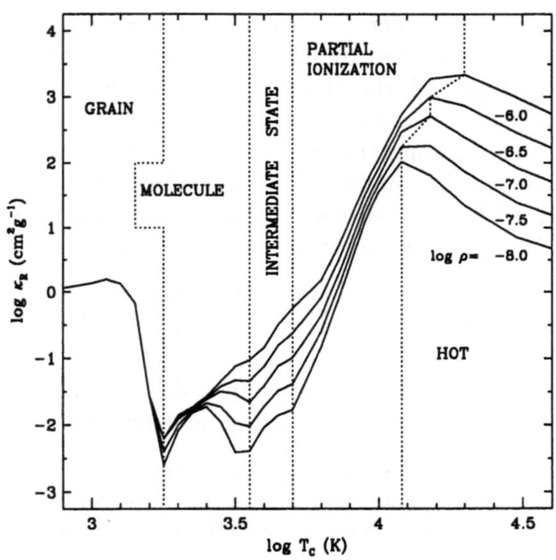

Fig. 6.5. Variation of disc opacity κ_R with temperature for different ρ. From [91].

Because temperature drops with disc radius, a simple condition[9] for the existence of "stable" (i.e. persistent) accretion discs is that the outer disc remains hotter than the HII recombination temperature, i.e. $T_d > T_H \simeq 10^4$ K, or equivalently in a Shakura & Sunyaev disc (see eq. 6.5)

$$\dot{M}_2 > \dot{M}_{\rm crit} \tag{6.13}$$

where $\dot{M}_2$ is the mass transfer rate from the companion star and $\dot{M}_{\rm crit} \propto R_d^3 T_H^4$ is the disc (critical) mass transfer rate which triggers a thermal instability in the outer disc.

These concepts were first applied to CVs by Smak [165] almost two decades ago. He obtained an approximated expression for $\dot{M}_{\rm crit}$ by assuming typical values $R_d \simeq 0.7 \times R_{L_1}$ and $M_2/M_1 \sim 0.1$ and making use of Paczynski's equation for the size of the primary's Roche lobe [138]

$$\frac{R_{L_1}}{a} \simeq 0.38 - 0.2 \log\left(\frac{M_2}{M_1}\right)$$

(valid for $M_1 > M_2$) together with Kepler's Third law $a^3 \propto P^2$. Smak's relation

$$\dot{M}_{\rm crit} \simeq 2.9 \times 10^{-9} \left(\frac{P}{3\ hr}\right)^2 M_\odot {\rm yr}^{-1} \tag{6.14}$$

shows a dependence of $\dot{M}_{\rm crit}$ on orbital period and, when applied to CVs, it was successful in separating the persistent novalikes ($\dot{M}_2 > \dot{M}_{\rm crit}$) from the transient dwarf novae ($\dot{M}_2 < \dot{M}_{\rm crit}$)[10]. However, van Paradijs [184] noted that this relation failed to accommodate the observations of LMXBs since $\dot{M}_{\rm crit}$ seemed to be a factor $\simeq 10^{-2}$ lower than predicted by eq. 6.14. He suggested that accretion discs in LMXBs are strongly affected by intense X-ray irradiation. The outer layers are efficiently heated up and, therefore, irradiated discs stabilize at much lower $\dot{M}_{\rm crit}$ than viscously heated discs (see Fig. 6.6).

He estimated the new temperature profile for irradiated discs (based on VR model with the disc opening angle $\gamma = 12°$ and albedo $\eta = 0.92$, as derived from model comparison with observations of a sample of LMXBs [43]) and derived a new relation for $\dot{M}_{\rm crit} - P$, which we reproduce here, adapted from [94],

$$\dot{M}_{\rm crit}^{\rm irr} \simeq 5 \times 10^{-11} \left(\frac{M_1}{M_\odot}\right)^{2/3} \left(\frac{P}{3\ hr}\right)^{4/3} M_\odot {\rm yr}^{-1}. \tag{6.15}$$

Note that mass ratios in LMXBs, as opposed to CVs, will vary significantly according to whether the compact object is a neutron star or a black hole. Thus,

---

[9] Strictly speaking, a viscous instability is also needed for an outburst to occur (i.e. the mass transfer rate is not constant through the disc) which leads to the concept of limit cycle (see [165]). Note, however, that thermal instability is always necessary and therefore our reasoning is still valid.

[10] The $\dot{M}_2$ values of individual systems were obtained by dividing the mass accreted during outburst by the observed recurrence time.

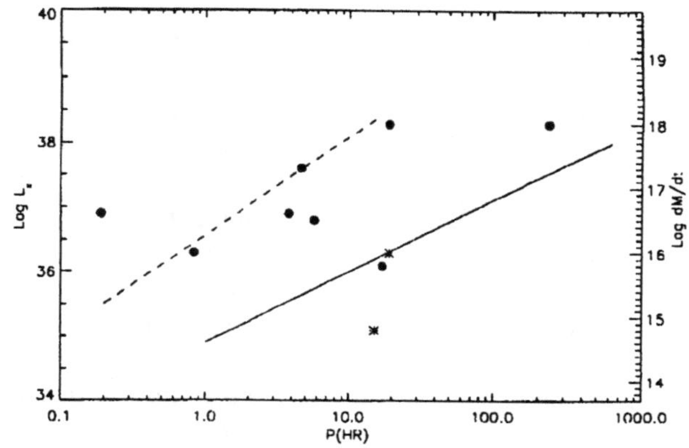

Fig. 6.6. Distribution of neutron star LMXBs in $\log L_X$ (or $\log \dot{M}$) vs orbital period P. Solid circles mark persistent LMXBs whereas stars indicate transients. Smak's and van Paradijs's relations for $\dot{M}_{crit}$ are drawn as dotted and continuous lines, respectively. Persistent systems should lie above the line and transients below.

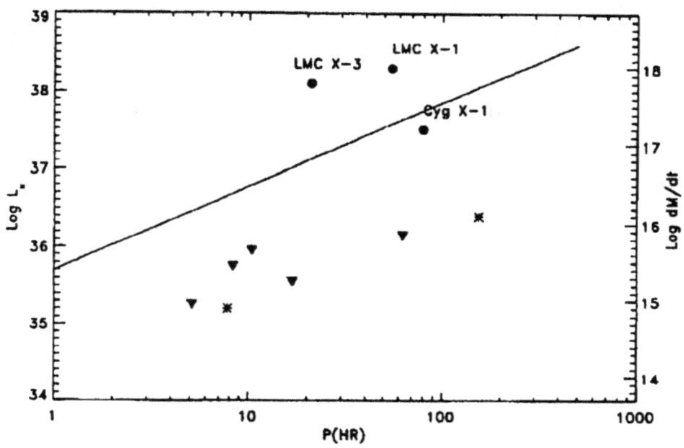

Fig. 6.7. Same as Fig. 6.6 but for X-ray binaries with black holes. Only the critical line for the irradiated disc case is shown.

instead of the approximation $M_2/M_1 \sim 0.1$ used in eq. 6.14 we now use $M_2 \sim 0.4$ $M_\odot$. This is why M_1 appears as an extra free parameter in eq. 6.15. The new expression provides a succesful division between persistent and transient LMXBs for both neutron star ($M_1 = 1.4$ $M_\odot$, Fig. 6.6) and black hole ($M_1 = 10$ $M_\odot$, Fig. 6.7) primaries.

6.2 Black Hole Candidates

The concept of a black hole dates back to the early works by Michell and Laplace at the turn of the 18th century, when they conceived the possible existence of very massive objects with escape velocities $v_e(=\sqrt{(2GM/R)})$ larger than the speed of light. This was still a classical physics concept of black hole, described as a large concentration of mass in the low density domain of normal stars (i.e. $\rho \propto R_s^{-2}$). The real breakthrough did not happen until 150 years later, with Einstein's General Relativity where black holes were found as singularities in Schwarzschild's solution to the field equations. But it has only been in the last three decades that X-ray astronomy has started to provide the first observational evidence for the existence of stellar objects with $R_s/R \sim 1$ in the Universe. These have been followed by the first mass function determination in Cyg X-1 [193],[9] and an intense (and controversial) debate on the possibility of deriving black hole signatures began (e.g. see [170] and included references).

On one hand, observational evidence for neutron stars (NS hereafter) is compelling through the detection of rapid X-ray pulsars (in HMXBs) and bursts (in LMXBs) both secure indications of a small solid surface (i.e. $R > R_s$). However, the demonstration of the presence of black holes (BH hereafter) remains elusive and indirect since it requires the proof of the existence of *event horizons*. Currently, the most solid evidence is based on the detection of accreting compact objects more massive than the maximum possible mass for a NS, M_{NS}^{max}, which ultimately depends on the assumed equation of state (EOS hereafter).

The EOS provides M-R relations for a fixed core density ρ_c, and each EOS has a maximum NS mass M_{NS}^{max} (Fig. 6.8). One can distinguish between *soft* EOS, with $R_{max} \sim 8 - 10$ km and $M_{NS}^{max} \leq 1.6$ M$_\odot$, and *stiff* EOS with $R_{max} \sim 15$

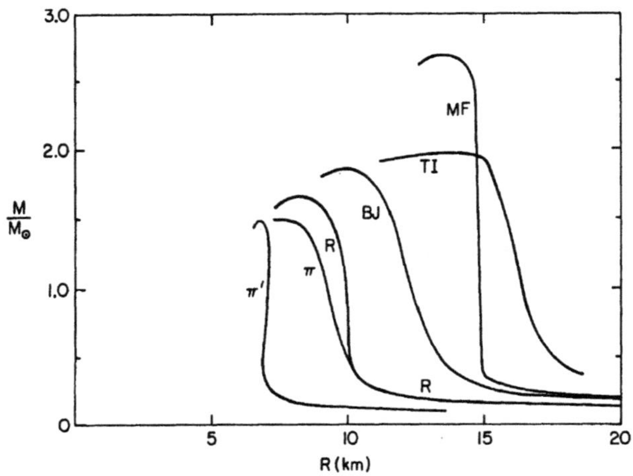

Fig. 6.8. Standard equations of state of neutron stars

km. For the stiffest EOS M_{NS}^{max} is found to be $\simeq$ 2.7 $M_\odot$. The effect of rapid rotation can raise this limit up to 20 %, yielding $M_{NS}^{max} \simeq 3.2$ $M_\odot$ [54]. A firm upper bound can also be established independently from the EOS selected, by assuming that General Relativity is the correct theory of gravity and causality holds inside NS i.e. the sound velocity is lower than the speed of light. This provides $M_{NS}^{max} \simeq 3.0$ $M_\odot$ [145]. However, it should be noted that in all cases where a neutron star is known to be present (through bursts or pulsations) and its mass can be accurately measured, those masses are all consistent with $M_{NS} < 2$ $M_\odot$ (see Sect. 5). The most secure way to demonstrate the existence of BHs is, therefore, by weighing the masses of compact stars, and this experiment can only be performed in X-ray binaries.

After the discovery of the first two black holes (Cyg X-1 and LMC X-3) involving HMXBs, almost all the newly found black hole candidates ($\simeq$22) belong to the class of Soft X-ray Transients (SXTs), or *X-ray novae*, a subgroup of LMXBs. The SXTs are ideal for such studies because of the low mass of their secondary stars. This allows firm lower limits to be set to the mass of the compact object, as the secondaries can be studied during the long periods of quiescence between outbursts (when the accretion disc is faint). In the next section we will focus on the techniques employed to measure masses in X-ray binaries and the impact of systematic effects on those masses. We will show how our final mass determination can be seriously compromised if these systematic effects are not properly accounted for.

6.3 Mass Determination

A complete determination of the component masses in a binary system requires the radial velocity curves of both stars and a knowledge of the inclination angle. Unfortunately, compact objects in LMXBs do not radiate or pulse (with Her X-1 as the only exception) and, consequently, their radial velocity curves are not available. The study of gas dynamics could, in principle, constrain the system parameters. In particular, the line wings originate in the vicinity of the compact object and hence they ought to share its orbital motion. Unfortunately, the emission profiles are broad, complex and highly variable (e.g. [24]). The "double-gaussian" method was proposed as a way of revealing the relevant velocity information by convolving the spectra with two Gaussian bandpasses[151]. This technique enables us to extract the centroid of the line regions separated by the distance a between the two Gaussian filters. Sine waves are subsequently fitted to the velocity points and the 3 free parameters (the systemic velocity γ, the velocity semiamplitude K_1 and phasing ϕ of the gas) are displayed as a function of a in the so-called "diagnostic diagram"[162].

K_1 will approach the true compact object's velocity at the extreme line wings and the choice of a is critical in the final determined value. However, the selection of a is rather arbitrary, and independent studies can sometimes yield inconsistent results (e.g. [73],[132]). In addition, systematic phase lags (relative to the

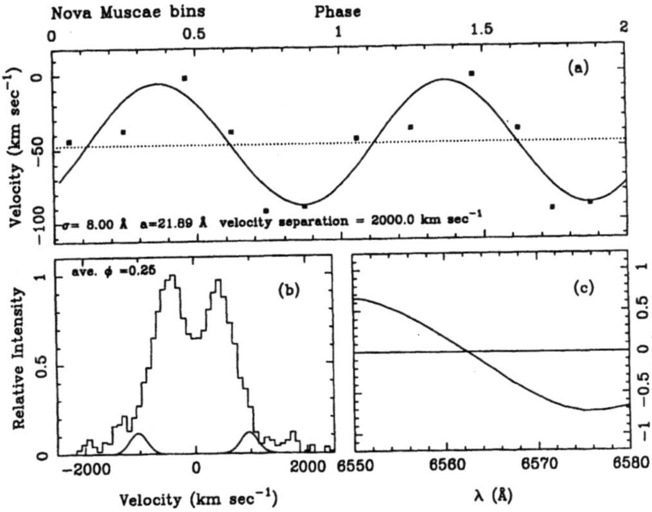

Fig. 6.9. Example of a radial velocity curve determined from the wings of Hα in N Mus 91 (=GU Mus), using the "double-gaussian" method with a bandpass separation a = 2000 km s^{-1} (a). Note the phase shift of -0.12 relative to the expected radial velocity curve of the compact object, which should be maximum at phase 0.5 in the phase convention used in this plot. Panel (b) sketchs the Gaussian bandpass used over a sample profile. Panel (c) shows the convolution function marking the velocity point. From [132].

expected motion of the primary star)[132] plus mean γ-velocities that are different from the absorption line γ-velocity [168] are commonly found (see Fig. 6.9). These results indicate that either the emission from the inner disc is non axisymmetric or the line wings are corrupted by other components (e.g. S-waves from the line core or winds). The "light-center" method was subsequently proposed as a strategy to correct for the line core contamination by extrapolating the "diagnostic diagram" towards phase 0 in velocity space (see [107] for details). However, it has been found that distortion emission from the hot-spot and other asymmetries are still present [109]. Furthermore, the uncertainties involved are dominated by large systematic errors which are very difficult to quantify. Therefore, the analysis of the emission line velocities is systematically biased and the results obtained through this technique are unreliable for accurate studies of binary system parameters.

Consequently, the determination of masses in LMXBs is usually reduced to a *single-line* spectroscopic binary problem where **all** the information has to be extracted from the optical star. A complete solution to the system parameters, however, can be obtained from three observational experiments involving high resolution optical spectroscopy and infrared photometry. This prescription has been summarized in several papers (e.g. [33], [22]). Here we present an updated version including illustrative examples based on recent results.

6.3.1 The Mass Function

This is a fundamental equation for the determination of binary system parameters which derives directly from Kepler's 2nd and 3rd laws. It relates the masses of the two stars (M_1, M_2) and the inclination angle (i) through two observable quantities which are readily measured from the radial velocity curve of the optical star: the orbital period $P_{\rm orb}$ and the radial velocity semiamplitude K_2. It has the form

$$f(M) = \frac{K_2^3 P_{\rm orb}}{2\pi G} = \frac{M_1^3 \sin^3 i}{(M_1 + M_2)^2} = \frac{M_1 \sin^3 i}{(1+q)^2} \tag{6.16}$$

where $q = M_2/M_1$ is the mass ratio. It is clear from eq. 6.16 that the mass function $f(M)$ represents an absolute lower limit to M_1, i.e. at the extreme case when the mass of the companion star is neglected (i.e. $q \to 0$) and the binary is seen edge on ($i = 90°$). Therefore, given our previous discussion on the EOS for NS, a mass function in excess of 3 $M_\odot$ is considered as a secure black hole diagnostic. This is why the work on V404 Cygni revolutionized the field of black hole searches: its mass function is the highest yet measured ($6.1 \pm 0.1 M_\odot$ [25],[27]; see Fig. 6.10) and, for the first time, placed an accreting compact object comfortably above the upper limit of maximally rotating neutron stars for any "standard" EOS assumed. Therefore, V404 Cygni is widely considered as the best evidence for a black hole, where no additional assumptions on i nor M_2 have to be invoked.

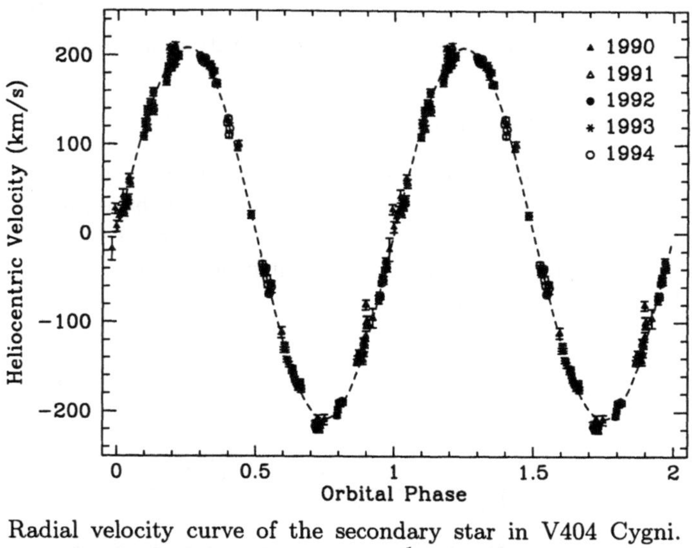

Fig. 6.10. Radial velocity curve of the secondary star in V404 Cygni. The radial velocity semiamplitude $K_2 = 209 \pm 1$ km s^{-1} combined with the orbital period $P = 6.4714 \pm 0.0001$ d yield $f(M) = 6.08 \pm 0.06 M_\odot$. Adapted from [25].

Note, however, that $f(M) \propto K_2^3$, and hence this limit is extremely sensitive to uncertainties in the radial velocity amplitude. Any non-uniform brightness distribution across the surface of the companion star will modify the radial velocity curve and affect our system parameter determination. There are several sources of systematic errors in K_2, the most important being X-ray irradiation which tends to supress the absorption lines near the L1 point. This will displace the center of light away from the companion's center of mass. Consequently, the radial velocity curve will be significantly eccentric and a sine-wave fit to it will give an observed K_{obs} which will be larger than the true K_2. The ratio K_2/K_{obs} is called the *K-correction* [192] and it depends on the X-ray luminosity and geometrical effects (q and shielding of the companion's inner face by the disc). The presence of the irradiation effect can be revealed through an elliptical fit to the radial velocity curve and by performing a test of significance (as compared to a simple sine-wave fit) [42].

Irradiation is considered to be negligible in quiescent X-ray transients (where $L_{\rm x} \leq 10^{33}$ erg s^{-1}) but it can be important in X-ray active states [3]. A dramatic example is offered by observations of GRO J1655-40 (=N Sco 94) during the decay of the 1994 outburst, when $L_{\rm x} = 1.4 \times 10^{37}$ erg s^{-1}. Orosz & Bailyn fitted the radial velocity curve with a simple sine-wave, obtaining $K_2 = 228.2 \pm 2.2$ km s^{-1}. This, combined with P=2.6d gave $f(M) = 3.24 \pm 0.09 M_\odot$ and hence was a strong case for a black hole [135]. The same data was subsequently fitted by Phillips et al. using an irradiation model, obtaining $K_2 = 192 - 214$ km s^{-1} which reduces the mass function to $f(M) = 1.93 - 2.67 M_\odot$ [140] (Fig. 11). This latter result was finally confirmed by observations in true quiescence which give $K_2 = 215.5 \pm 2.4$ km s^{-1} and $f(M) = 2.73 \pm 0.09 M_\odot$ [160]. The new mass function is 16 % lower than the first reported value, enough to disclaim GRO J1655-40 as a secure black hole candidate (based purely on dynamical information).

6.3.2 Rotational Broadening

The companion stars transfer matter onto their compact objects and hence they must be filling their Roche lobes with their sizes given by eq 6.3. In addition, the short orbital periods ($\sim$ hr) and old ages ($> 10^7$ yr) of these close binaries suggest that the companion stars must be synchronised, i.e.

$$\omega_{\rm s} = \omega_{\rm orb}$$

where $\omega_{\rm s}$ and $\omega_{\rm orb}$ are the stellar and orbital angular velocities respectively. The projected linear velocities would then be

$$\frac{V_{rot} \sin i}{R_2} = \frac{K_2 + K_1}{a} = \frac{K_2(1+q)}{a} \tag{6.17}$$

where $q = M_2/M_1 = K_1/K_2$. Combining eq 6.3 and 6.17 one obtains [192]

$$V_{rot} \ \sin i = 0.46 \ K_2 \ q^{1/3}(1+q)^{2/3} \tag{6.18}$$

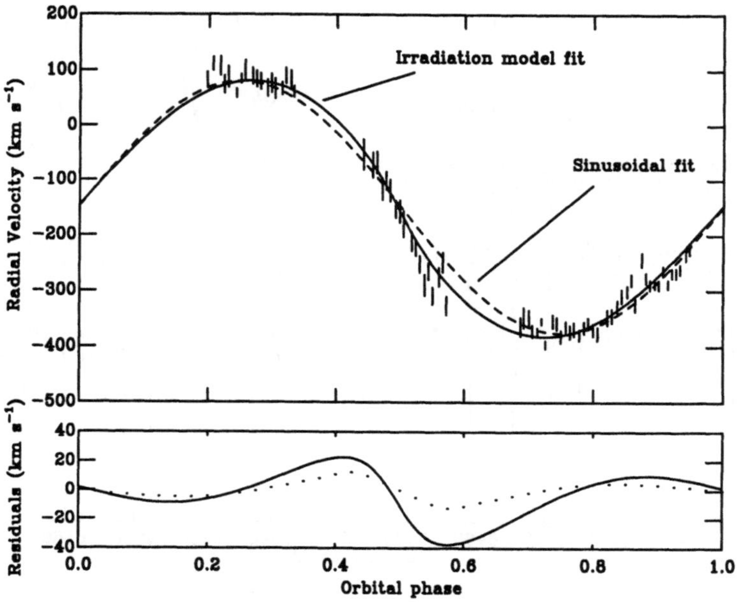

Fig. 6.11. Influence of irradiation effects in the radial velocity curve. From [140].

and hence the mass ratio q can be measured directly from the radial velocity curve and the observed rotational broadening ($V_{rot} \sin i$) of the secondary's absorption lines [59]. The rotational velocity is determined by comparing our target with broadened versions of spectral type templates (e.g. through a χ^2 minimization technique [109]; see Fig. 6.12). The rotationally broadened versions of the template are computed by convolution with a Gray profile of the form [62]

$$G(x) = \frac{6(1-\epsilon)}{\pi(3-\epsilon)}\left(1-x^2\right)^{1/2} + \frac{3\epsilon}{2(3-\epsilon)}\left(1-x^2\right) \tag{6.19}$$

where ϵ is the limb darkening coefficient, which takes into account the attenuation of the radiation field I_λ towards the stellar limb, $x = v/V_{rot} \sin\ i$ and v is the radial velocity of a surface element of the star. The Gray profile is computed under the assumption of a spherical geometry and a linear approximation of the limb darkening effect i.e. $I_\lambda \propto 1 - \epsilon + \epsilon \cos\theta$, with θ being the angle between the line of sight and the normal to the surface. The limb darkening coefficient depends smoothly on $T_{\rm eff}$ and λ and is usually taken as constant over the line profile. More details about the treatment of limb darkening can be found in [40].

Absorption features are usually shallower in the target than in the broadened template due to dilution by the residual accretion disc continuum or *veiling*, which is found to contribute $\leq 40\%$ percent in the R band. This analysis is also sensitive to the spectral type of the template, thus enabling us to constrain this simultaneously to $V_{rot} \sin i$ and the veiling (e.g. see [31]). Note that the template spectra must be observed with the same instrumental setup as the target in

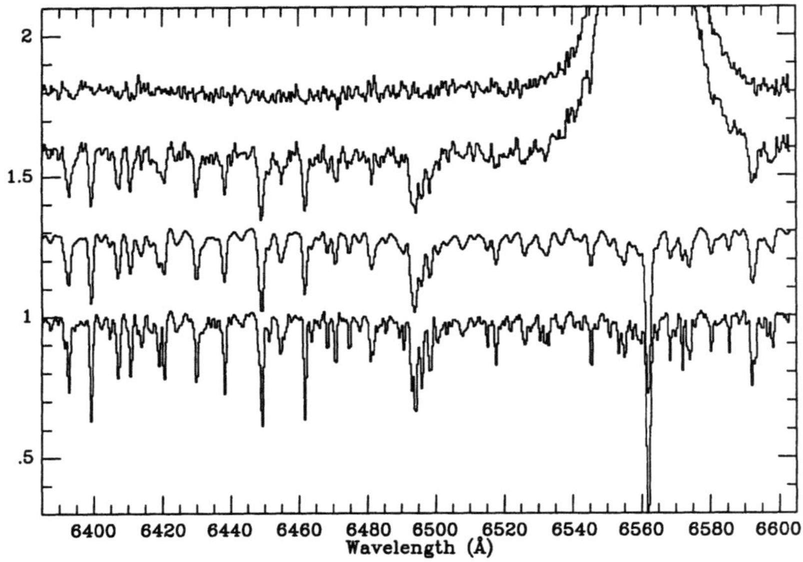

Fig. 6.12. Example of rotational broadening analysis of V404 Cygni. From bottom to top: a KOIV template, the template broadened by $V_{rot} \sin i = 39kms^{-1}$, the spectrum of V404 Cyg in the rest frame of the secondary star and the residual after subtraction of the broadened template. A formal uncertainty of 1 km s^{-1} (2%) in $V_{rot} \sin i$ was achieved. From [27].

order to cancel out instrumental resolution effects. For the typically low q values of LMXBs $\delta(q)/q \simeq 2.6 \times \delta(V_{rot} \sin\ i)/V_{rot} \sin\ i$ and hence a 6 % accuracy in $V_{rot} \sin i$ is needed to yield a 15 % determination of q. This experiment is technically challenging as typical values of $V_{rot} \sin i$ are 40-100 km s^{-1}, so high spectral resolution ($\leq 1\AA$) is needed. Given the faintness of these objects (V$\geq$18), large telescopes are required even for the brightest SXTs.

Two main sources of systematic errors are at play in this calculation. To start with, the limb darkening coefficients (e.g. as listed in [1]–[191]) have been computed using atmospheric stellar models and they are appropriate for the continuum. Absorption lines are formed higher up in the stellar atmosphere and thus one would expect limb darkening to be smaller, especially in the line center [41]. A conservative estimate of the systematic uncertainty introduced by this approximation can be obtained by computing $V_{rot} \sin i$ using the extreme possible values $\epsilon = 0, 1$ and this has been shown to be $\sim$ 14% [195].

Companion stars in LMXBs are clearly non-spherical and hence the use of a Gray profile introduces a second source of systematic error. In fact, the rotational profile changes with orbital phase because of the varying aspects of the Roche potential (see Fig. 6.13) and it is not clear that the orbitally averaged profile necessarily coincides with the spherical Gray profile. Marsh et al. [109] explored this effect by computing the phase-averaged rotational profile in A0620-00 through properly integrating the velocity field in a Roche geometry (but assuming a continuum limb-darkening law). They concluded that the use of eq 6.19 can result in a 5% underestimation of q. This is at (or smaller than) the level of the current statistical errors achieved and so the results are not seriously compromised. A correct treatment of the problem would require integrating the rotational profile in a Roche geometry including a properly computed limb darkening law varying within the line profile, and this has not yet been solved. In the meantime, the use of the Gray profile seems an acceptable approximation which introduces a systematic uncertainty of typically $2\sigma \sim 14$ % due to our ignorance of the "averaged" limb darkening law for the lines. This is demonstrated in the mass ratio determination in J1655-40 where the $V_{rot} \sin i$ analysis provides $q = 0.39 \pm 0.05$ [160] whereas a completely independent calculation, based on the analysis of the light curve (see Fig. 6.15) yields $q = 0.33 \pm 0.01$ [135].

Whereas the phase averaged $V_{rot} \sin i$ depends mainly on q, its orbital variation is a strong function of i and could be compared to synthetic models to derive this important parameter. The details of the modulation also depend on the gravity darkening coefficient β (especially around phase 0.5 where the local gravity changes more rapidly) and ϵ, to a smaller extent. This way of determining i has the advantage over the ellipsoidal modulation (see below) of not being affected by any residual veiling, although the effect is very subtle ($\leq 10\%$). Resolving a few percent variation in $V_{rot} \sin i$ requires excellent quality spectra of

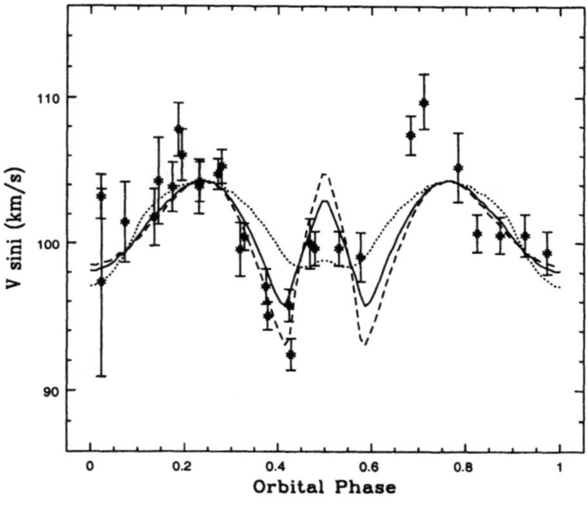

Fig. 6.13. Orbital modulation of $V_{rot} \sin i$ in AE Aqr compared to synthetic models computed for $\beta = 0.08$, $\epsilon = 0.65$ and three inclinations: $i = 40°$ (dotted line), $i = 58°$ (solid line) and $i = 70°$ (dashed line). The modulation is a direct confirmation of the geometrical distortion of the companion star. From [30].

signal-to-noise ratio $\sim$ 100, obtained at typically $R = \lambda/\delta\lambda \geq 10000$ and time resolution better than $\sim 5/R \times P/K_2$ to avoid orbital smearing. Figure 6.13 shows the first application of this technique to the brightest cataclysmic variable AE Aqr [30], and has yielded $i = 58 \pm 6°$ in good agreement with other independent estimates [195]. This work clearly shows that the technique is feasible even though the actual i value may be affected by systematics (in particular, the use of the spherical approximation in the determination of the $V_{rot} \sin i$ values [158]). The correct approach would involve a direct fit of unblended lines with simulated profiles to determine all free parameters simultaneously (i, q, β and ϵ). A new generation of larger telescopes are, however, required before the orbital variations of $V_{rot} \sin i$ are detected in the fainter SXTs and used to derive the binary orbital inclination.

6.3.3 Ellipsoidal Modulation

The lightcurves of companion stars in contact binaries display the characteristic double-humped variation on the orbital period, with amplitudes $\leq$10% (Fig. 6.14). This modulation is a consequence of the tidal distortion of the secondary star and the non-uniform distribution of the surface brightness, due to a combination of limb and gravity darkening. In the case of SXTs and LMXBs in general, the gravity darkening effect is treated in a modified version of von Zeipel's theorem [205] $T_{eff} \propto g^{\beta}$ with $\beta = 0.08$ appropriate for stars with convective envelopes [104]. The dependence of T_{eff} on the local surface gravity g implies that the inner hemisphere of the secondary (facing the compact object)

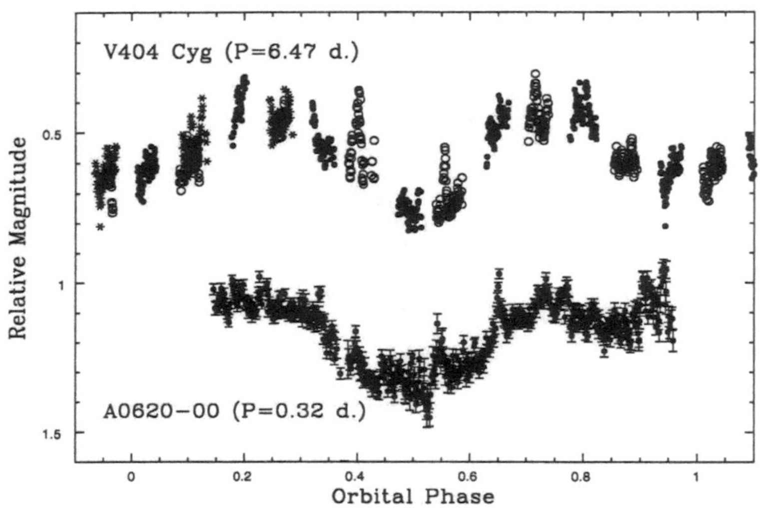

Fig. 6.14. Optical light curves of the two brightest SXTs showing evidence for contaminating fast variability superposed on the classical ellipsoidal modulation.

will be cooler and, therefore, the phase 0.5 minimum becomes deeper than the minimum at phase 0.

Model calculations show that the shape and amplitude of the ellipsoidal modulation are functions of q and i [2]. In particular, the amplitude is a strong (increasing) function of i, but is insensitive to q if $q \leq 0.1$. This is normally the case in SXTs (see Table 6.2) and therefore model fits to the ellipsoidal modulation can be used to determine i directly. Unfortunately, there is additional complicating structure in the optical light curves [76], and it is likely that they are contaminated by the residual continuum from the accretion disc and/or possible flares or starspots on the surface of the secondary (cf. RS CVn-type activity). In particular, the ellipsoidal modulation in V404 Cyg has shown a superimposed variability with a characteristic timescale of ~6hr (e.g. [26],[139]), which presumably originates in the accretion disc. This type of "short" time-scale variability has also been detected in A0620-00 [74] and seems to scale as a fixed fraction of the orbital period (i.e. $\sim 1/25 P_{orb}$). It may be a common property of quiescent SXTs which has escaped detection in fainter sources given the long integrations used (see Fig. 6.14).

Long-term (a few % longer than P_{orb}) distortion waves have also been detected in the light curves of quiescent SXTs (e.g. [36],[75]). These are identified as an oscillation in the height of the two maxima and it has been proposed that they could be produced either by large spots in a non-synchronized secondary or a precessing accretion disc, like the "superhump" modulation seen during outbursts [130] when the disc exceeds the 3:1 resonance radius [199].

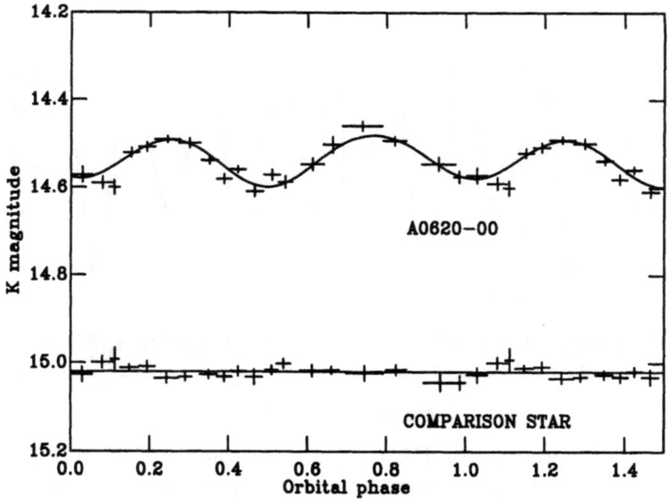

Fig. 6.15. Ellipsoidal fits to the IR light curves of A0620-00. From [154].

The effect of these contaminating continuum sources will be to dilute the total amplitude of the optical lightcurves and, consequently, ellipsodal model fits will only provide a *lower limit* to the inclination. In [139], the short-time scale variability was removed by fitting ellipsoidal models to the lower envelope of the lightcurve, which should define the minimum contamination level. Alternatively, if the veiling is known (e.g. through comparing the depth of the absorption lines with a broadened template) then a reasonable i determination can also be provided.

The spectral distribution of the veiling in the UV/Optical has been measured in a few systems (A0620-00 [131],[118], Cen X-4 [35],[118], V404 Cyg [26]) and it consistently drops towards longer wavelengths. Moreover, since the companion stars in SXTs all have late spectral types they are expected to dominate completely at IR wavelengths. This assumption allows us to fit lightcurves of SXTs in the IR with pure ellipsoidal models, yielding the most realistic i determinations (e.g. [153],[155]). Furthermore, models are less affected by uncertainties in the limb and gravity darkening coefficients than in the optical. An example of an ellipsoidal fit to a K band lightcurve is shown in Fig. 6.15.

Subsequent IR photometry [148] has shown, however, that short-time scale contamination is still present in H-band observations of V404 Cyg. This result undermines the key assumption of negligible IR veiling and suggests that accretion discs in SXTs, which may be optically thick in their cool outer parts, also contribute significantly to the IR flux. Therefore, ellipsoidal fits to IR lightcurves would strictly provide only lower limits to i and thus upper limits to M_1. Attempts to quantify the veiling in the K band using medium resolution spectroscopy have yielded upper limits (<14% in V404 Cyg [156] and <27% in A0620-00 [159]) leaving the matter still unresolved.

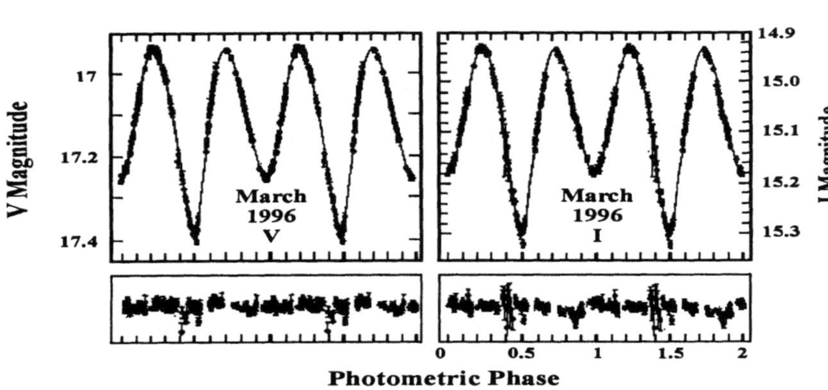

Fig. 6.16. Orbital lightcurves of J1655-40 in quiescence and model fits including ellipsoidal modulation and grazing eclipse of the companion star by the accretion disc at phase 0.5. The disc contribution to the optical light is completely negligible in this case. Adapted from [135].

N Sco 94 (=J1655-40) and 4U1543-47 are atypical SXTs containing early type companions (A2V and F6IV) which completely dominate the observed flux at optical wavelengths. The veiling is therefore negligible and the results of the ellipsoidal fits can then be taken as unbiased i determinations. The mass ratio for N Sco lies in the range where ellipsoidal models are sensitive to q ($q > 0.1$) so this parameter can also be measured. In addition, during the outburst phase the binary was found to show grazing eclipses between the star and the accretion disc [4] (but not of the X-ray source) which set tight constraints on the inclination. Detailed ellipsoidal model fits to lightcurves in quiescence, including the phase 0.5 eclipse by the dim accretion disc, have yielded $i = 70 \pm 1°$ and $q = 0.33 \pm 0.01$ [135] (see Fig. 6.16). These are the most accurate determinations in any SXT and provide the most precise component masses yet measured (see also [79] for an independent result, with a more conservative treatment of the error bars).

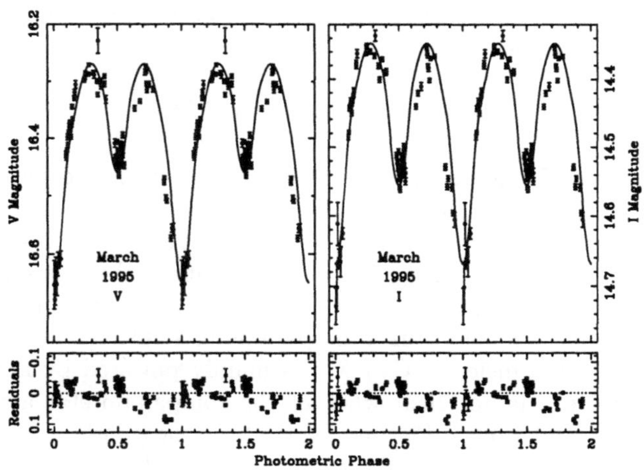

Fig. 6.17. Orbital lightcurves of J1655-40 during outburst and best model fits showing the effect of X-ray irradiation. The best fits were obtained by fixing i and q to the values derived in Fig. 6.16 and leaving the disc radius, temperature law and flared angle as free parameters. Only irradiation of the companion star was included. (see [135] for details)

Moderate X-ray irradiation is known to affect the thermal structure of the star's outer layers thereby changing the brightness distribution of the Roche lobe and the shape of the ellipsoidal modulation. The main effect will be to fill-in the phase 0.5 minimum which, under severe X-ray heating, can even reverse into a single maximum (e.g. Fig. 6.4). The influence of irradiation shows up clearly in the light curves presented in Fig. 6.17, where the deepest minimum has now shifted to phase 0. Model fits were presented in [78] which included both mutal eclipses and irradiation effects on the disc (according to VR model) and the secondary (allowing for the X-ray shadow cast by a flared disc). They yielded $i = 65° - 76°$ in good agreement with the quiescent fits. The mass ratio is, however, poorly constrained to $q = 0.18 - 0.26$ and is only marginally consistent with the more precise fits to the clean ellipsoidal light curve in quiescence.

6.4 Dynamical Black Holes

The combination of these three techniques (i.e. optical mass function, rotational broadening and ellipsoidal modulation) has rendered the first realistic compact object masses and the evidence that most SXTs harbour black holes. A first application of this prescription was already presented in [59] for the case of the HMXB Cyg X-1. I will now summarize the results of this work as a general example of the methodology for weighing masses in X-ray binaries.

TCyg X-1 was the first black hole candidate discovered, but the low value of its optical mass function

$$M_1 \sin i^3 (1+q)^{-2} = 0.25 M_\odot, \tag{6.20}$$

required estimates of i and q in order to place any sensible constraints on M_1. Restrictions on the inclination came from the the absence of X-ray eclipses i.e. $i \leq 60°$ which, combined with the expected mass for a "normal" O9.7Iab ($M_2 = 33 M_\odot$), yield $M_1 > 4 M_\odot$ [10]. Unfortunately, there is compelling evidence for optical companions to X-ray binaries to be severely undermassive for their spectral types [178] and thus a full parameter solution was required. Being a HMXB, the mass transfer in Cyg X-1 is mainly powered by a strong stellar wind and therefore it is not clear whether the companion star actually fills its Roche lobe. Gies and Bolton [59] modelled the optical lightcurves of Cyg X-1 using an extension of the ellipsoidal models discussed in Sect 3.3 allowing for a fill-out factor $\rho \leq 1$. For a given inclination, the amplitude of the ellipsoidal modulation will always be lower and thus $\rho = 1$ provides an absolute lower limit to the true inclination. Also, the star may not be synchronised and, therefore, eq 6.18 has to adopt the more general expression

$$V_{rot} \sin i = \Omega \; \rho \; 0.462 \; K_2 \; q^{1/3} (1+q)^{2/3}, \tag{6.21}$$

where Ω is the *corotation factor* which parameterizes the degree of synchronism i.e. $\Omega = \omega_s / \omega_{orb}$. They computed ellipsoidal models and synthetic line profiles, in a Roche potential geometry, which were simultaneously fitted to the data to reproduce the observed $V_{rot} \sin i$ and amplitude of the ellipsoidal modulation. Figure 6.18 presents the fitted solutions in the $M_1 - M_2$ plane as defined by eqs 6.20 and 6.21 for a set of ρ and Ω values and a measured $V_{rot} \sin i \simeq 96$ km s^{-1}. The allowed region in the parameter space is constrained by the inclinations $i \leq 60°$ (i.e. no X-ray eclipses; line *E*), $i \geq 33°$ (i.e. $\rho = 1$) and $M_2 \leq 50 M_\odot$, based on evolutionary arguments applied to a O9.7Iab star[11]. Adding the extra free parameters ρ and Ω means that a wide range of (i, M_1, M_2) solutions can equally fit the data and then these too need to be constrained. Actually, ρ and Ω are limitted by two theoretical restrictions:

- The absence of statistically significant eccentricity in the radial velocity curve indicates that the orbit is circularised [58] which, together with the advanced evolutionary state of the optical star suggest $\Omega \simeq 1$ [149].
- Stellar wind model fits to the HeII λ4686 emission indicate large filling-out factors in the range $\rho \simeq 0.9 - 1.0$ [60]. These new constraints lead to the solutions $M_1 = 16 \pm 5 M_\odot$ and $M_2 = 33 \pm 9 M_\odot$[12].

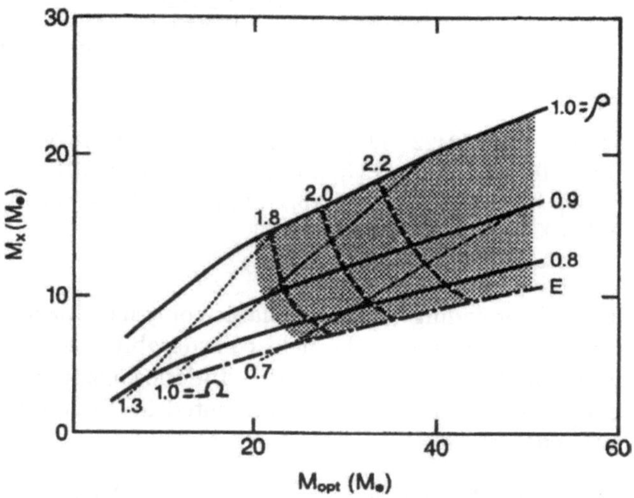

Fig. 6.18. Constraints on the masses of the two components in Cyg X-1. Allowed solutions are contained in the shaded region. See text for details. From [58].

SXTs present cleaner examples of the application of this prescription since the optical companions are low-mass stars tidally locked ($\rho = 1$) in short period, circularised orbits which transfer mass through Roche lobe overflow ($\Omega = 1$). Therefore, the determination of the optical mass function, $V_{rot} \sin i$ and model fits to the ellipsoidal light curve (corrected for veiling) provide a full solution to the system parameters. For a decade, 4-m telescopes like WHT in La Palma, UKIRT in Hawaii or CTIO in Chile, have enabled us to test this method with the brighter transients. Systems like A0620-00, V404 Cyg and GU Mus could be

11 A further constraint is set by a lower limit to the distance $d > 1.9$ kpc [202] which, combined with the observed luminosity and the Roche geometry leads to a lower limit on M_2.

12 Note that these numbers have been refined to $M_1 = 4 - 15 M_\odot$ and $M_2 = 12 - 19 M_\odot$ by an independent technique which employs detailed atmospheric modelling of the OBI star to yield $\log g$ and $T_{\rm eff}$ [77].

Table 6.2. Optical properties of soft X-ray transients

Source	P (d)	γ (km/s)	K_2 (km/s)	$f(M)$ ($M_\odot$)	q (M_2/M_1)	i (deg.)	M_1 ($M_\odot$)	Sp. Type	$\overline{\rho}/\overline{\rho_{MS}}$	LiI
4U1543-47	1.12	-87	124	0.22 ± 0.02	$\sim$0.5	20-40	$\sim$ 2.7-7.5	A2V	0.28	
J0422+32	0.21	9	381	1.21 ± 0.06	$0.116^{+0.079}_{-0.071}$	20-40	10 ± 5	M2V±2	1.19	?
V616 Mon	0.32	10	433	2.72 ± 0.06	0.067 ± 0.010	37 ± 5	10 ± 5	K3-5V	0.62	Y
J1655-40	2.62	-142	216	2.73 ± 0.09	0.39 ± 0.05	67 ± 3	6.7 ± 1.2	F6IV	0.03	Y
GU Mus	0.43	16	406	3.01 ± 0.15	0.128 ± 0.04	54 ± 20	6^{+4}_{-2}	K3-4V	0.35	Y
N. Vel 93	0.29	30	475	3.17 ± 0.12	0.137 ?	$\sim$ 37-51	$\sim$ 3-5	K8V±2	0.77	
N. Oph 77	0.52	-54	441	4.65 ± 0.21	$0.014^{+0.019}_{-0.012}$	70 ± 10	6 ± 2	K5V±2	0.23	
QZ Vul	0.34	19	520	5.01 ± 0.12	0.042 ± 0.012	56 ± 15	10 ± 4	K3-5V	0.55	Y
V404 Cyg	6.47	-0.4	208	6.08 ± 0.06	0.067 ± 0.005	55 ± 4	12 ± 2	K0IV	0.002	Y
J2123-058	0.25					73±4				
Cen X-4	0.63	186	146	0.21 ± 0.08	$0.16^{+0.04}_{-0.05}$	43 ± 11	1.4 ± 0.5	K5-7IV	0.16	Y
Aql X-1	0.79					>36		K6-M0	0.10	

References: 4U1543-47: [136]; J0422+32: [133], [28], [49], [71]; V616 Mon: [118], [109], [154]; J1655-40: [135], [160]; GU Mus: [134], [31], [157]; N. Vel 93: [51]; N. Oph 77: [50], [70]; QZ Vul: [29], [48], [69], [6]; V404 Cyg: [25], [27], [155]; J2123-058: [206]; Cen X-4: [117], [153], [141]; Aql X-1: [37], [38], [196].

intensively scrutinized, providing the first reliable compact object masses. The advent of 10-m Keck has expanded the field considerably towards the fainter members of the class and has opened the door to statistical comparison of system parameters. A compilation of the results is presented in Table 6.2. Four main trends can be outlined:

a) **Most SXTs contain BHs.** Compact object masses derived using the previous prescription yield 9 dynamical black holes with $M_1 > 3.2 M_\odot$ and one neutron star, Cen X-4, with $M_1 = 1.4 \pm 0.5 M_\odot$. The detection of X-ray bursts in 5 additional systems makes a total of 6 NS out of 28 SXTs (21%) whereas the remaining 79% are likely to contain BHs. On the other hand, almost all persistent LMXBs are known to contain NS. In Sect 1.6 we saw how X-ray irradiation can stabilize accretion discs, i.e. preventing the development of instabilities if $T_d^{irr} > T_H$ or, equivalently, $\dot{M}_2 > \dot{M}_{crit}^{irr} \propto P^2$. Therefore, the high impact of BHs in SXTs indicates that BH systems tend to have lower $\dot{M}_2$. This question has been intensively studied in the last four years by King and colleagues demonstrating that the reason behind this behaviour rests on evolutionary considerations [94],[95],[96][97]. These can be summarized as follows:

LMXBs are a particular phase in binary evolution right after the common envelope episode and the explosion of the more massive star in a supernova [188]. The output will be a detached binary containing either a NS or a BH and a low mass main sequence star in a $\sim$ 0.5-10 day orbit. The system will come into contact because of two processes: the nuclear expansion of the secondary star and the shrinkage of the orbit caused by angular momentum loss. The characteristic physical timescales of these two mechanisms are t_{MS} and t_{AML}. The binary evolution and subsequent activity (i.e. *transient/ persistent*) will depend on the interplay between t_{MS} and t_{AML} and we have two possible scenarios:

(1) $t_{MS} \ll t_{AML}$**:** The secondary star evolves off the Main Sequence before angular momentum loss bring the system into contact, and this occurs for $P \geq 2$ days. The secondary star will be a subgiant or *stripped giant* with M_2 and L_2 being controlled by the mass of the degenerate helium core [169]-[194]. The secondary star will transfer mass on a nuclear timescale at a rate $\dot{M}_2$ growing with P according to the equation

$$\dot{M}_2 \simeq 4.0 \times 10^{-10} P_d^{0.93} m_2^{1.47} M_\odot yr^{-1}$$

where $m_2 = M_2/M_\odot$ and $P_d = P/day$ [92]. Note that $\dot{M}^{irr}_{crit} \simeq 1.9 \times 10^{-7} P_d^2$ $M_\odot$ yr^{-1} and, because $P \geq 2$ d, $\dot{M}_2 < \dot{M}^{irr}_{crit}$ and the systems will be transients. An example is V404 Cyg [93].

(2) $t_{MS} \geq t_{AML}$**:** angular momentum loss shrinks the orbit before the secondary evolves off the Main Sequence and this occurs for $P \leq 2$ days. If $t_{MS} \sim t_{AML}$ the secondary star can be significantly evolved before the onset of the mass transfer phase. This is the situation for most SXTs and the mass transfer rate is given by the expression

$$\dot{M}_2 = 2 \times 10^{-9} m_1^{-2/3} \overline{m}_2^{7/3} \left(\frac{P}{3hr}\right)^{5/3} + 7.6 \times 10^{-11} m_1^{2/3} \overline{m}_2^{2} \left(\frac{P}{3hr}\right)^{-2/3}, \tag{6.22}$$

where $m_1 = M_1/M_\odot$, $\overline{m}_2 = M_2/M_2(MS)$ and $M_2(MS) \simeq 0.33(P/3hr)M_\odot$ is the main sequence mass. The first term in eq 6.22 grows with P and accounts for the angular momentum loss due to magnetic braking. The second term, reflects the angular momentum loss by gravitational radiation. It decreases with P and hence only dominates for orbital periods $\leq 3hr$. Note the inverse dependence of the first term with m_1 which gives massive compact objects a higher chance to be transients. Figure 6.19 plots eq 6.22 for the case of an unevolved secondary star (i.e. $\overline{m}_2 = 1$) and a range of compact object masses. The dash-dotted line represents $\dot{M}_{crit}$ whereas the dotted line indicates the relevant case $\dot{M}^{irr}_{crit}$.

Neutron star binaries with main sequence companions have $\dot{M}_2 > \dot{M}^{irr}_{crit}$ at all orbital periods and they must be persistent. Therefore, the few transient neutron star binaries (e.g. Cen X-4) require highly evolved companions (i.e. $\overline{m}_2 \ll 1$) in order to trigger disc instabilities. According to Fig. 6.19, BH binaries with main sequence companions would also be persistent and evolutionary models predict a significant population of BH binaries with unevolved companions. However, there is evidence of only one possible persistent BH binary (GX339-4) in $\sim$ 200 LMXBs. A solution to this puzzle arose in [95]

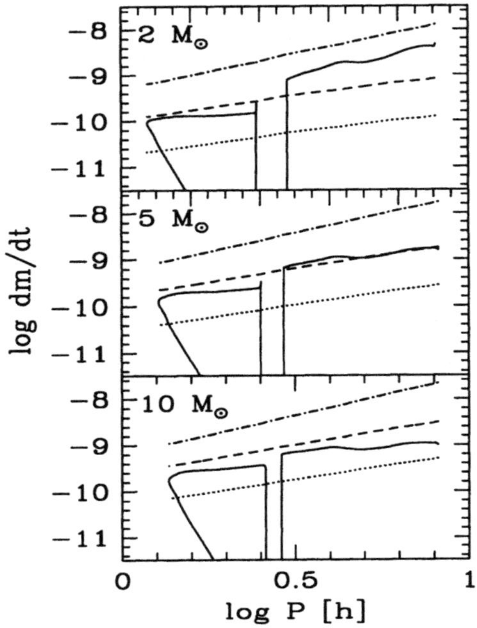

Fig. 6.19. $\dot{M}_2$ vs P for different compact object masses. Companion stars are assumed to be main sequence.

where they proposed disc irradiation must be less efficient in BHs than in NS binaries, the reason simply being the lack of a hard surface in BHs which leaves the inner disc region as the likely X-ray source. This lies at the orbital plane and subtends a solid angle (as seen by the outer disc) a factor $\sim H/R$ smaller than the compact object's surface, with H, R being the outer disc height and radius. X-ray irradiation of the outer disc will consequently be smaller and hence $\dot{M}_{crit}^{irr}$ will rise by a factor ~ 6 (dashed line in Fig. 6.19). This explains why the short period BH binary J0422+32, which contains a main sequence M2 companion, is also a SXT[13].

b) **SXTs contain late-type companions** and hence their mass ratios are very extreme ($q \leq 0.15$). There are two exceptions, 4U1543-47 and J1655-40, with intermediate mass donors (2-3 $M_\odot$) and slightly different evolutionary history (see [86]). The companions are also considerably **undermassive** for their spectral types. This is clearly seen in column 10 of Table 6.2 which provides the star's mean density ($\overline{\rho}$) normalised to the density of a main

[13] The real scenario is somehow more complex since another stability limit holds for cold quiescent accretion discs at lower mass transfer rates ($\dot{M}_{cold}$) [67] which in turn defines an instability band in Fig. 6.19, at $\dot{M}_{cold} < \dot{M}_2 < \dot{M}_{crit}^{irr}$. $\dot{M}_{cold}$ depends critically on the inner disc radius and the long recurrence times of SXT outbursts suggest that accretion discs may be truncated in their inner regions [120], rendering support to ADAF models.

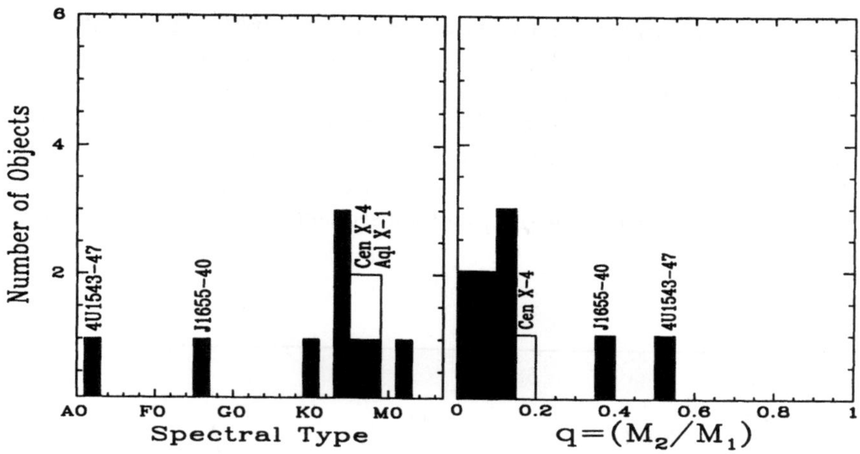

Fig. 6.20. Distribution of spectral types of companion stars and mass ratios in SXTs. Neutron star SXTs are marked in white.

sequence star (ρ_{MS}) with spectral type determined from spectroscopy (i.e. column 9). Since the companion star fills its Roche lobe its size is given by Paczynski's equation (eq.3) which, combined with Kepler's Third law yield

$$R_2 = 0.234 P_{hrs}^{2/3} M_2^{1/3}$$

or equivalently

$$\bar{\rho}\ [gr/cm^3] = 110 P_{hr}^{-2}.$$

Therefore, the Roche lobe density depends entirely on the orbital period P_{hr}. $\bar{\rho}/\rho_{MS}$ is equivalent to the fill-out fraction of a normal mainn sequence star and is always < 1 (except for J0422+32) meaning that **secondary stars in SXTs are evolved** (see also [167]). The most extreme cases are V404 Cyg and J1655-40. In the first case the companion is a stripped-giant or subgiant whereas, in the second case, it is crossing the Hertzsprung gap, just before ascending the giant branch [100],[99].

c) **BH binaries have significantly lower systemic velocities** than NS binaries, suggesting the lack of a recoil velocity during the collapse of the progenitor (see [198]). Here again 4U1543-47 and J1655-40 are the exceptions with space velocities comparable to neutron star binaries (e.g. Cen X-4). The most likely explanation for the high space velocity in these two binaries is a *delayed BH formation* i.e. a neutron star is formed in the initial supernova collapse which shortly afterwards will become a BH because of subsequent accretion of fall-back matter or a phase transition [12]. However, there is an alternative view where the velocity dispersion in *all* BH binaries can be explained by variable mass ejection during the SN explosion and hence no quick velocities would be required in BH formation[198],[126].

d) **Companion stars tend to show strong ^{7}Li λ6708 lines** with EWs of typically $\sim$ 300 mÅ (see Fig. 6.21). This was first noted during the analysis of the first observations to determine the spectral type of the secondary star in V404 Cyg [110] where the Li feature stood out as the one significant difference between V404 Cyg and the K0IV template star with which it was fitted. Later, it was also detected in the high resolution high S/N spectra of V616 Mon [109], Cen X-4 [111] and subsequently in GU Mus [114] and QZ Vul [48].

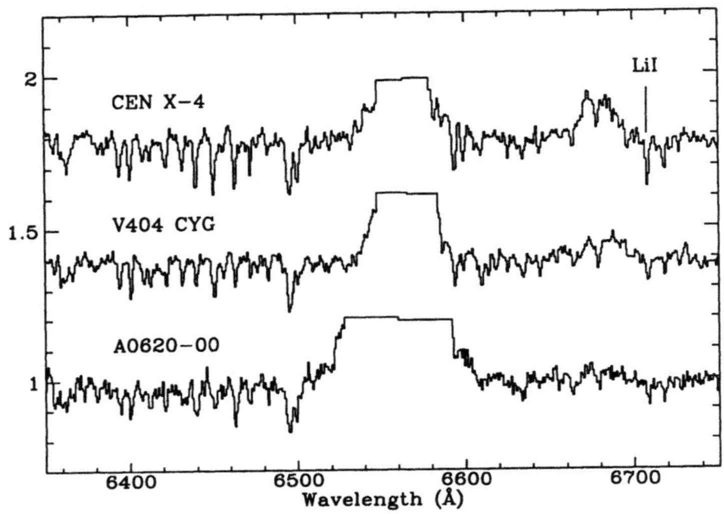

Fig. 6.21. AAT and WHT spectra of Cen X-4, V404 Cyg and V616 Mon showing the presence of LiI λ6708. From [34].

The EWs have been converted into abundances through non-LTE line model fits after correcting for the amount of veiling measured in the spectra (e.g. see [111]). Table 6.3 list the EWs and derived abundances. They are in the range $\log N_{Li} = 2-3$, consistent with the primordial cosmic abundance (see Fig. 6.22). This is a striking result as Li is found with this abundance only in young stars, where convective mixing (that leads to Li destruction) has not yet had time to substantially reduce the initial Li content; e.g. observations of K stars in the Pleiades and Hyades indicate typical depletion times of $10^7 - 10^8$ yrs. And this must be an upper limit since extra depletion of a factor 10 is expected from steady mass transfer at typically $\sim 10^{-9}$ M$_\odot$ yr^{-1}. It is *very* unlikely that all these SXTs are young ($\leq 10^7$yrs), so the conclusion must be that Li is being created at these sites.
This group of SXTs have a wide range of orbital periods (0.32–6.5 days), four are black holes while one is a neutron star and the secondaries are of signifi-

Table 6.3. Li abundances in companion stars to SXTs

Name	Sp. T.	T_{eff} K	Veiling	EW (LiI) mÅ	log N_{Li} NLTE
Cen X-4	K5-K7 V	4250	25%	480±30	3.1± 0.4
V404 Cyg	G9-K1 IV	4750	5%	290±30	2.7± 0.4
V616 Mon	K3-K5 V	4500	6%	235±40	2.1± 0.4
GU Mus	K3-K5 V	4500	12%	420±60	3.0± 0.5
GS2000+25	K5 V	4400	15%	250 ± 40	2.2 ± 0.5
J0422+32	M2 V	3700	40%	$\leq$ 480	$\leq$ 2.0
N. Oph 77	K3-7 V	$\sim$ 4400	32%	$\leq$ 120	
J1655-40	F6-7 IV	6400	$\leq$ 2%	55 ± 8	

cantly different size. The *only* property they share is the massive X-ray outburst that recurs every few decades, and it has been proposed [111],[114],[112] that CNO spallation by α particles and $\alpha - \alpha$ collisions during these outbursts result in the production of Li. With large mass outflows seen during the outbursts [23] the Li can thereby be transferred to the secondary star.

Alternative spallation scenarios have also been proposed involving the high ion energies ($\sim$ MeV) arising in advection dominated flows around compact objects. In [203] Li is produced in the accretion flow by α particles and subsequently expelled, thereby contaminating the companion star. In [63] CNO spallation is directly produced by neutron flux onto the companion star. Recently, it has been argued that the Li overabundance observed in SXTs is consistent with observations of RS CVn systems, i.e. chromospherically active K stars in close binaries, [8] and hence the same mechanism might be at work (see Fig. 6.22). Independent evidence of possible coronal activity is the detection of Hα [31],[70] and IR CaII [21] emission from the companion star in Doppler images.

Since the orbital periods and atmospheric structure of the secondaries in CVs are comparable to SXTs, they might also show strong Li lines. However, a follow-up campaign has given only upper limits [113]. A clear implication of this result is that the high Li abundances in SXTs are not a consequence of inhibiting depletion mechanisms or chromospheric activity; it also tells us that white dwarf primaries do not provide sufficient accretion energy to power the nucleosynthesis reactions.

It has been suggested that a possible test for Li production mechanisms through spallation is related to the observed γ-ray line at 476 keV in Nova Mus 1991 [111]. This was originally interpreted as a gravitationally-redshifted e^--e^+ annihilation line, double-peaked due to Keplerian rotation of the disc [61]. Instead, this feature can be associated with the $^7Li^*$ de-excitation, which also provides a more natural explanation of the width of the line and

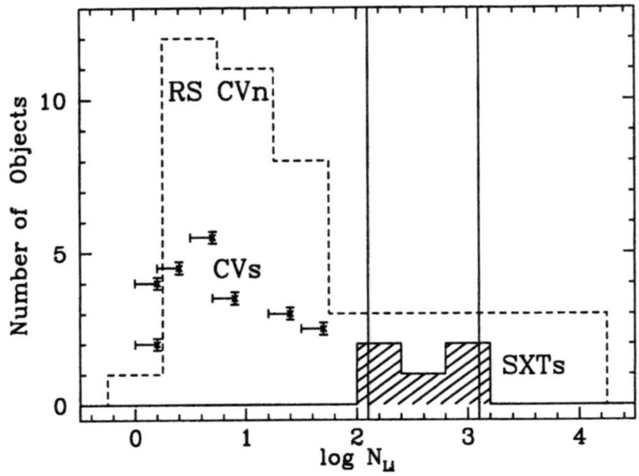

Fig. 6.22. Number distribution of ^{7}Li abundance in the companion stars of SXTs and CVs (upper limits). Vertical lines mark the Li abundance in young cluster stars ($\log N_{Li} = 3.1$) and in old stars of the galactic halo ($\log N_{Li} = 2.1$). The primordial cosmic abundance is constrained between these two lines. The dashed-style histogram represents the distribution observed in RS CVn binaries.

the time delay of the event. The recent discovery of Li in the companion star of N Mus 1991 lends strong support to this scenario [111]. Other γ-ray lines of interest are the 429 keV, resulting from the de-excitation of ^{7}Be*, and the 3.561 MeV line from the ^{6}Li* de-excitation, both nuclei are produced in the course of $\alpha - \alpha$ collisions. Alternatively, the presence of the 2.224 MeV line would indicate neutron capture onto protons, favouring [63]. Possible tests of spallation mechanisms in the optical are the detection of a high abundance of side-elements (D, Be, B) and a large isotopic ratio $^7Li/^6Li \sim 5$, much higher than in the interstellar medium. However, these observations are challenging. The Li doublet separation is only ~ 0.1 Å, much lower than the rotational broadening of the companion star features (typically ≥ 40 km s^{-1}. Also D, Be and B lines are located in the near UV, where the stellar spectrum is strongly hampered by a combination of interstellar reddening and the large veiling of the accretion disc.

6.5 Mass Distribution of Collapsed Objects

The analysis of the mass distribution of collapsed objects has a crucial impact on the most fundamental areas of modern physics, namely Particle Physics and General Relativity. In particular, the distribution of neutron star masses provides important constraints on the equation of state of condensed matter (see Sect. 2). A fundamental experiment, for instance, would be to discover a neutron star with a mass larger than 1.6 M$_\odot$ since this will automatically rule out *soft equations of state* as in [13]. Very precise mass determinations (e.g. [173],[127],[174]) are available from time delays in millisecond radio-pulsars and are all consistent with a narrow Gaussian distribution with $M_1 = 1.35 \pm 0.04\,\mathrm{M}_{\odot}$ (Fig. 6.23). Finn [52], on the other hand, has examined the observed distribution using a statistical Bayesian analysis which assumes that neutron star masses are uniformly distributed between a lower and upper bound. He finds that neutron star masses in radio pulsars are bound by the limits 1.01-1.34 and 1.43-1.6, just consistent with soft EOS.

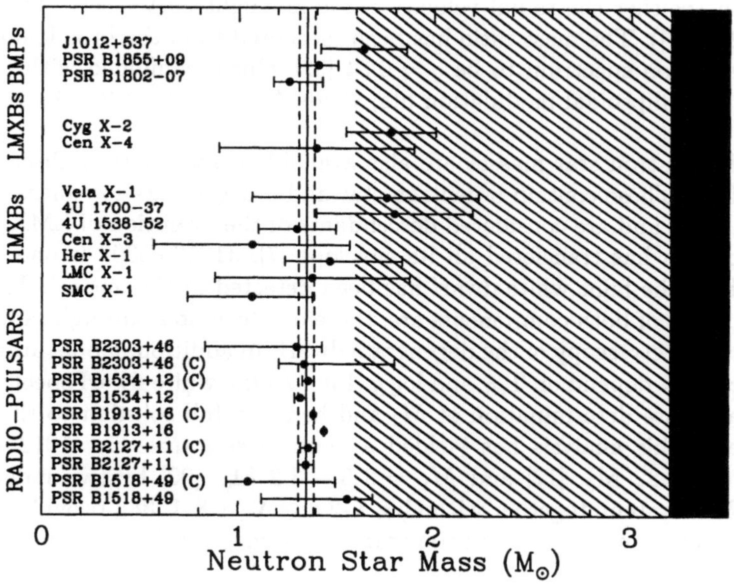

Fig. 6.23. Distribution of neutron star masses. The shaded region is forbidden by soft equations of state, while the black region marks the black hole domain. Vertical lines indicate the mean mass of the Gaussian distribution for radio pulsars ($1.35 \pm 0.04\ M_{\odot}$).

Dynamical masses are also available from *pulsing* neutron stars in six eclipsing HMXBs and they have all been found to lie in the range 1.0–$1.9\,\mathrm{M}_{\odot}$ (see [89] and references therein). Unfortunately, the uncertainties involved are significantly larger than in radio pulsars, the reason being non-Keplerian perturbations in the radial velocity curves. These are driven by stellar wind contamination (which becomes very important in the early type companions to HMXBs) and X-ray heating effects.

Quiescent SXTs are free from these systematic effects although there is only a handful containing neutron stars. The brightest of all is Cen X-4 and its neutron star mass has been measured as $M_1 = 1.4 \pm 0.5\ \mathrm{M}_{\odot}$ [153],[141]. The unsatisfactory large error is governed by the uncertainty in the inclination angle. Persistent LMXBs, on the other hand, are suspected to contain *relatively* "massive" neutron stars since steady near-Eddington accretion rates can supply a few extra 0.1 $M_\odot$ in the course of their life-times ($\sim 10^8$ yr [181]). However, mass determinations are usually hampered by the difficulty to see the optical companion, normally overwhelmed by X-ray reprocessed radiation. In a few exceptions (e.g. Her X-1, Cyg X-2), the companion is a giant star or sufficiently evolved to dominate the optical spectrum, like Cyg X-2 which contains a F0III in a 9.8d orbit. Despite Cyg X-2 being a strong X-ray source, the irradiation flux onto the companion's surface is unimportant because of its early spectral type and the large orbital separation (neither the radial velocity curve nor light curves seem to be influenced [32],[137]). The mass ratio has been measured from the broadening of the F0III absorption spectrum to be $q = 0.34$ [32] which, combined with the mass function and ellipsoidal modulation fits, yields $M_1 = 1.78 \pm 0.23$ $M_\odot$ [137].

Finally, neutron star masses are also available from binary millisecond pulsars (BMPs). These contain a neutron star and a degenerate low-mass star in a detached orbit and are considered descendants of the evolution of LMXBs, after the companion has transfered all its envelope [7]. J1012+5307 is one of a few BMPs where the faint white dwarf has been detected in the optical. The radial velocity curves of both components are, therefore, available through radio pulses and optical spectroscopy of the degenerate dwarf. In addition, model atmosphere fits (which only depend on $T_{\rm eff}$ and $\log g$ [146]) to the white dwarf spectrum renders an accurate determination of M_2 and hence, a full solution to the system parameters. This analysis has yielded a neutron star mass in J1012+5307 of $M_1 = 1.6 \pm 0.22$ $M_\odot$ [17] and $M_1 = 1.5 - 3.2$ $M_\odot$ [90] by two independent studies. This system, together with Cyg X-2 and Vela X-1 [88], provide the most promising cases for *massive* neutron stars which may invalidate soft EOS.

Figure 6.24 presents the observed distribution of black hole masses as given in Table 6.2. Despite the large errorbars (mainly dominated by uncertainties in the inclination), Bailyn *et al.* [5] have attempted to study the statistical properties of seven systems (excluding 4U1543-47 and N Vel 93) following a Bayesian analysis identical to that by Finn and found:

- (1) All determinations, except for V404 Cyg, are consistent with a narrow range of black hole masses around ~ 7 $M_\odot$. This suggests that V404 may belong to a different population of parent binaries and also a preference in nature to form black holes with a particular mass.
- (2) There is a gap between 3-5 $M_\odot$ in the distribution, implying that these black holes have not been formed through accretion induced collapse of neutron stars since at least $\sim 3 M_\odot$ would have been accreted.

However, these conclusions were obtained before 4U1543-47 and N Vel 93 were discovered and their preliminary mass estimates are in open contradiction with (2). In addition, it is evident that (1) is completely dominated by the small error in the J1655-40 mass determination. Clearly more accurate determinations are needed for other systems before drawing any statistically significant conclusion on the distribution of BH masses. In particular, a factor 2 improvement in

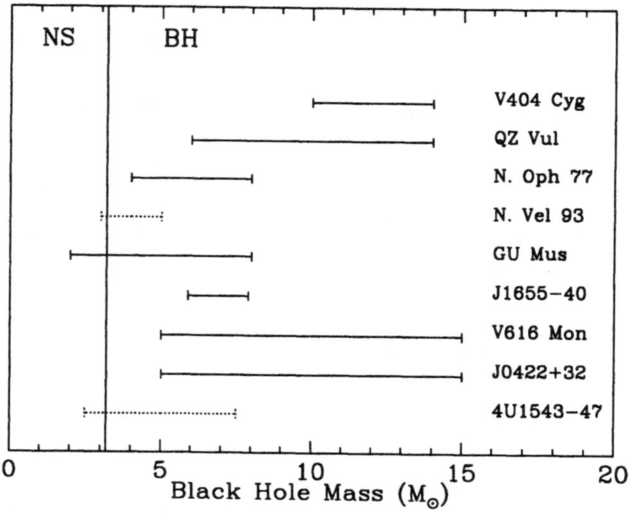

Fig. 6.24. Observed distribution of black hole masses. Errorbars in dashed style indicate preliminary mass estimates.

the error bars of J0422+32, V616 Mon, and QZ Vul will be a major progress. At least for V616 Mon, this should be feasible with the better sensitivity of current IR detectors.

6.6 Comparison with Supernovae Models

The theoretical black-hole and neutron star mass distributions depend critically on the EOS (which establishes the critical mass dividing NS and BH formation) and the current understanding of the late evolution stages of massive stars and supernovae models. The latter contain several theoretical uncertainties which completely dominate the final distribution, namely:

- The treatment of *convective mixing* in massive stars, which conditions the amount of fresh material being brought from the radiative atmosphere into the convective nucleus, and hence the size of He to Fe cores.
- The *mass-cut* or fraction of the stellar core to become the compact remnant (the remaining will be expelled). This determines the size of the collapsing nucleus.
- The *amount of fallback* or fraction of envelope swallowed by the remnant. This depends critically on the budget between the binding energy of the star (calculated by integrating the stellar mass profile [55]) and the kinetic energy E_k of the mantle at infinity (typically $\sim 1.2 \times 10^{51}$ erg as determined from fits to supernova lightcurves [46]). E_k is a fraction of the total available explosive energy and the actual fraction is highly uncertain as it depends on

the detailed physics of the supernova explosion such as the neutrino cross sections, EOS, the angular momentum of the progenitor, etc... Supernova models computed in [175], for instance, suggest that progenitors more massive than ~ 30 M$_\odot$ will swallow most of the envelopes leaving ≥ 10 M$_\odot$ black hole remnants.

- The *mass-loss* of envelopes through strong winds during the Wolf-Rayet phase of stars more massive than $\sim 35-40$ M$_\odot$ (e.g. see [44]). This defines the size of the He core prior to the Type Ib supernova and is usually computed from the expression [200]

$$\dot{M}_{WR} = 5 \times 10^{-8} \left(\frac{M_{WR}}{M_\odot}\right)^{2.6} M_\odot yr^{-1}$$

  Since the mass-loss rate increases with M_{WR} the final He cores will converge to $\sim 3-5$ M$_\odot$ for a wide range of initial masses $\sim$35-80 M$_\odot$. However, see [68] for evidence of reduced mass-loss rates.
- The influence of *binarity* in the evolution of the massive star. Particularly, the massive star can transfer its hydrogen envelope during the common envelope phase leaving a "naked" He core which will behave like a Wolf-Rayet star experiencing large mass-loss through a wind [14]. The collapsed remnant can thereby end up with much lower mass than if it had evolved from a single star.
- The *initial mass function IMF* of the progenitors. There is observational evidence that $IMF \propto M_{\rm prog}^{-\gamma}$ with $\gamma \sim 1.8-3.1$ in our mass range of interest i.e. $\geq 8-10 M_\odot$ [105].

Figure 6.25 presents an example of remnant distributions computed in [176] using massive star evolution (up to 40 M$_\odot$) and explosive models from [201]. They choose a Salpeter IMF (i.e. $\gamma = 1.35$) and fixed the kinetic energy $E_k = 1.2 \times 10^{51}$ ergs for all masses but dismissed the effect of fallback. Massive stars were exploded either before or after the H envelope has been transfered through Roche lobe overflow (i.e. Type II or Type Ib supernovae, respectively). In the second scenario, only the "naked" He nucleus is exploded and two close peaks at ~ 1.32 M$_\odot$ are produced (bottom panel). This seems to be in good agreement with the observed distribution of neutron stars in MSPs and most HMXBs, except for Vela X-1. The first scenario (top panel) predicts a bimodal birth function with peak masses at 1.28 and 1.73 M$_\odot$ which can now also accommodate a population of "massive" NS, such as Cyg X-2 and J1012+537, or low-mass black holes (if a soft EOS is assumed). However, the BH population observed in SXTs, with masses in the range $\sim 3-14 M_\odot$ cannot be explained.

High-mass BHs (≥ 3 M$_\odot$) are difficult to obtain in close binaries because the companion will uncover the He core during the common envelope stage, allowing for strong winds to drastically reduce its mass. The effect is important when Roche lobe overflow takes place during *Case A* (ZAMS) and *Case B* (H burning in shell) mass transfer. However, if the H envelope is removed after the

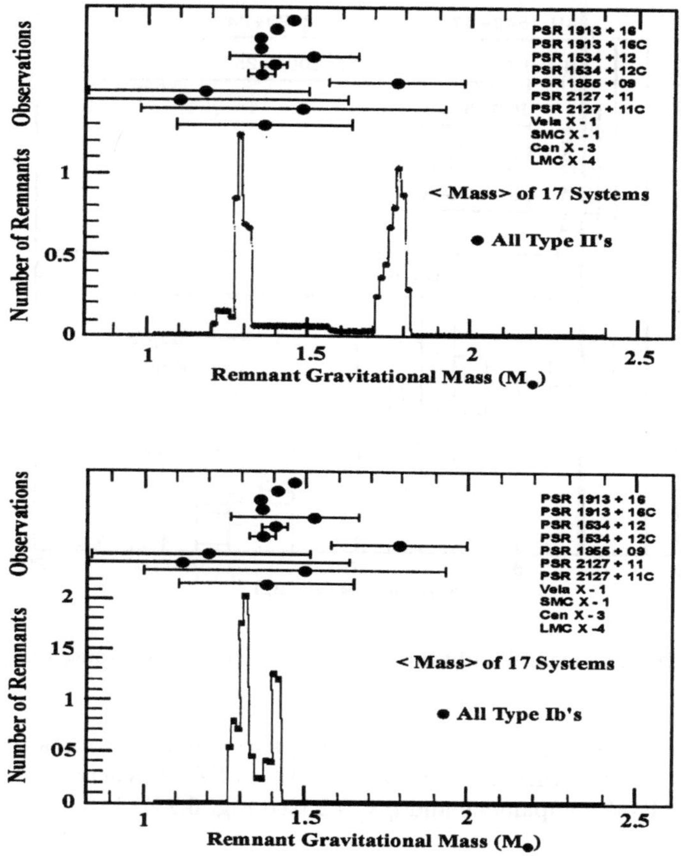

Fig. 6.25. Remnant distribution function computed in [176] compared to the observed distribution of neutron star masses in MSPs and HMXBs. Stars were assumed to undergo either a Type II supernova explosion (top) or a Type Ib supernova (bottom).

massive star has finished He core burning (*Case C* mass transfer), it will be in the short-lived ($\sim 10^4$ yrs) supergiant stage with shell He burning and the effects of mass-loss will be small [15]. The massive star has essentially evolved as a single star and the resulting core masses can be obtained from [201]

$$M_{He} \simeq 0.08\,(M_{ZAMS})^{1.4} . \tag{6.23}$$

The He core masses left by 20-35 $M_\odot$ ZAMS will lie in the range 5-12 $M_\odot$ and will collapse into $\sim 3 - 10 M_\odot$ BHs. This scenario will only take place if the binary orbit is wide enough to avoid *Case B* mass transfer but tight enough to go through *Case C* evolution and it has been estimated to be in a very narrow range $\sim 1600 - 1800 R_\odot$ [87]. In addition, the large orbital shrinkage needed to expel the envelope of the massive star during the common envelope stage

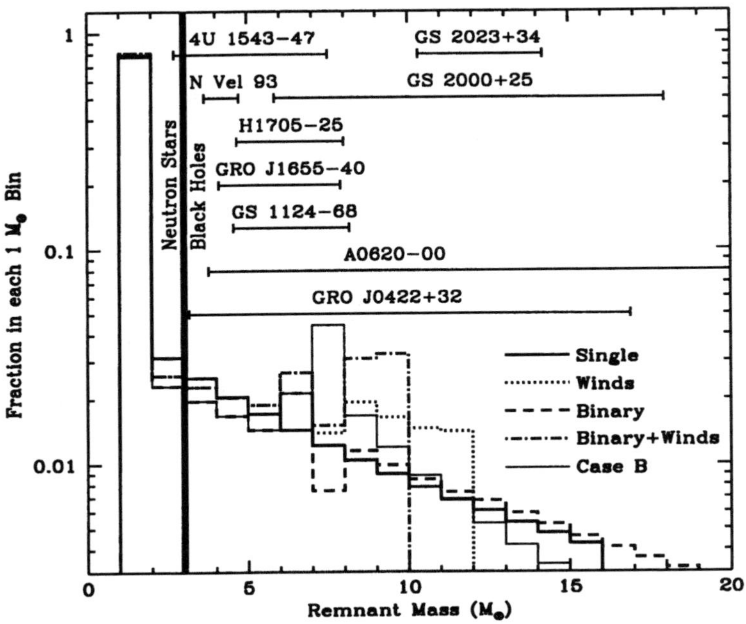

Fig. 6.26. The mass distribution of compact remnants computed in [56].

favours low-mass companion stars and hence, extreme binary mass ratios. This is in good agreement with observations of BHs in SXTs, where $\sim$ 75 % of the binaries have K-M companions and $q < 0.2$ (see Fig. 6.23).

Figure 6.26 presents remnant distributions computed in [56] using a 2D core-collapse simulation and a Scalo IMF (γ=2.7), compared to the observed distribution of BHs in SXTs. Different line styles demonstrate the effect of wind-loss, binarity and mass transfer through *Case B* in the final distribution. These show continuous quasi-exponential functions from NS up to BH with no evidence for gaps and some indication of upper mass cut-off at $\sim 10 - 15 M_\odot$. Neutron stars outnumber ($\sim$ 90 %) BHs and the fraction of BH masses in the range 3-5 $M_\odot$ is very sensitive to the power-law index γ of the IMF while the fraction of masses in the range 5-10 $M_\odot$ depends on the assumptions about binaries and winds. The observed distribution of BHs in SXTs qualitatively agrees with the theoretical function including binary effects and mass loss by winds in the standard treatment (dashed-dotted line). Only V404 Cyg, with a 12 $M_\odot$ remnant, remains outside the theoretical distribution and which may indicate an overestimation of the mass-loss rates.

Companion stars in X-ray binaries have survived a supernova explosion and hence they might have been contaminated by nucleosynthesis products of the ejecta. Therefore, the study of possible abundance anomalies in their atmospheres could be used to set further constraints on supernova models. This new approach has been applied, for the first time, by Israelian et al. [84] to observations of J1655-40, the BH SXT with the most precise mass determination (Table 6.2). Their high resolution spectrum (Fig. 6.27) shows a clear excess of OI (λ8446 and the triplet at λ7771-5) and other α-elements (e.g. Mg, Si, Ti) relative to F6III-IV template stars. LTE model atmospheres were used to compute synthetic spectra for a normal F6III star (i.e. $T_{eff} = 6400$ K, $\log g = 3.7$, microturbulence $\xi = 2$ km s^{-1} and $[Fe/H] = 0$) and the abundance of α-elements was iterated until a good fit to the J1655-40 spectrum was achieved. This analysis shows that α-elements are overabundant by a factor $\simeq$ 6-10. They cannot be synthesized in the core of the $\simeq 2.3 M_{\odot}$ companion and therefore must come from the ejecta of the supernova which formed the BH.

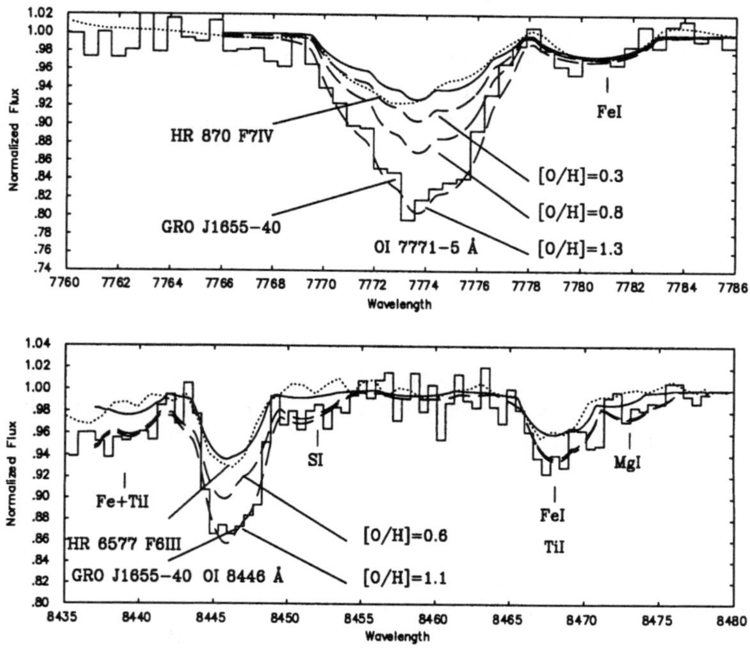

Fig. 6.27. Spectral analysis of the secondary star in J1655-40. Dotted lines correspond to spectra of F6 templates whereas dashed lines show synthetic spectra for a set of α-element abundances. Note that the abundance of element X is defined as $[X/H] = \log [N(X)/N(H)]_{star} - \log [N(X)/N(H)]_{\odot}$, where $N(H)$ is the number density of atoms. Figure adapted from [84].

An upper limit to the He core mass which underwent a Type Ib supernova is set by the remnant's mass, i.e. $M_{He} \geq 6 - 8 M_{\odot}$, whereas the binary survival condition [11] yields $M_{He} \leq 2M_{BH} + M_2 \simeq 10 - 16 M_{\odot}$. Standard supernovae models (e.g. using eq. 6.23) require massive stars in the range $\simeq 25 - 40 M_{\odot}$

to leave $\simeq 5 - 16 M_\odot$ remnants and they also predict the ejection of typically $\geq 1 - 2 M_\odot$ of matter enriched with nucleosynthesis products, 70 % of which is Oxygen (e.g. [128]). In order to explain the observed abundance of α-elements one requires an enrichment of $\sim 10^{-3} M_\odot$ and Israelian et al. show this is plausible since it is significantly lower than the fraction intercepted by the companion in a spherically symmetric explosion[14]. The measured S abundance, however, does not agree with nucleosynthesis yields in standard supernova models since very little of this element is predicted [128]. New models have been recently proposed by Nomoto et al. [129] to fit the lightcurves of SN 1998bw and SN 1997ef which require a factor 10 higher kinetic energies than in ordinary supernovae. These are called *Hypernovae* and nucleosynthesis yields computations are in good agreement with the observations of J1655-40, strongly suggesting that the precursor was a hypernova. Further evidence of the hypernova nature of J1655-40 has been presented in [16].

Detailed chemical analysis of the atmospheres of companion stars can, therefore, be used to extract "fosil" information on supernova explosions in X-ray binaries and estimate the precursor's masses. The combination of this new technique with new accurate dynamical masses of remnants, is expected to provide fundamental constraints on current supernovae models, the theory of massive star and close binary evolution.

Acknowledgments

I thank Phil Charles and Garik Israelian for helpul comments on the manuscript. The ING telescopes are operated on the island of La Palma by the Royal Greenwich Observatory in the Spanish Observatorio del Roque de Los Muchachos of the Instituto de Astrofísica de Canarias.

References

1. H.M. Al-Naimy, *Astrophys. Space Sci.* **53**, 181 (1978).
2. Y. Avni and J.N. Bahcall, *ApJ* **197**, 675 (1975).
3. J.N. Bahcall, P.C. Joss and Y. Avni, *ApJ* **191**, 211 (1974).
4. C.D. Bailyn, J.A. Orosz, J.E. McClintock and R.A. Remillard, *Nature* **378**, 157 (1995).
5. C.D. Bailyn, R.K. Jain, P. Coppi and J.A. Orosz, *ApJ* **499**, 375 (1998).
6. G. Beekman, T. Shahbaz, T. Naylor and P.A. Charles, *MNRAS* **281**, L1 (1996).
7. D. Bhattacharya and E.P.J. van den Heuvel, *Phys. Rep.* **203** 1 (1991).
8. L. Bildsten and R. Rutledge, *ApJ*, astro-ph/9912304.
9. C.T. Bolton, *Nature* **235**, 271 (1972).
10. C.T. Bolton, *ApJ* **200**, 269 (1975).
11. W.N. Brandt and Ph Podsiadlowski, *MNRAS* **274**, 461 (1995).
12. W.N. Brandt, Ph Podsiadlowski and S. Sirgursson, *MNRAS* **277**, L35 (1995).

[14] Note that this was assuming mixing with the matter in the convection zone, which is very shallow in an F6 subgiant.

13. G.E. Brown and H.A. Bethe, *ApJ* **423**, 659 (1994).
14. G.E. Brown, J.C. Weingartner and R.A.M.J. Wijers, *ApJ* **463**, 297 (1996).
15. G.E. Brown, C.-H. Lee and H.A. Bethe, *ApJ* **astro-ph/9807221**.
16. G.E. Brown *et al.*, *NewA* **astro-ph/0003361**.
17. P.J. Callanan, P.M. Garnavich and D. Koester, *MNRAS* **298**, 207 (1998).
18. J.K. Cannizzo, *ApJ* **435**, 389 (1994).
19. J.K. Cannizzo, W.. Chen and M. Livio *ApJ* **454**, 880 (1995).
20. J.K. Cannizzo, *ApJ* **494**, 366 (1998).
21. J. Casares, in *Cataclysmic Variables and Related Objects*, IAU Coll. 158, eds. A. Evans and J.H. Wood, Kluwer., p395 (1996).
22. J. Casares, in *Relativistic Astrophysics and Cosmolgy*, Proc. of the Spanish Relativity Meeting, ed. J. Buitrago, E. Mediavilla and A. Oscoz, World Scientific, p.3 (1997).
23. J. Casares *et al.*, *MNRAS* **250**, 712 (1991).
24. J. Casares and P.A. Charles, *MNRAS* **255**, 7 (1992).
25. J. Casares, P.A. Charles and T. Naylor, *Nature* **355**, 614 (1992).
26. J. Casares, P.A. Charles, T. Naylor and E.P. Pavlenko, *MNRAS* **265**, 834 (1993).
27. J. Casares and P.A. Charles, *MNRAS* **271**, L5 (1994).
28. J. Casares *et al.*, *MNRAS* **276**, L35 (1995).
29. J. Casares, P.A. Charles and T.R. Marsh, *MNRAS* **277**, L45 (1995).
30. J. Casares, M. Mouchet, I.G. Martínez-Pais and E.T. Harlaftis, *MNRAS* **282**, 182 (1996).
31. J. Casares, E.L. Martín, P.A. Charles, P.A. Molaro and R. Rebolo, *NewA* **1**, 299 (1997).
32. J. Casares, P.A. Charles and E. Kuulkers, *ApJ* **493**, L39 (1998).
33. P.A. Charles, in *Compact stars in binaries*, IAU Symp. 165, ed. J. van Paradijs, E.P.J. van den Heuvel, E. Kuulkers., Kluwer, Dordrecht, p331 (1996).
34. P.A. Charles, J. Casares, E.L. Martín and R. Rebolo, in *The Evolution of X-ray Binaries*, eds. S.S. Holt and C.S. Day (AIP 308), p.371 (1994).
35. C. Chevalier, S.A. Ilovaisky, J. van Paradijs, H. Pedersen and M. van der Klis, *A&A* **210**, 114 (1989).
36. C. Chevalier and S.A. Ilovaisky, *A&A* **269**, 301 (1993).
37. C. Chevalier and S.A. Ilovaisky, *IAU CIRC.* 6806 (1998).
38. C. Chevalier, S.A. Ilovaisky, P. Leisy and F. Patat, *A&A* **347**, L51 (1999).
39. W. Chen, C. Shrader and M. Livio, *ApJ* **491**, 312 (1997).
40. A. Claret and A. Giménez **this volume** (2001).
41. G.W. Collins and R.J. Truax, *ApJ* **439**, 860 (1995).
42. S. Davey and R.C. Smith, *MNRAS* **257**, 476 (1992).
43. J. A. de Jong, J. van Paradijs and T. Augusteijn, *A&A* **314**, 484 (1996).
44. C. de Loore, **this volume** (2001).
45. G. Dubus *et al.*, *MNRAS* **303**, 139 (1999).
46. L.M. Ensman and S.E. Woosley, *ApJ* **333**, 754 (1988).
47. A.A. Esin, J.E. McClintock and R. Narayan, *ApJ* **489**, 865 (1997).
48. A.V. Filippenko, T. Matheson and A.J. Barth, *ApJ* **455**, L139 (1995).
49. A.V. Filippenko, T. Matheson and L.C. Ho, *ApJ* **455**, L614 (1995).
50. A.V. Filippenko *et al.*, *PASP* **109**, 461 (1997).
51. A.V. Filippenko *et al.*, *PASP* **111**, 969 (1999).
52. L.S. Finn, *Phys. Rev. Lett.*, **73**, 1878 (1994).
53. J. Frank, A.R. King and D.J. Raine: *Accretion Power in Astrophysics* (Cambridge University Press, Cambridge 1992).

54. J.L. Friedman, J.R. Ipser and L. Parker, *ApJ* **304**, 115 (1986).
55. C.L. Fryer, *ApJ*, **522** 413 (1999).
56. C.L. Fryer and V. Kalogera, *ApJ*, astro-ph/9911312.
57. R. Giacconi *et al.*, *ApJ* **148**, 129 (1967).
58. D.R. Gies and C.T. Bolton, *ApJ* **260**, 240 (1982).
59. D.R. Gies and C.T. Bolton, *ApJ* **304**, 371 (1986).
60. D.R. Gies and C.T. Bolton, *ApJ* **304**, 389 (1986).
61. A. Goldwurm *et al.*, *ApJ* **389**, L79 (1992).
62. D.F. Gray, *The Observation and Analysis of Stellar Atmospheres,* Vol 56, *Ed. Wiley-Interscience, New York, p.398* (1976).
63. N. Guessoum and D. Kazanas, *ApJ* **512**, 332 (1999).
64. J.-M. Hameury, A.R. King and J.P. Lasota, *A&A* **162**, 71 (1986).
65. J.-M. Hameury, A.R. King and J.P. Lasota, *A&A* **171**, 140 (1987).
66. J.-M. Hameury, A.R. King and J.P. Lasota, *A&A* **353**, 585 (1990).
67. J.M. Hameury, K. Menou, G. Dubus, J.-P. Lasota and J.-H. Huré, *MNRAS* **298**, 1048 (1998).
68. W.-R. Hamann and L. Koesterke, Journal*A&A*33510031998.
69. E.T. Harlaftis, K. Horne and A.V. Filippenko, *PASP* **108**, 762 (1996).
70. E.T. Harlaftis, D. Steeghs, K. Horne and A.V. Filippenko, *AJ* **114**, 1170 (1997).
71. E.T. Harlaftis, S.J. Collier, K. Horne and A.V. Filippenko, *A&A* **341**, 491 (1999).
72. J.V. Hartman, A&A **322**, 127 (1997).
73. C.A. Haswell and A. W. Shafter, *ApJ* **359**, L47 (1990).
74. C.A. Haswell, *Ph.D. Thesis,* Univ. Texas (1992).
75. C.A. Haswell, E.L. Robinson, K. Horne, R.F. Stiening and T.M.C. Abbott, *ApJ* **411**, 802 (1993).
76. C.A. Haswell, in *Compact stars in binaries,* IAU Symp. 165, ed. J. van Paradijs, E.P.J. van den Heuvel, E. Kuulkers., Kluwer, Dordrecht, p351 (1996).
77. A. Herrero, R.P. Kudritzki, R. Gabler, J.M. Vilchez and A. Gabler, A&A **297**, 556 (1995).
78. F. van der Hooft *et al.*, *MNRAS* **286**, L43 (1997).
79. F. van der Hooft, M.H.M. Heemskerk, F. Alberts and J.van Paradijs, *A&A* **329**, 538 (1998).
80. K. Horne *et al.*, *ApJ* **367**, L5 (1991).
81. M. Huang and J.C. Wheeler, *ApJ* **343**, 229 (1989).
82. R.I. Hynes, K. O'Brien, K. Horne, W. Chen ans C.A. Haswell, *MNRAS* **299**, L37 (1998)
83. I.Jr. Iben and A.V. Tutukov, *ApJ Supp. Series* **100**, 217 (1995).
84. G. Israelian, R. Rebolo, G. Basri, J. Casares and E.L. Martin, *Nature* **401**, 142 (1999).
85. H.M. Johnston, S.R. Kulkarni and J.B. Oke, *ApJ* **345**, 492 (1989).
86. V. Kalogera, *ApJ* **521**, 723 (1999).
87. V. Kalogera and R.F. Webbink, *ApJ* **493**, 351 (1998).
88. M.H. van Kerkwijk *et al.* *A&A* **303**, 483 (1995).
89. M.H. van Kerkwijk, J. van Paradijs and E.J. Zuiderwijk, *A&A* **303**, 497 (1995).
90. M.H. van Kerkwijk, P. Bergeron and S.R. Kulkarni, *ApJ* **467**, L89 (1996).
91. S-W. Kim, J.C. Wheeler and S. Mineshige, *PASP* **51**, 393 (1999).
92. A.R. King, *QJRAS* **29**, 1 (1988).
93. A.R. King, *MNRAS* **260**, L5 (1993).
94. A.R. King, U. Kolb and L. Burderi, *ApJ* **464**, L127 (1996).
95. A.R. King, J. Frank, U. Kolb and H. Ritter, *ApJ* **482**, 919 (1997).

96. A.R. King, J. Frank, U. Kolb and H. Ritter, *ApJ* **484**, 844 (1997).
97. A.R. King, U. Kolb and E. Szuszkiewicz, *ApJ* **488**, 89 (1997).
98. A.R. King and H. Ritter, *MNRAS* **293**, 42 (1998).
99. U. Kolb, *MNRAS* **297**, 419 (1998).
100. U. Kolb, A.R. King, H. Ritter and J. Frank, *MNRAS* **485**, L33 (1997).
101. E. Kuulkers, *NewAR* **42**, 1 (1998).
102. W.H.G. Lewin, in *X-Ray Binaries*, eds. W.H.G. Lewin, J. van Paradijs and E.P.J. van den Heuvel (CUP 26, Cambridge), p.175 (1995).
103. R. London, R. McCray and L.H. Auer, *ApJ* **243**, 970 (1981).
104. L.B. Lucy, *Z. Astrophys.*, **65**, 89 (1967).
105. P. Massey and A.B. Thompson, *AJ* **101**, 1408 (1991).
106. L. Maraschi, A. Treves and E.P.J. van den Heuvel, *Nature* **259**, 292 (1976).
107. T.R. Marsh, *MNRAS* **231**, 1117 (1988).
108. T.R. Marsh, **this volume** (2001).
109. T.R. Marsh, E.L. Robinson and J.H. Wood, *MNRAS* **266**, 137 (1994).
110. E.L. Martín, R. Rebolo, J. Casares and P.A. Charles, *Nature* **358**, 129 (1992).
111. E.L. Martín, R. Rebolo, J. Casares and P.A. Charles, *ApJ* **435**, 791 (1994).
112. E.L. Martín, H.C. Spruit and J. van Paradijs, *A&A* **291**, L43 (1994).
113. E.L. Martín, J. Casares, P.A. Charles and R. Rebolo, *A&A* **303**, 785 (1995).
114. E.L. Martín, J. Casares, P. Molaro, R. Rebolo and P.A. Charles, *NewA* **1**, 197 (1996).
115. J.E. McClintock, C.R. Canizares and C. B. Tarter, *ApJ* **198**, 641 (1975).
116. J.E. McClintock and R.A. Remillard, *ApJ* **308**, 110 (1986).
117. J.E. McClintock and R.A. Remillard, *ApJ* **350**, 386 (1990).
118. J.E. McClintock and R. Remillard, *ApJ* **531**, 956 (2000).
119. K. Menou *et al.*, *ApJ* **520**, 276 (1999).
120. K. Menou, J.-M. Hameury, J.-P. Lasota and R. Narayan, *MNRAS* **314**, 498 (2000).
121. M. Milgrom, *A&A* **67**, 25 (1978).
122. S. Mineshige, *ApJ* **431**, 99 (1994).
123. R. Narayan, J.E. and I. Yi, *ApJ* **444**, 231 (1995).
124. R. Narayan, J.E. McClintock and I. Yi, *ApJ* **457**, 821 (1996).
125. R. Narayan, D. Barret and J.E. McClintock, *ApJ* **482**, 448 (1997).
126. G. Nelemans, T.M. Tauris and E.P.J. van den Heuvel, *A&A* **352**, 87 (1999).
127. D.J. Nice, R.W. Sayer and J.H. Taylor, *ApJ* **466**, L87 (1996).
128. K. Nomoto *et al.*, *Nuclear Phys.*, **A616** 79 (1997).
129. K. Nomoto *et al.*, *in "The Greatest Explosions since the Big Bang: Supernovae and Gamma-Ray Bursts"* **astro-ph/0003077**.
130. D. O'Donoghue and P.A. Charles, *MNRAS* **282**, 191 (1996)
131. J.B. Oke, *ApJ* **217**, 181 (1977).
132. J.A. Orosz, C.D. Bailyn, R.A. Remillard, J.E. McClintock and C.B. Foltz, *ApJ* **436**, 848 (1994).
133. J.A. Orosz and C.D. Bailyn, *ApJ* **446**, L59 (1995).
134. J.A. Orosz, C.D. Bailyn, J.E. McClintock and R.A. Remillard *ApJ* **468**, 380 (1996).
135. J.A. Orosz and C.D. Bailyn, *ApJ* **477**, 876 (1997).
136. J.A. Orosz, R.K. Jain, C.D. Bailyn, J.E. McClintock and R.A. Remillard, *ApJ* **499**, 375 (1998).
137. J. A. Orosz and E. Kuulkers, *MNRAS* **305**, 1320 (1999).
138. B. Paczynski, *Ann. Rev. Astron. Astrophys.* **9**, 183 (1971).

139. E.P. Pavlenko, A.C. Martin, J. Casares, P.A. Charles and N.A. Ketsaris, *MNRAS* **281**, 1094 (1996).
140. S.N. Phillips, T. Shahbaz and Ph. Podsiadlowski, *MNRAS* **304**, 839 (1999).
141. M. Pérez-Torres, J. Casares, I.G. Martínez-Pais and P.A. Charles, *MNRAS* , in preparation (2000).
142. J.A. Petterson, *ApJ* **216**, 827 (1977).
143. J.E. Pringle, *MNRAS* **281**, 357 (1996).
144. R. Ramachandran and D. Bhattacharya, *MNRAS* **288**, 565 (1997).
145. C.E. Rhoades and R. Ruffini, *Phys. Rev. Lett.* **32**, 324 (1974).
146. I.N. Reid, *AJ* **111**, 2000 (1996).
147. A. Sandage *et al.*, *ApJ* **146**, 316 (1966).
148. D. Sanwal *et al.*, *ApJ* **460**, 437 (1996).
149. G.J. Savonjie and J.C.B. Papaloizou, *MNRAS* **207**, 685 (1984).
150. J. Schachter, A.V. Filippenko and S.M. Kahn, *ApJ* **340**, 1049 (1989).
151. D.P. Schneider and P. Young, *ApJ* **238**, 946 (1980).
152. E. Schreier *et al.*, *ApJ* **172**, 79 (1972).
153. T. Shahbaz, T. Naylor and P.A. Charles, *MNRAS* **265**, 655 (1993).
154. T. Shahbaz, T. Naylor and P.A. Charles, *MNRAS* **268**, 756 (1994).
155. T. Shahbaz *et al.*, *MNRAS* **271**, L10 (1994).
156. T. Shahbaz, R. Bandyopadhyay, T. Naylor and P.A. Charles, *MNRAS* **282**, 977 (1996).
157. T. Shahbaz, T. Naylor and P.A. Charles, *MNRAS* **285**, 607 (1997).
158. T. Shahbaz, *MNRAS* **298**, 153 (1998).
159. T. Shahbaz, R. Bandyopadhyay and P.A. Charles, *MNRAS* **346**, 82 (1999).
160. T. Shahbaz, F. van der Hooft, J. Casares, P.A. Charles and J. van Paradijs, *MNRAS* **306**, 89 (1999).
161. T. Shahbaz, P.A. Charles and A.R. King, *MNRAS* **301**, 382 (1998).
162. A.W. Shafter, P. Szkody and J.R. Thortensen, *ApJ* **308**, 765 (1986).
163. N.I. Shakura and R.A. Sunyaev, Å **24**, 337 (1973).
164. I.S. Shklovskii, *Astron. Zhur.* **44**, 930 (1967).
165. J. Smak, *ApJ* **272**, 234 (1983).
166. J. Smak, **this volume** (2001).
167. D.A. Smith and V.S. Dhillon, *MNRAS* **301**, 767 (1998).
168. R. Soria, D.T. Wickramasinghe, R.W. Hunstead and K. Wu, *ApJ* **495**, L45 (1998).
169. R.E. Taam, *ApJ* **270**, 694 (1983).
170. Y. Tanaka and W.H.G. Lewin, in *X-Ray Binaries*, eds. W.H.G. Lewin, J. van Paradijs and E.P.J. van den Heuvel (CUP 26, Cambridge), p.126 (1995).
171. Y. Tanaka and N. Shibazaki, *Ann. Rev. Astron. Astrophys.* **34**, 607 (1996).
172. J.L. Terman, R.E. Taam and C.O. Savage, *MNRAS* **281**, 552 (1996).
173. S.E. Thorsett, Z. Arzoumanian, M.M. McKinnon and J.H. Taylor, *ApJ* **405**, L29 (1993).
174. S.E. Thorsett and D. Chakrabarty, *ApJ* **512**, 288 (1999).
175. F.X. Timmes, S.E. Woosley and T.A. Weaver, *ApJ Supp. Series* **98**, 617 (1995).
176. F.X. Timmes, S.E. Woosley and T.A. Weaver, *ApJ* **457**, 834 (1996).
177. J. Truemper *et al.*, *Sp. Sci. Rev.***40**2551985.
178. E.P.J. van den Heuvel, in *Accretion-Driven Stellar X-ray Sources*, eds. W.H.G. Lewin and E.P.J. Van den Heuvel, C.U.P., p303 (1983).
179. E.P.J. van den Heuvel and R.E. Taam, *Nature* **309**, 235 (1984).
180. E.P.J. van den Heuvel and S. Rappaport, *in "Physics of Be Stars"*, C.U.P., p. 291 (1987).

181. E.P.J. van den Heuvel and O. Bitzaraki, A&A **297**, L41 (1995).
182. E.P.J. van den Heuvel and J. van Paradijs, *ApJ* **483**, 399 (1997).
183. E.P.J. van den Heuvel, S.F. Portegies, D. Bhattacharya and L. Kaper, *A&A***astro-ph/0005245**.
184. J. van Paradijs, *ApJ* **464**, L139 (1996).
185. J. van Paradijs, in *Accretion-Driven Stellar X-ray Sources*, eds. W.H.G. Lewin and E.P.J. Van den Heuvel, C.U.P., p189 (1983).
186. J. van Paradijs and N. White, *ApJ* **447**, 33 (1995).
187. J. van Paradijs and J.E. McClintock, in *X-Ray Binaries*, eds. W.H.G. Lewin, J. van Paradijs and E.P.J. van den Heuvel (CUP 26, Cambridge), p58 (1995).
188. F. Verbunt and E.P.J. van den Heuvel, in *X-Ray Binaries*, eds. W.H.G. Lewin, J. van Paradijs and E.P.J. van den Heuvel (CUP 26, Cambridge), p.457 (1995).
189. S.D. Vrtilek *et al.*, Å **235**, 162 (1990).
190. S.D. Vrtilek *et al.*, Å **376**, 278 (1991).
191. R.A. Wade and S.M. Rucinski, *ApJ Supp. Series* **60**, 471 (1985).
192. R.A. Wade and K. Horne, *ApJ* **324**, 411 (1988).
193. B.L. Webster and P. Murdin, *Nature* **235**, 37 (1972).
194. R.F. Weebink, S. Rappaport and G.J. Savonije, *ApJ* **270**, 678 (1983).
195. W.F. Welsh, K. Horne and R. Gomer, *ApJ* **410**, L39 (1993).
196. W.F. Welsh, E.L. Robinson and P. Young *AJ*, **astro-ph/0004344**.
197. N.E. White, in *The Evolution of X-ray Binaries*, ed. S.S. Holt and C.S. Day (AIP 308, 1994), p.53.
198. N. E. White and J. van Paradijs, *ApJ* **473**, L25 (1996).
199. R. Whitehurst, *MNRAS* **266**, 35 (1994).
200. S.E. Woosley, N. Langer and T.A. Weaver, *ApJ* **411**, 823 (1993).
201. S.E. Woosley and T.A. Weaver, *ApJ Supp. Series* **101**, 181 (1995).
202. C.-C. Wu *et al.*, *PASP* **94**, 149 (1982).
203. I. Yi and R. Narayan, *ApJ* **486**, 363 (1997).
204. I. Yi, R. Narayan, D. Barret and J.E. McClintock, *A&A Supp.* **120**, 187 (1996).
205. H. von Zeipel, *MNRAS* **84**, 665 (1924).
206. C. Zurita *et al.*, *MNRAS*, **astro-ph/0003092**.

Lecture Notes in Physics

For information about Vols. 1–531
please contact your bookseller or Springer-Verlag

Vol. 532: S. C. Müller, J. Parisi, W. Zimmermann (Eds.), Transport and Structure. Their Competitive Roles in Biophysics and Chemistry. XII, 400 pages. 1999.

Vol. 533: K. Hutter, Y. Wang, H. Beer (Eds.), Advances in Cold-Region Thermal Engineering and Sciences. Proceedings, 1999. XIV, 608 pages. 1999.

Vol. 534: F. Moreno, F. González (Eds.), Light Scattering from Microstructures. Proceedings, 1998. XII, 300 pages. 2000

Vol. 535: H. Dreyssé (Ed.), Electronic Structure and Physical Properties of Solids: The Uses of the LMTO Method. Proceedings, 1998. XIV, 458 pages. 2000.

Vol. 536: T. Passot, P.-L. Sulem (Eds.), Nonlinear MHD Waves and Turbulence. Proceedings, 1998. X, 385 pages. 1999.

Vol. 537: S. Cotsakis, G. W. Gibbons (Eds.), Mathematical and Quantum Aspects of Relativity and Cosmology. Proceedings, 1998. XII, 251 pages. 1999.

Vol. 538: Ph. Blanchard, D. Giulini, E. Joos, C. Kiefer, I.-O. Stamatescu (Eds.), Decoherence: Theoretical, Experimental, and Conceptual Problems. Proceedings, 1998. XII, 345 pages. 2000.

Vol. 539: A. Borowiec, W. Cegła, B. Jancewicz, W. Karwowski (Eds.), Theoretical Physics. Fin de Siècle. Proceedings, 1998. XX, 319 pages. 2000.

Vol. 540: B. G. Schmidt (Ed.), Einstein's Field Equations and Their Physical Implications. Selected Essays. 1999. XIII, 429 pages. 2000.

Vol. 541: J. Kowalski-Glikman (Ed.), Towards Quantum Gravity. Proceedings, 1999. XII, 376 pages. 2000.

Vol. 542: P. L. Christiansen, M. P. Sørensen, A. C. Scott (Eds.), Nonlinear Science at the Dawn of the 21st Century. Proceedings, 1998. XXVI, 458 pages. 2000.

Vol. 543: H. Gausterer, H. Grosse, L. Pittner (Eds.), Geometry and Quantum Physics. Proceedings, 1999. VIII, 408 pages. 2000.

Vol. 544: T. Brandes (Ed.), Low-Dimensional Systems. Interactions and Transport Properties. Proceedings, 1999. VIII, 219 pages. 2000

Vol. 545: J. Klamut, B. W. Veal, B. M. Dabrowski, P. W. Klamut, M. Kazimierski (Eds.), New Developments in High-Temperature Superconductivity. Proceedings, 1998. VIII, 275 pages. 2000.

Vol. 546: G. Grindhammer, B. A. Kniehl, G. Kramer (Eds.), New Trends in HERA Physics 1999. Proceedings, 1999. XIV, 460 pages. 2000.

Vol. 547: D. Reguera, G. Platero, L. L. Bonilla, J. M. Rubí(Eds.), Statistical and Dynamical Aspects of Mesoscopic Systems. Proceedings, 1999. XII, 357 pages. 2000.

Vol. 548: D. Lemke, M. Stickel, K. Wilke (Eds.), ISO Surveys of a Dusty Universe. Proceedings, 1999. XIV, 432 pages. 2000.

Vol. 549: C. Egbers, G. Pfister (Eds.), Physics of Rotating Fluids. Selected Topics, 1999. XVIII, 437 pages. 2000.

Vol. 550: M. Planat (Ed.), Noise, Oscillators and Algebraic Randomness. Proceedings, 1999. VIII, 417 pages. 2000.

Vol. 551: B. Brogliato (Ed.), Impacts in Mechanical Systems. Analysis and Modelling. Lectures, 1999. IX, 273 pages. 2000.

Vol. 552: Z. Chen, R. E. Ewing, Z.-C. Shi (Eds.), Numerical Treatment of Multiphase Flows in Porous Media. Proceedings, 1999. XXI, 445 pages. 2000.

Vol. 553: J.-P. Rozelot, L. Klein, J.-C. Vial Eds.), Transport of Energy Conversion in the Heliosphere. Proceedings, 1998. IX, 214 pages. 2000.

Vol. 554: K. R. Mecke, D. Stoyan (Eds.), Statistical Physics and Spatial Statistics. The Art of Analyzing and Modeling Spatial Structures and Pattern Formation. Proceedings, 1999. XII, 415 pages. 2000.

Vol. 555: A. Maurel, P. Petitjeans (Eds.), Vortex Structure and Dynamics. Proceedings, 1999. XII, 319 pages. 2000.

Vol. 556: D. Page, J. G. Hirsch (Eds.), From the Sun to the Great Attractor. X, 330 pages. 2000.

Vol. 557: J. A. Freund, T. Pöschel (Eds.), Stochastic Processes in Physics, Chemistry, and Biology. X, 330 pages. 2000.

Vol. 558: P. Breitenlohner, D. Maison (Eds.), Quantum Field Theory. Proceedings, 1998. VIII, 323 pages. 2000

Vol. 559: H.-P. Breuer, F. Petruccione (Eds.), Relativistic Quantum Measurement and Decoherence. Proceedings, 1999. X, 140 pages. 2000.

Vol. 560: S. Abe, Y. Okamoto (Eds.), Nonextensive Statistical Mechanics and Its Applications. IX, 272 pages. 2001.

Vol. 561: H. J. Carmichael, R. J. Glauber, M. O. Scully (Eds.), Directions in Quantum Optics. XVII, 369 pages. 2001.

Vol. 562: C. Lämmerzahl, C. W. F. Everitt, F. W. Hehl (Eds.), Gyros, Clocks, Interferometers...: Testing Relativistic Gravity in Space. XVII,507 pages. 2001.

Vol. 563: F. C. Lázaro, M. J. Arévalo (Eds.), Binary Stars. Selected Topics on Observations and Physical Processes. 1999.IX, 327 pages. 2001.

Vol. 564: T. Pöschel, S. Luding (Eds.), Granular Gases. VIII, 457 pages. 2001.

Vol. 565: E. Beaurepaire, F. Scheurer, G. Krill, J.-P. Kappler (Eds.), Magnetism and Synchrotron Radiation. XIV, 396 pages. 2001.

Vol. 566: J. L. Lumley (Ed.), Fluid Mechanics and the Environment: Dynamical Approaches. VIII, 412 pages. 2001.

Vol. 567: D. Reguera, L. L. Bonilla, J. M. Rubí (Eds.), Coherent Structures in Complex Systems. IX, 465 pages. 2001.

Vol. 568: P. A. Vermeer, S. Diebels, W. Ehlers, H. J. Herrmann, S. Luding, E. Ramm (Eds.), Continuous and Discontinuous Modelling of Cohesive-Frictional Materials. XIV, 307 pages. 2001.

Monographs
For information about Vols. 1–23 please contact your bookseller or Springer-Verlag

Vol. m 24: A. Ern, V. Giovangigli, Multicomponent Transport Algorithms. XIV, 427 pages. 1994.

Vol. m 25: A. V. Bogdanov, G. V. Dubrovskiy, M. P. Krutikov, D. V. Kulginov, V. M. Strelchenya, Interaction of Gases with Surfaces. XIV, 132 pages. 1995.

Vol. m 26: M. Dineykhan, G. V. Efimov, G. Ganbold, S. N. Nedelko, Oscillator Representation in Quantum Physics. IX, 279 pages. 1995.

Vol. m 27: J. T. Ottesen, Infinite Dimensional Groups and Algebras in Quantum Physics. IX, 218 pages. 1995.

Vol. m 28: O. Piguet, S. P. Sorella, Algebraic Renormalization. IX, 134 pages. 1995.

Vol. m 29: C. Bendjaballah, Introduction to Photon Communication. VII, 193 pages. 1995.

Vol. m 30: A. J. Greer, W. J. Kossler, Low Magnetic Fields in Anisotropic Superconductors. VII, 161 pages. 1995.

Vol. m 31 (Corr. Second Printing): P. Busch, M. Grabowski, P.J. Lahti, Operational Quantum Physics. XII, 230 pages. 1997.

Vol. m 32: L. de Broglie, Diverses questions de mécanique et de thermodynamique classiques et relativistes. XII, 198 pages. 1995.

Vol. m 33: R. Alkofer, H. Reinhardt, Chiral Quark Dynamics. VIII, 115 pages. 1995.

Vol. m 34: R. Jost, Das Märchen vom Elfenbeinernen Turm. VIII, 286 pages. 1995.

Vol. m 35: E. Elizalde, Ten Physical Applications of Spectral Zeta Functions. XIV, 224 pages. 1995.

Vol. m 36: G. Dunne, Self-Dual Chern-Simons Theories. X, 217 pages. 1995.

Vol. m 37: S. Childress, A.D. Gilbert, Stretch, Twist, Fold: The Fast Dynamo. XI, 406 pages. 1995.

Vol. m 38: J. González, M. A. Martín-Delgado, G. Sierra, A. H. Vozmediano, Quantum Electron Liquids and High-Tc Superconductivity. X, 299 pages. 1995.

Vol. m 39: L. Pittner, Algebraic Foundations of Non-Com-mutative Differential Geometry and Quantum Groups. XII, 469 pages. 1996.

Vol. m 40: H.-J. Borchers, Translation Group and Particle Representations in Quantum Field Theory. VII, 131 pages. 1996.

Vol. m 41: B. K. Chakrabarti, A. Dutta, P. Sen, Quantum Ising Phases and Transitions in Transverse Ising Models. X, 204 pages. 1996.

Vol. m 42: P. Bouwknegt, J. McCarthy, K. Pilch, The W3 Algebra. Modules, Semi-infinite Cohomology and BV Algebras. XI, 204 pages. 1996.

Vol. m 43: M. Schottenloher, A Mathematical Introduction to Conformal Field Theory. VIII, 142 pages. 1997.

Vol. m 44: A. Bach, Indistinguishable Classical Particles. VIII, 157 pages. 1997.

Vol. m 45: M. Ferrari, V. T. Granik, A. Imam, J. C. Nadeau (Eds.), Advances in Doublet Mechanics. XVI, 214 pages. 1997.

Vol. m 46: M. Camenzind, Les noyaux actifs de galaxies. XVIII, 218 pages. 1997.

Vol. m 47: L. M. Zubov, Nonlinear Theory of Dislocations and Disclinations in Elastic Body. VI, 205 pages. 1997.

Vol. m 48: P. Kopietz, Bosonization of Interacting Fermions in Arbitrary Dimensions. XII, 259 pages. 1997.

Vol. m 49: M. Zak, J. B. Zbilut, R. E. Meyers, From Instability to Intelligence. Complexity and Predictability in Nonlinear Dynamics. XIV, 552 pages. 1997.

Vol. m 50: J. Ambjørn, M. Carfora, A. Marzuoli, The Geometry of Dynamical Triangulations. VI, 197 pages. 1997.

Vol. m 51: G. Landi, An Introduction to Noncommutative Spaces and Their Geometries. XI, 200 pages. 1997.

Vol. m 52: M. Hénon, Generating Families in the Restricted Three-Body Problem. XI, 278 pages. 1997.

Vol. m 53: M. Gad-el-Hak, A. Pollard, J.-P. Bonnet (Eds.), Flow Control. Fundamentals and Practices. XII, 527 pages. 1998.

Vol. m 54: Y. Suzuki, K. Varga, Stochastic Variational Approach to Quantum-Mechanical Few-Body Problems. XIV, 324 pages. 1998.

Vol. m 55: F. Busse, S. C. Müller, Evolution of Spontaneous Structures in Dissipative Continuous Systems. X, 559 pages. 1998.

Vol. m 56: R. Haussmann, Self-consistent Quantum Field Theory and Bosonization for Strongly Correlated Electron Systems. VIII, 173 pages. 1999.

Vol. m 57: G. Cicogna, G. Gaeta, Symmetry and Perturbation Theory in Nonlinear Dynamics. XI, 208 pages. 1999.

Vol. m 58: J. Daillant, A. Gibaud (Eds.), X-Ray and Neutron Reflectivity: Principles and Applications. XVIII, 331 pages. 1999.

Vol. m 59: M. Kriele, Spacetime. Foundations of General Relativity and Differential Geometry. XV, 432 pages. 1999.

Vol. m 60: J. T. Londergan, J. P. Carini, D. P. Murdock, Binding and Scattering in Two-Dimensional Systems. Applications to Quantum Wires, Waveguides and Photonic Crystals. X, 222 pages. 1999.

Vol. m 61: V. Perlick, Ray Optics, Fermat's Principle, and Applications to General Relativity. X, 220 pages. 2000.

Vol. m 62: J. Berger, J. Rubinstein, Connectivity and Superconductivity. XI, 246 pages. 2000.

Vol. m 63: R. J. Szabo, Ray Optics, Equivariant Cohomology and Localization of Path Integrals. XII, 315 pages. 2000.

Vol. m 64: I. G. Avramidi, Heat Kernel and Quantum Gravity. X, 143 pages. 2000.

Vol. m 65: M. Hénon, Generating Families in the Restricted Three-Body Problem. Quantitative Study of Bifurcations. XII, 301 pages. 2001.

Vol. m 66: F. Calogero, Classical Many-Body Problems Amenable to Exact Treatments. XIX, 749 pages. 2001.